GEOTECHNICAL NO. 101

SLOPE STABILITY 2000

PROCEEDINGS OF SESSIONS OF GEO-DENVER 2000

SPONSORED BY
The Geo-Institute of the American Society of Civil Engineers

August 5-8, 2000
Denver, Colorado

EDITED BY
D.V. Griffiths
Gordon A. Fenton
Timothy R. Martin

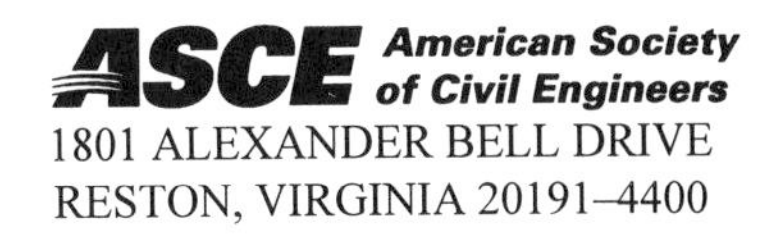

Abstract: Slope stability analysis is one of the oldest geotechnical engineering subjects, yet as we enter the 21st century it remains one of the most active areas of study for geotechnical practitioners and researchers. The 26 state-of-the-art papers contained within this publication entitled *Slope Stability 2000* cover many facets of the subject, including case histories of both natural and constructed slopes, slope stabilization methods, rock slope analysis, shear strength evaluation, centrifuge testing, limit analysis, 3-D analyses, finite element and finite difference methods, progressive failure analyses, and probabilistic methods using Bayesian and random field approaches. It is hoped that readers of *Slope Stability 2000* will be stimulated and inspired by this wide range of quality papers written by a distinguished group of national and international authors.

Library of Congress Cataloging-in-Publication Data

Geo-Denver 2000 (2000 : Denver, Colo.)

Slope stability 2000 : proceedings of sessions of Geo-Denver 2000 : August 5-8, 2000, Denver, Colorado / sponsored by the Geo-Institute of the American Society of Civil Engineers ; edited by D.V. Griffiths, Gordon A. Fenton, Timothy R. Martin.

p. cm. (Geotechnical special publication ; no. 101)

Includes bibliographical references and index.

ISBN 0-7844-0512-3

1. Slopes (Soil mechanics)—Congresses. 2. Soil stabilization—Congresses. 3. Engineering geology—Mathematical models—Congresses. I. Griffiths, D. V. II. Fenton, Gordon A. III. Martin, Timothy R. VI. American Society of Civil Engineers. Geo-Institute. V. Title. VI. Series.

TA710.A1 G34 2000a
624.1'51363--dc21 00-042139

Library of Congress Catalog Card No: 00-042139 ISBN 0-7844-0512-3
Manufactured in the United States of America.

Geotechnical Special Publications

1 *Terzaghi Lectures*
2 *Geotechnical Aspects of Stiff and Hard Clays*
3 *Landslide Dams: Processes, Risk, and Mitigation*
4 *Tiebacks for Bulkheads*
5 *Settlement of Shallow Foundation on Cohesionless Soils: Design and Performance*
6 *Use of In Situ Tests in Geotechnical Engineering*
7 *Timber Bulkheads*
8 *Foundations for Transmission Line Towers*
9 *Foundations & Excavations in Decomposed Rock of the Piedmont Province*
10 *Engineering Aspects of Soil Erosion, Dispersive Clays and Loess*
11 *Dynamic Response of Pile Foundations–Experiment, Analysis and Observation*
12 *Soil Improvement: A Ten Year Update*
13 *Geotechnical Practice for Solid Waste Disposal '87*
14 *Geotechnical Aspects of Karst Terrains*
15 *Measured Performance Shallow Foundations*
16 *Special Topics in Foundations*
17 *Soil Properties Evaluation from Centrifugal Models*
18 *Geosynthetics for Soil Improvement*
19 *Mine Induced Subsidence: Effects on Engineered Structures*
20 *Earthquake Engineering & Soil Dynamics II*
21 *Hydraulic Fill Structures*
22 *Foundation Engineering*
23 *Predicted and Observed Axial Behavior of Piles*
24 *Resilient Moduli of Soils: Laboratory Conditions*
25 *Design and Performance of Earth Retaining Structures*
26 *Waste Containment Systems: Construction, Regulation, and Performance*
27 *Geotechnical Engineering Congress*
28 *Detection of and Construction at the Soil/Rock Interface*
29 *Recent Advances in Instrumentation, Data Acquisition and Testing in Soil Dynamics*
30 *Grouting, Soil Improvement and Geosynthetics*
31 *Stability and Performance of Slopes and Embankments II*
32 *Embankment of Dams–James L. Sherard Contributions*
33 *Excavation and Support for the Urban Infrastructure*
34 *Piles Under Dynamic Loads*
35 *Geotechnical Practice in Dam Rehabilitation*
36 *Fly Ash for Soil Improvement*
37 *Advances in Site Characterization: Data Acquisition, Data Management and Data Interpretation*
38 *Design and Performance of Deep Foundations: Piles and Piers in Soil and Soft Rock*
39 *Unsaturated Soils*
40 *Vertical and Horizontal Deformations of Foundations and Embankments*
41 *Predicted and Measured Behavior of Five Spread Footings on Sand*
42 *Serviceability of Earth Retaining Structures*
43 *Fracture Mechanics Applied to Geotechnical Engineering*
44 *Ground Failures Under Seismic Conditions*
45 *In Situ Deep Soil Improvement*
46 *Geoenvironment 2000*
47 *Geo-Environmental Issues Facing the Americas*
48 *Soil Suction Applications in Geotechnical Engineering*
49 *Soil Improvement for Earthquake Hazard Mitigation*
50 *Foundation Upgrading and Repair for Infrastructure Improvement*
51 *Performance of Deep Foundations Under Seismic Loading*
52 *Landslides Under Static and Dynamic Conditions–Analysis, Monitoring, and Mitigation*
53 *Landfill Closures–Environmental Protection and Land Recovery*
54 *Earthquake Design and Performance of Solid Waste Landfills*
55 *Earthquake-Induced Movements and Seismic Remediation of Existing Foundations and Abutments*
56 *Static and Dynamic Properties of Gravelly Soils*
57 *Verification of Geotechnical Grouting*
58 *Uncertainty in the Geologic Environment*
59 *Engineered Contaminated Soils and Interaction of Soil Geomembranes*
60 *Analysis and Design of Retaining Structures Against Earthquakes*

61 *Measuring and Modeling Time Dependent Soil Behavior*
62 *Case Histories of Geophysics Applied to Civil Engineering and Public Policy*
63 *Design with Residual Materials: Geotechnical and Construction Considerations*
64 *Observation and Modeling in Numerical Analysis and Model Tests in Dynamic Soil-Structure Interaction Problems*
65 *Dredging and Management of Dredged Material*
66 *Grouting: Compaction, Remediation and Testing*
67 *Spatial Analysis in Soil Dynamics and Earthquake Engineering*
68 *Unsaturated Soil Engineering Practice*
69 *Ground Improvement, Ground Reinforcement, Ground Treatment: Developments 1987-1997*
70 *Seismic Analysis and Design for Soil-Pile-Structure Interactions*
71 *In Situ Remediation of the Geoenvironment*
72 *Degradation of Natural Building Stone*
73 *Innovative Design and Construction for Foundations and Substructures Subject to Freezing and Frost*
74 *Guidelines of Engineering Practice for Braced and Tied-Back Excavations*
75 *Geotechnical Earthquake Engineering and Soil Dynamics III*
76 *Geosynthetics in Foundation Reinforcement and Erosion Control Systems*
77 *Stability of Natural Slopes in the Coastal Plain*
78 *Filtration and Drainage in Geotechnical/Geoenvironmental Engineering*
79 *Recycled Materials in Geotechnical Applications*
80 *Grouts and Grouting: A Potpourri of Projects*
81 *Soil Improvement for Big Digs*
82 *Risk-Based Corrective Action and Brownfields Restorations*
83 *Design and Construction of Earth Retaining Systems*
84 *Effects of Construction on Structures*
85 *Application of Geotechnical Principles in Pavement Engineering*
86 *Big Digs Around the World*
87 *Jacked Tunnel Design and Construction*
88 *Analysis, Design, Construction, and Testing of Deep Foundations*
89 *Recent Advances in the Characterization of Transportation Geo-Materials*
90 *Geo-Engineering for Underground Facilities*
91 *Special Geotechnical Testing: Central Artery/Tunnel Project in Boston, Massachusetts*
92 *Behavioral Characteristics of Residual Soils*
93 *National Geotechnical Experimentation Sites*
94 *Performance Confirmation of Constructed Geotechnical Facilities*
95 *Soil-Cement and Other Construction Practices in Geotechnical Engineering*
96 *Numerical Methods in Geotechnical Engineering: Recent Developments*
97 *Innovations and Applications in Geotechnical Site Characterization*
98 *Pavement Subgrade, Unbound Materials, and Nondestructive Testing*
99 *Advances in Unsaturated Geotechnics*
100 *New Technologies and Design Developments in Deep Foundations*
101 *Slope Stability 2000*

Preface

Slope stability analysis is one of the oldest geotechnical engineering subjects, yet as we enter the 21st century it remains one of the most active areas of study for geotechnical practitioners and researchers. The 26 refereed papers contained within this publication entitled *Slope Stability 2000* cover many facets of the subject, ranging from case histories to advanced probabilistic studies and numerical analysis. The broad range of topics covered in this publication is due in part to a decision by the editors to merge two overlapping technical areas for which papers had been originally solicited, namely:

Slope Stability

- Laboratory vs. Field Shear Strength Assessment of Slope Stability
- Geology/Geophysics
- Numerical Methods
- Three Dimensional Analysis
- Innovative Mitigation Measures

and

Risk Assessment

- Dam Safety
- Probabilistic Slope Stability

A common objective of all the papers in this publication however, is a desire to use modern analysis and instrumentation tools to better define the factors both causing and mitigating stability problems in slopes. Essentially, an improved understanding of both the numerator *and* denominator in the Factor of Safety equation! The generic title *Slope Stability 2000*} was chosen to capture this spirit, and to draw attention to this state-of-the-art publication on slope stability in analysis, design and practice as we enter the new millennium.

The split between "practical" and "analysis" papers was fairly even, and the Editors were tempted to divide the contents into more descriptive sub-categories such as Case Histories, Limit Analysis, Probabilistic Geotechnics etc.. In the event this idea was abandoned, as it soon became apparent that this could lead to artificial and misleading distinctions between papers that did not lie entirely within one sub-heading. We have therefore resorted to the safety of simple alphabetic order according to names of the first authors. It is hoped that readers of *Slope Stability 2000* will be stimulated and inspired by this wide range of quality papers written by a distinguished group of national and international authors.

All the papers accepted for this ASCE Geotechnical Specialty Publication were peer reviewed and went through at least two iterations before reaching the publication stage. The papers are eligible for discussion in the *Journal of Geotechnical and Geoenvironmental Engineering* and for ASCE award nominations.

Finally, a publication of this quality would not be possible without the dedication and professionalism of the paper reviewers. At the risk of leaving some deserving names off our list, we wish to recognize, Tiffany Adams, Scott Anderson, Rex Baum, Brian Collins, Imran Gillani, Daniel L. Harpstead, Russell Jernigan, Ning Lu, Ray E. Martin, Gordon M. Matheson, William J. Murphy, Ugur Ozbay, Stein Sture, and Dobroslav Znidarcic.

D.V. Griffiths, P.E., Colorado School of Mines
Gordon A. Fenton, P.Eng., Dalhousie University
Timothy R. Martin, P.E., Powell-Harpstead, Inc.

4/24/00

Contents

Rapid Reactivation of a Large Composite Earth Slide - Earth Flow

Scott A. Anderson, Ph.D., P.E., [1]
T. Samuel Holder, P.E., [2]
Gene Dodd, P.E., [3]

Abstract

A dormant earth flow crossed by Forest Highway 78 in Montrose County, Colorado, reactivated as an earth slide - earth flow in June 1997. The earth slide - earth flow moved at rates of up to a few meters per day and accumulated more than 150 meters of displacement during the summer of 1997, causing considerable damage to the highway and temporary road closures. An irrigation ditch crossing the earth slide - earth flow was also damaged and required reconstruction several times during the summer. Movement slowed rapidly and has essentially stopped. This paper presents the investigation into the cause of the reactivation, the rapid cessation of movement, and the probability of a similar period of activity occurring in the future. The investigation includes study of subsurface conditions, sliding history in the area, climate, and manmade impacts to the slide, such as the road and irrigation ditch.

Introduction

The Cimarron Valley is south of U.S. 50 and east of U.S. 550, near the town of Montrose, in southwest Colorado. There are numerous historic and prehistoric landslides in the valley. A pre-existing earth flow, part of a landslide complex referred to herein as the Prehistoric Wells Basin Landslide (Figure 1) had been exhibiting slow creep for years. The earth flow has had an unlined irrigation ditch crossing its upper half since the early 1900's and, prior to 1996, maintenance efforts required of the irrigation ditch company, the electric utility that has a

[1] Landslide Technology Leader, URS Greiner Woodward Clyde,
4582 S. Ulster St., Denver, CO 80237
[2] Project Manager, Federal Highway Administration - CFLHD,
Lakewood, CO 80228
[3] Geotechnical Engineer Federal Highway Administration - CFLHD,
Lakewood, CO 80228

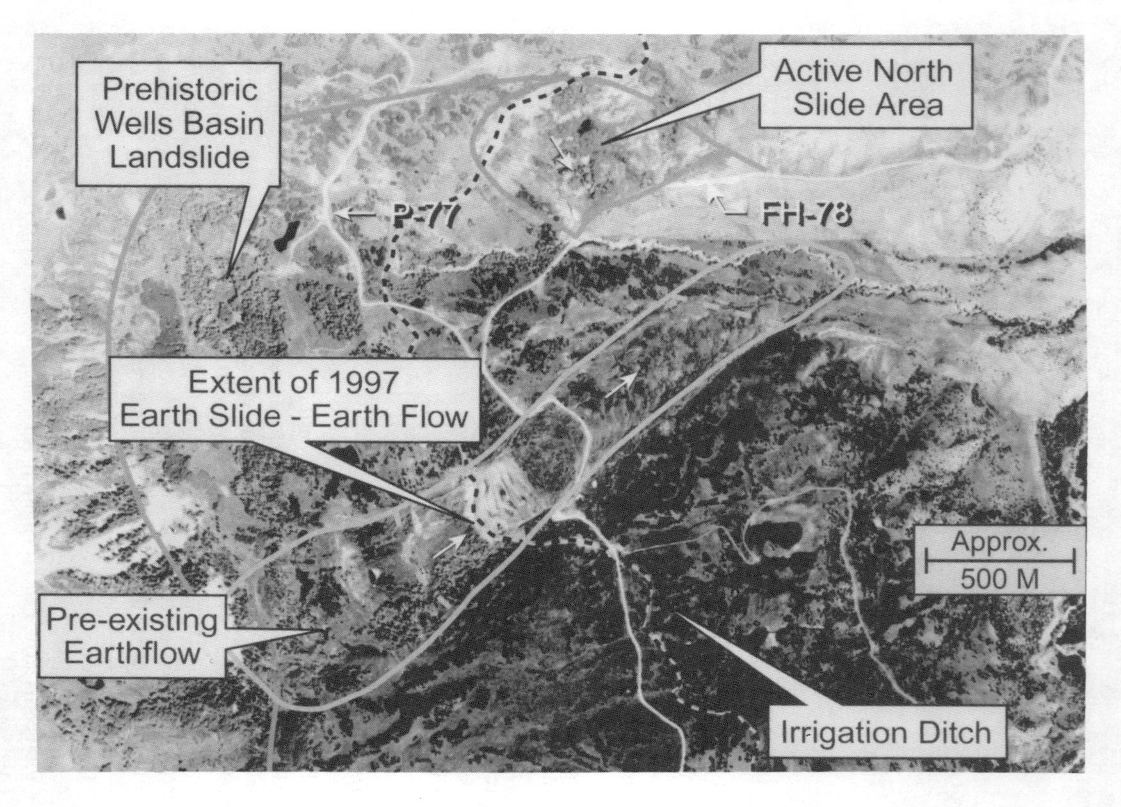

Figure 1. Wells Basin Landslide in 1998. Arrows show direction of movement.

power line crossing the slide, and road maintenance crews were minor. By 1996, however, the ditch width has reached about 50m, probably through gradual movement and erosion. In 1996 the ditch company reconstructed the ditch to its narrower section (Autrey, 1998) and the Forest Service filled a depression that had developed in Forest Highway 78 where it crosses the earth flow below the ditch (Burke, 1998). No other actions were taken in 1996 and no unusual movement was observed.

The spring of 1997 was wetter than average and in June 1997 part of the earth flow started to move rapidly, reaching rates of up to a few meters per day. More than 150 m of horizontal movement accumulated during the summer and the rapid movement stopped by November. This paper discusses the conditions leading to the 1997 reactivation of the earth flow and the rather spectacular amount of movement that occurred. Using observations of movement, site reconnaissance, and results of investigation on the slide mass, hypotheses are presented to explain the rapid remobilization.

Research for this work led to uncovery of descriptive reports of similar landsliding in the valley more than 100 years ago (Cross, 1886). Because Cross's descriptions are very vivid, they are very valuable in the understanding of this type of landsliding. For this reason and because Cross's publication is somewhat obscure, we have included several quotes here.

Site Description

The Cimarron River flows northward from the Uncompahgre Plateau to the Gunnison River, about 30 km east of the town of Montrose, in southwest Colorado. Colorado Forest Highway 78 provides access from U.S. 50, through the Cimarron River valley to the Uncompahgre National Forest. The landslide site is located in the Cimarron River valley, about 9.5 km south of U.S. 50, in an area known as Wells Basin. The site is crossed by Forest Highway 78, as shown on Figure 1.

The Cimarron River valley elevation is between about 2,100 m and 2,800 m above sea level. The lower valley slopes are vegetated with grass and sagebrush, and limited other woody vegetation. The upper slopes contain aspen and conifer trees. The lower valley slopes are gently sloped and undulating, and the upper slopes are notably steeper, in many places consisting of bedrock outcrop.

The geology of the area has been mapped by Yeend (1961). The primary surficial deposit on the valley slopes is Wisconsin till. In many places the till has been moved by landsliding, and Yeend has mapped distinct areas of recent slides, recent mudflows, and Pleistocene landslides and slumps. So pervasive is the landsliding that Colton et al. (1975) mapped essentially the entire valley as landslide deposits. The lower valley slopes are underlain by Mancos shale, which does outcrop in a number of locations. The upper slopes are capped by sedimentary and volcanic rocks that form the mesas and ridges east and west of the valley. These rocks include the Alboroto rhyolite, San Juan tuff, Telluride conglomerate, and Mesa Verde sandstone (which structurally overlies the Mancos shale).

Historic Landsliding

Cimarron Earth Flow

The historic record of landsliding in the Cimarron River valley dates back more than 100 years to an earth flow that occurred in 1886. The disturbance was first thought to have been caused by an earthquake and an earthquake was described in newspaper articles in Montrose and in Denver (Cross, 1886). The earth flow occurred between July 18 and July 25, the period between subsequent visits to the site by ranchers. Cross, who was from the U.S. Geological Survey in Denver, visited the site in August 1886 and described the discovery of the site as follows:

> "On the 25th of July Mr. Samuel Scheldt was engaged in looking after his cattle that ranged over these slopes, and found that a considerable area of hill-side had been terribly convulsed and very much changed in appearance. The trees, here growing quite luxuriantly, had been overturned, the earth fissured, and large masses of soil with the trees growing upon them had been broken off from the steeper slopes. In one jungle of fallen timber a bunch of cattle were imprisoned, unable to make their way out. He at once procured assistance, and the terrified cattle when liberated ran for miles before stopping. On examining the region, Mr. Scheldt and his companions found the detailed features over an area of several hundred acres very much changed. A small lake they had previously known could not be found, and apparently on its site was a ridge. Near by another sharp ridge had been thrown up some 25 feet, and one side had again sunk down, leaving a vertical wall. Everywhere the trees had been disturbed, and in some small areas none were left standing. The ground was traversed by innumerable fissures which were all shallow, owing to the soft, crumbling nature of the soil. The news of the phenomenon spread rapidly and the spot was visited by many. The explanation uniformly adopted was that the convulsions were caused by an earthquake."

Following his site reconnaissance, Cross described the nature of the movement as follows:

> "The movement was a downward sliding of the whole surface, unequal in different places, apparently greatest in the upper part, and dying out gradually as distance from the upper line increased. The upper limit of movement runs along the steep mesa slope at a present elevation of 50-125 feet above the basin floor. A steep surface of freshly exposed earth and shaley rock marks the line. Above are undisturbed trees, turf or debris. At

the foot of this surface is a tangle of overturned trees and bushes, half buried in loose soil and rocks. Upon the slide surface lies a few uprooted trees, or a small patch of earth which has caught, half way down. Along the upper edge are partly detached sections, with their trees inclined at various angles.

The shoulder mentioned as projecting from the mesa out into the upper part of the basin has suffered on all its steep sides, as did the above slope; and the entire mass is divided into sections by fissures, so that it seems strange that all did not slip, piece by piece, to the basin below.

Along the northern side, near the base of the bounding ridge, runs a more or less continuous line, which is nothing less than an anticlinal fold or plication of the surface soil or turf, caused by the lateral pressure of the downward moving mass. On the outer or northern side of this fold the bushes and trees, where such exist, are simply tipped from the vertical position, corresponding to the sharpness of the plication. They are not uprooted, and in many places this side of the little ridge is unbroken, while on the basin side the downward movement has torn away nearly all of that half. A mile below the head of the slide this lateral movement is manifested very plainly by the cracking of a grassy surface, the turf from the basin side being simply shoved sideways a foot or more over the undisturbed part on the ridge side."

Cross drew several conclusions as to the cause of sliding and its significance in the formation of the valley. These conclusions include the following:

1. There was no earthquake. The movement is attributed to landsliding as a result of heavy rain. Local ranchers said that although little rain fell where they were it appeared to be raining all week up in the valley.
2. This type of landsliding has occurred elsewhere in the valley and apparently at this very site in the past. Observations from vantage points within the valley suggest such landsliding is an important valley forming process.
3. Sag ponds provide some of the water necessary for such fluid movement of the earth, but other contributing sources would be required.
4. The level to which sections of the slide are disturbed varies considerably.
5. A small slip may have started the movement and been followed by progressive failure up the slope, described by Cross as follows:

   "A small slip may have started the movement and, by removing the resistance which held another mass in place, have paved the way to a successive slipping of section after section until the higher bounding grounds were reached.

> Such a theory would allow a slight movement in the lower portions to lead to much greater displacements on the upper limit, and such seems to be in fact the case. That the slipping did occur in sections is shown by the appearance or ridges here and there in the midst of the area, which were plainly formed as was the one on the northern limit, described above."

Aside from the general reference to landslide features such as hummocky ground, closed drainage basins, sag ponds, and dead trees covering a large part of the valley, Cross made no reference of movement at the site of the pre-existing Wells Basin earth flow. Based on observations of Yeend (1961), the initial earth flow episode at Wells Basin apparently predates the 1886 Cimarron earth flow visited by Cross. Given the thoroughness of Cross's observations the fact that he would have practically had to traverse the Wells Basin site to get to the Cimarron earth flow, and the lack of reference to the Wells Basin site, it seems probable that the pre-existing earth flow at Wells Basin was not active during 1886. The close proximity of the Wells Basin earth flow and the Cimarron earth flow (same site as the 1997 earth slide-earth flow) is shown on Figure 2.

Wells Basin Landslide

Wells Basin developed its basin-like morphology as the result of prehistoric landsliding, which is shown on Figures 1 and 2. As identified on Figure 1, the most recent landslide activity is along the lateral margins of the prehistoric landslide. The North Slide Area has experienced some movement in recent years, requiring maintenance to the irrigation ditch that crosses near the headscarp. There has been no recent impact to the Forest Highway (FH-78) that passes along the toe of the slide area.

The Pre-existing Earth Flow identified on Figure 1 along the southern margin of the prehistoric landslide is apparently not active over its full length. However, the distinct earth flow morphology of this feature suggests it is more recent than other inactive parts of the prehistoric landslide. The central 1200 m portion of the older earth flow is the area that reactivated in 1997; the reactivation included the entire width but not the entire length of the older earth flow.

1997 Reactivated Earth Slide - Earth Flow

The characteristics of the movement and the landslide morphology lead to the classification of the landslide as an earth slide - earth flow (Cruden and Varnes, 1996). Herein, the generic term, landslide, is used. The reactivated landslide is approximately 300 - 400 m wide and 1200 m long and the overall slope angle is about 6 - 7 degrees.

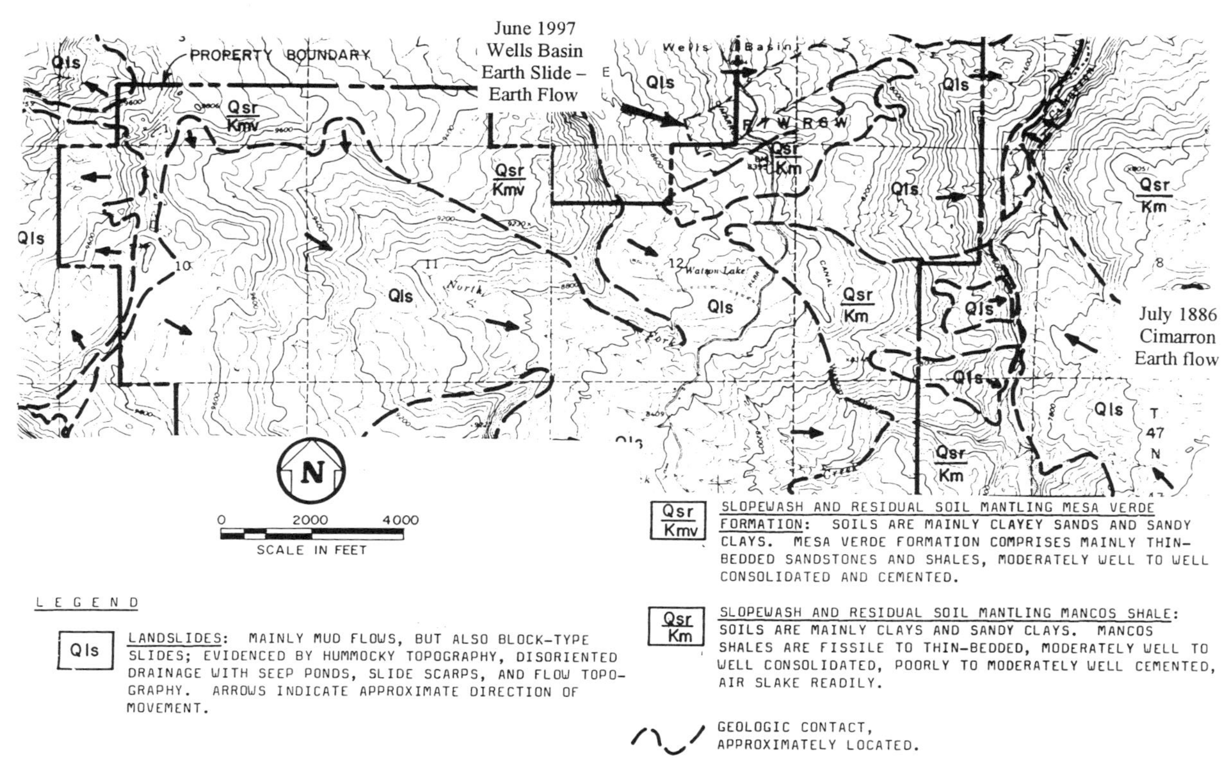

Figure 2. Geologic Map of part of the Cimarron River Valley showing the Cimarron earth flow and Wells Basin earth slide - earth flow (URSGWC, 1973).

Conditions Leading to Reactivation

The landslide had been experiencing slow movement for at least 18 to 30 years based on the records of the Delta Montrose Electric Association - the owner of a power line across the site (Blowers, 1998), and Montrose County Maintenance (Jeffries, 1998). The landslide is crossed by an unlined irrigation ditch and according to the ditch owner, the ditch has been there since the early 1900's and has not required significant maintenance in this area; the North Slide area has been more problematic (Autrey, 1998). Nevertheless, by 1996, the ditch width had increased from probably about 5 m to approximately 50 meters where it crossed the pre-existing earth flow.

A "sink hole" in the road was repaired with fill in the summer of 1996 (Burke, 1998) and the ditch was moved about 50 meters in the upslope direction at approximately the same time (Autrey, 1998). The specific amount or nature of earthwork conducted in 1996 is not known but, given the large size of the reactivated landslide, it seems unlikely that the earthwork itself would have significantly impacted its global stability.

The rapid movement started on June 16, 1997 and it was apparently preceded by somewhat wetter than average weather. The 1997 precipitation totals were exceeded only one to four times in the previous 45 years. Because some precipitation probably fell as snow at the site, mean temperature was evaluated as a potential indicator of unusual snowmelt; the temperature for April was about normal, May was warmer than any previous year, and June was above average. The warm temperatures during the snowmelt season suggest snowmelt may have been more rapid than average.

Flood frequency analysis performed on the stream gage data from a gage farther up the valley indicate the peak flow, which occurred on June 5, has a return interval of 25 years. This peak apparently occurred 5 days after one period of heavy rain and before a second period of heavy rain (June 8-10), as indicated by newspaper reports on precipitation in Montrose.

In summary, none of the precipitation or stream gage data is for the site itself. The data and reports from Montrose indicate regional precipitation was greater than average. The lack of precise agreement of the peak flows to the heavy rain indicates there is some uncertainty with respect to how much the precipitation and infiltration amounts were above normal at the site itself.

Reactivation

The reactivated landslide at Wells Basin has an appearance very similar to the Cimarron earth flow described by Cross (1886), with open cracks, grabens, and horsts, in some areas, pressure ridges in some areas, and considerably less evidence of movement elsewhere. Sag ponds near the head of the pre-existing earth flow (see Figure 1) appear well established in 1973 aerial photos, and were not impacted by the 1997 reactivation. The landslide moved at a peak rate of a few meters per day and accumulated more than 150 m of displacement during the summer.

The active movement of the reactivated landslide caused considerable maintenance difficulties for Montrose County and the ditch company. The County provided maintenance for Forest Highway 78 and, during the period of the peak rate of movement, this required stationing equipment and an operator continuously at the site. Since the traffic volume is low, the road would undergo considerable distress between vehicles, and grading was required for nearly every vehicle. The ditch company had to reconstruct the irrigation ditch five times during the summer because of downslope movement of the landslide. Figure 3 shows the road crossing the landslide and the abandoned ditch segments. Observations from these maintenance efforts provide the only record of movement rates during the summer. The greatest displacement occurred near the irrigation ditch and Forest Highway.

Cessation of Rapid Movement

By November 1997 landslide movement had essentially stopped and it was judged safe to drill through the slide. An inclinometer installed in November showed movement was continuing. There was 25 mm of movement at 22.5 m depth between November and February, and the inclinometer casing was sheared by May 1998. The inclinometer showed sliding (after November 1997) was occurring within a discrete shear zone at the top of the Mancos shale. The investigation and analysis continued over the winter because it was uncertain what would happen the following spring. As it turned out, the spring of 1998 was relatively dry and movement, as observed at the surface, was not notable.

In contrast to 1998, the spring of 1999 was wet, and in many aspects similar to the conditions of 1997, when the reactivation occurred. Records from the Cimarron weather station are compared in Table 1. Despite the apparently similar precipitation histories near the site, no movement was noted in 1999.

Table 1
Precipitation Records from the Cimarron Weather Station

Month	1997	% Normal	1999	% Normal
March	22 mm	90	2 mm	7
April	58 mm	251	59 mm	257
May	63 mm	263	51 mm	213
June	46 mm	244	49 mm	256
July	17 mm	46	34 mm	95
Total for Period	206 mm	163	195 mm	154

Figure 3. Portion of reactivated earth slide - earth flow showing multiple abandoned canal sections, cracking, and sag ponds. Note 18-wheel truck on road for scale. Movement is from top right to bottom left of photograph.

Investigation and Analysis

As a result of the 1997 reactivation, Montrose County had a road that was passable, but not to design specifications, and surrounded by unstable ground with fresh open cracks and, at times, ponded water. The road needed to be restored because it provided important access to the National Forest, permanent residences, and Silver Jack dam, a 42 m high U.S. Bureau of Reclamation dam. Montrose County looked for assistance in evaluating the preferred alternative of either rebuilding across the landslide on the original alignment, or reconstructing the road on a new alignment to avoid the landslide and, because the road provided National Forest access, the county was eligible to receive federal funding assistance through the Federal Highway Administration - Central Federal Lands Highway Division (FHWA). The FHWA contracted URS Greiner Woodward

Clyde (URSGWC) and URSGWC provided assistance through its Landslide Technology Group.

Investigation

A drilling investigation of the landslide was conducted by the FHWA in the fall and winter of 1997, once movement had slowed. Two holes were advanced through the landslide and one hole was advanced just below its toe. The results of the investigation show moist to wet, brown to olive- gray clay and silt to a depth of about 22.5 meters. Based on Standard Penetration Test N-values, the consistency of this material is stiff to hard. At about 22.5 m depth a soft, wet clay layer was encountered. No N-values were measured, but a 75 mm diameter Shelby tube was pushed 550 mm until claystone bedrock was encountered at 23.1 m depth. The soft clay zone was interpreted to be the basal shear zone; bedrock was hard and dry below this zone.

In both borings on the landslide, water was first encountered at the basal shear zone and, once the shear zone was penetrated, the water rose 4 to 4.5 meters in the drill holes. The rise of water in the holes indicates water is confined at the shear zone and suggests pore-water pressure cannot be simply approximated as hydrostatic.

Samples from the basal shear zone indicate a liquid limit of about 60, a plasticity index of about 40, and about 50% silt and 50% clay-size particles. Unconsolidated-undrained triaxial test results indicate an undrained strength of about 130 kPa. No testing of residual friction angle (ϕ'_r) was available, but index tests suggest a value of 10 to 15 degrees.

Analysis

Limit equilibrium slope stability analyses were conducted using Spencer's method and the UTEXAS3 (Wright, 1991) computer software program. Limit equilibrium analyses assume that strength is mobilized equally on the entire shear surface and that the failure mass behaves as a rigid block. This assumption is not true for the reactivated landslide because surface observations indicate about 150 meters of movement near the road and irrigation ditch and essentially no movement near the head and toe. Nevertheless, limit equilibrium analysis is probably the best practical tool for analyzing the conditions required to initiate movement, the current stability of the landslide, and the merits of potential mitigation measures.

The shear strength mobilized on the shear surface at failure was backcalculated using the geometry shown on Figure 4 and a piezometric level extrapolated from that observed in November 1997, in FH 78B-1. The results for the active zone indicate a friction angle, ϕ'_r, of 6.0 degrees corresponds to a factor of safety = 1.0 (Figure 4). The index properties and material descriptions provided do not

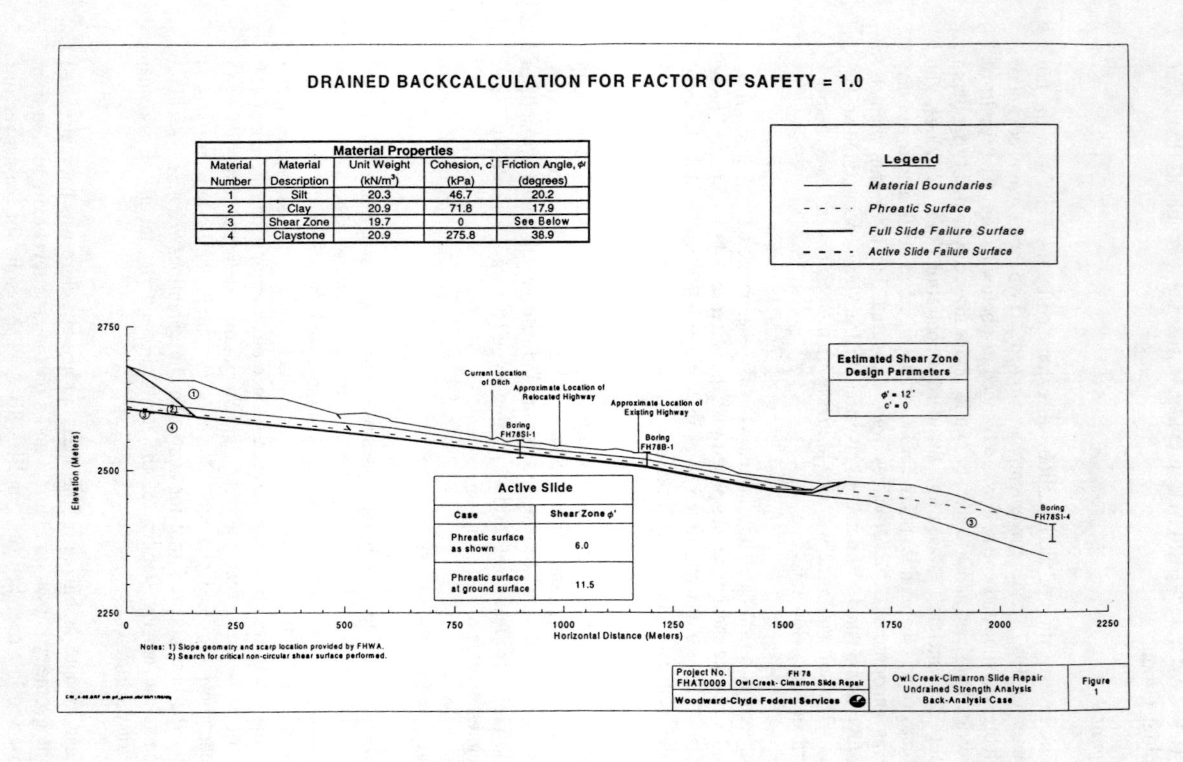

Figure 4. Maximum section of earth slide - earth flow.

support a ϕ'_r as low as 6.0 degrees, which implies the modeled pore-water pressure is too low. Considering that when movement occurred in the spring of 1997 the water pressure in the slope was probably much higher than in November, analyses were performed assuming the piezometric line to be effectively at the ground surface. This assumption of pore-water is believed to be realistic; however, the reference to the ground surface is somewhat arbitrary because the pore-water is confined below the shear zone and, therefore, not limited by the ground surface elevation.

The backcalculated ϕ'_r for this assumption of pore-water pressure is 11.5 degrees, which corresponds well with the plasticity and grain size measured from samples of the slide mass and basal shear zone. Thus, it is likely high pore-water pressure at the basal shear zone contributed to the initial movement of the reactivated landslide.

Drained strength parameters (ϕ'_r and c'=o) are appropriate for initiation of failure caused by changing pore-water pressure within the slope. Once the soil begins to strain, however, strain-induced pore-water pressure change occurs and shearing of the soft clay encountered in the basal shear zone elevates pore-water pressure. The low permeability of the high plasticity clay soil prohibits rapid dissipation of the pore-water pressure, so the soil becomes undrained, and undrained strength parameters are the most practical way of characterizing the shear strength until drainage occurs. Duncan (1996) has suggested the time for drainage to occur can be estimated using the principles of the time rate of consolidation. Using this approach, a coefficient of consolidation of 0.03 mm^2/s for soil with a liquid limit of 60 (Lambe and Whitman, 1969), and assuming a 10 m drainage path length, drainage of strain-induced pore-water pressure will take approximately 1 year.

The estimated time for drainage is very sensitive to the drainage path length, which is uncertain and variable along the shear surface. Nevertheless, this analysis shows that it is improbable that during the rapid movement in 1997 drainage was occurring in the shear zones. Back-analysis shows that an undrained shear strength of 36 kPa (ϕ=0) on the basal shear zone corresponds to a factor of safety of 1.0 for the reactivated landslide. This is a reasonable value for soft clay, as encountered, and similar to the undrained shear strength of granular soils involved in flow slides following liquefaction. Drainage occurs rapidly in granular flow slides, the soil strengthens and the slide comes to rest. The same process can be expected of this reactivated landslide, only the time for drainage and cessation of movement is much longer.

The investigation and analysis lead to the following understanding of the observed reactivation:

1. There is a zone of soft clay at the base of the landslide. The zone is approximately 0.5 m thick where sampled.
2. Groundwater is confined beneath this basal zone.
3. The drained strength was exceeded as a result of increased water pressure and/or the previous year's construction on the road and ditch. Shear strain occurred within the basal shear zone and, possibly, other shear zones.
4. The shear strain was significant enough that in soft clay zones excess pore-water pressure was developed.
5. The drainage rate was constrained by the low permeability of the clay and the length of drainage path.
6. Movement slowed as partial drainage occurred.

Risk Assessment

The stability analyses and the landslide history at the site were used to evaluate the probability of future reactivation and the risk of impact to the road and public safety. Probability was judged based on our understanding of the history of landsliding at the site, which includes the following:

1. The pre-existing earth flow at Wells Basin predates the Cimarron earth flow and is more than 115 years old.
2. Natural causes, probably heavy precipitation, caused the 1886 Cimarron earth flow.
3. The Wells Basin earth slide - earth flow near the road and irrigation ditch crept for many years, even decades before there was a reactivation of rapid movement.
4. The 1997 reactivation occurred during the wet season following localized earthwork on the site.
5. A source of surface water is present in the form of closed drainages, sag ponds and, at Wells Basin, an unlined irrigation ditch.
6. The surface morphology at both sites suggest that earth flow or relatively rapid movement had occurred prior to the studied event. Thus, movement is episodic.
7. The Cimarron earth flow did not reactivate when the Wells Basin earth slide - earth flow did, and visa versa.
8. There are some parts of the valley that are apparently more stable than others (areas not labeled Q1s on Figure 2).

Given the investigation, analyses, and understanding of landslide history at the site described here, a risk assessment was performed to evaluate alternatives for addressing the site. The risk assessment was conducted using a logic tree methodology and considering the following initiating events: no action, rebuild across the slide with minimal grading and drainage improvements, rebuild across the slide with mitigation to achieve a factor of safety ≥1.25 (despite impractical cost), and relocating on more stable ground within the valley. For each initiating event three modes of potential failure were considered and the probability of the following three events estimated: sliding, impact to Forest Highway 78, and safety hazard, for a 50-year design life. Probability estimates were made independently by a panel of experts, and averaged.

The results show that the risk of impact or safety hazard to Forest Highway 78 is less if it is reconstructed on an alternate alignment. However, because of the generally marginally stable nature of the valley slopes, the risk assessment shows the risk reduction by realigning the road is modest.

Conclusions

The Cimarron valley of southwestern Colorado has gentle valley slopes comprised of till overlying Mancos shale. Landslides and earth flows have been a significant valley forming process and the valley has a long history of landslide study, including Cross (1886), Yeend (1961), URSGWC (1973), and this study from 1997 to 1999. Earth flows have moved episodically, apparently being active for periods of days to months. During active periods, rates of movement can exceed 1.5 m/day, based on the Wells Basin earth slide - earth flow, and probably several times faster than that, based on descriptions of the Cimarron earth flow. During periods of quiescence, creep occurs and periodic maintenance to facilities, such as irrigation ditches, roads, and power lines is required.

An undisputed trigger of the reactivation has not been identified. The presence of the unlined irrigation ditch across the landslide would at first appear to be very significant, but when the long history of its presence on the site and the number of natural sag ponds and depressions is factored in, we come to the same conclusion as Cross (1886), that the presence of surface water cannot fully explain a water content of the failing materials high enough to cause them to flow. It appears that exceptionally high precipitation or a minor perturbation to the slope in conjunction with precipitation well above average is required to reactivate the earth flows.

The 1997 reactivation could have occurred in the manner described by Cross (1886). A small slide could have initiated under drained conditions in the spring of 1997 near the recently relocated irrigation ditch and repaired road. This slide would have unloaded the slope above and loaded the slope below, causing the factor of safety against sliding to drop below one, and slide movement to propagate upslope and/or downslope. The small slide movement could have caused undrained loading of soft clay shear zones, causing excess pore-water pressure development and reducing strength to the undrained residual valve. Such a hypothesis of mobilization of undrained failure from changes in drained loading conditions caused by rainfall-induced pore-water pressure changes is supported by the work of Anderson and Sitar (1995).

The 1997 reactivation could also be the episodic result of years of creep and an unusually wet spring season. The soft clay in shear zones is the result of drained creep, dilation, and the introduction of more water. If confined groundwater conditions develop, as on the basal shear zone, the effective confining pressure will become less than geostatic and a void ratio higher than the critical void ratio for geostatic stresses will develop. By this method, soil very susceptible to undrained failure could develop through drained creep. This process could take several years. The unusually wet conditions of 1997 could have caused the drained strength to be exceeded and caused undrained loading of the soft clay. The lack of time for soft clay to develop between 1997 and 1999 could explain why, despite a very wet spring, there was no reactivation in 1999.

At this time, the exact cause of the reactivation remains subject to speculation. Montrose County has used the risk assessment described herein and other considerations to make the decision to reconstruct the road on its existing alignment. The realignment alternative offer less risk to the road and public but not sufficiently less to justify the costs, including right-of-way acquisition, and environmental impacts of the proposed work. Currently, plans are being finalized for reconstructing on the existing alignment and the new road and associated work will provide continued opportunity to observe the impact of construction on the site, and the nature and cures of future movement.

Acknowledgements

We would like to acknowledge the assistance of Dr. Robert Schuster in researching the history of landsliding in the valley and assisting with the risk assessment. Montrose County and U.S. Forest Service personnel also provided valuable assistance.

References

Anderson, S.A. and Sitar, N., 1995. Analysis of Rainfall Induced Debris Flows, Journal of Geotechnical Engineering, Vol. 121, No. 7, ASCE, p. 544-552.

Autrey, G., 1998. Personal communication with Gordon Autrey, Bostwick Park Water Conservancy, May 1998.

Blowers, D., 1998. Personal communication with Don Blowers of Delta Montrose Electric Association, May 1998.

Burke, M., 1998. Personal communication with Michael Burke of the U.S. Forest Service, San Juan National Forest.

Colton, R.B., Patterson, P.E., Holligan, J.A., and Anderson, L.W., 1975. Preliminary Map of Landslide Deposits, Montrose 1° x 2° Quadrangle, Colorado, Miscellaneous Field Studies Map MF-702.

Cross, W., 1886. The Cimarron Land-Slide, July 1886, Proceedings of the Colorado Scientific Society, Vol. II, Part II, Denver, Colorado, p. 116-p.126.

Cruden, D.M. and Varnes, D.J., 1996. Landslide Types and Processes, in Transportation Research Board Special Report 247: Landslides - Investigation and Mitigation, A.K. Turner and R.L. Schuster, editors, p. 36-75.

Duncan, J.M., 1996. Soil Slope Stability Analysis, in Transportation Research Board Special Report 247: Landslides - Investigation and Mitigation, A.K. Turner and R.L. Schuster, editors, p. 337 - 371.

Jeffries, J., 1998. Personal communication with Jeff Jeffries of Montrose County, Maintenance Department.

Lambe, T.W. and Whitman, R.E., 1969. Soil Mechanics, John Wiley & Sons, New York, p. 553.

URS Greiner Woodward Clyde (formerly Woodward-Clevenger & Associates), 1973. Engineering Geologic Investigation, Storm King Project, Montrose County, Colorado. Project No. 17434.

Wright, S.G., 1991. UTEXAS3 - A Computer Program for Slope Stability Calculations, Shinoak Software, Austin, Texas.

Yeend, W.E., 1961. Geology of the Cimarron Ridge – Cimarron Creek Area, San Juan Mountains, Colorado. Masters degree thesis, Department of Geology, University of Colorado.

Importance of Three-Dimensional Slope Stability Analyses in Practice

David Arellano[1], Student Member and Timothy D. Stark[2], Associate Member

Abstract

This paper focuses on the importance of three-dimensional (3D) slope stability analyses in practice. Commercially available 3D slope stability software does not consider the shear resistance along the two sides of the slide mass that parallel the direction of movement in calculating the 3D factor of safety (Stark and Eid 1998). Consequently, the 3D factor of safety may be underestimated and the back-calculated shear strengths may be overestimated or unconservative. A method for incorporating the shear resistance along the two sides of a slide mass in existing 3D software is presented. A parametric study is used to investigate the importance of 3D end effects by providing a comparison of two-dimensional (2D) and 3D analyses for various slide mass geometries and shear strengths along the failure surface. A field case history is used to illustrate the use of the parametric study results and the importance of conducting a 3D analysis in practice.

Introduction

Two-dimensional (2D) limit equilibrium methods are based on a plane-strain condition. It is assumed that the failure surface is infinitely wide such that three-dimensional (3D) effects are negligible compared to the overall driving and resisting forces. A 2D analysis yields a conservative estimate of the factor of safety because the shear resistance along the two sides of the slide mass or end effects are not included in the 2D estimate of the factor of safety. In general, a 2D analysis is

----

[1]Graduate Research Assistant, Dept. of Civ. and Env. Eng., Univ. of Illinois, 205 N. Mathews Ave., Urbana, IL 61801-2352.

[2]Prof. of Civ. and Env. Eng., Univ. of Illinois, 205 N. Mathews Ave., Urbana, IL 61801-2352.

appropriate for slope design because it yields a conservative estimate of the factor of safety (Duncan 1992). A 3D analysis is recommended for back-analysis of slope failures so the back-calculated shear strength reflects the 3D end effects (Stark and Eid 1998). The back-calculated shear strength then can be used in remedial measures for failed slopes or slope design at sites with similar conditions. If the 3D end effects are not included, the back-calculated shear strengths may be too high or unconservative. A 3D analysis may also be useful to analyze slopes with a complicated topography, large differences in shear strength between the foundation materials and/or overlying materials, and/or a complex pore-water pressure condition because a 3D analysis can incorporate the spatial variation of each of these effects in the calculation of the 3D factor of safety.

Commercially available 3D slope stability software does not consider the shear resistance along the two sides of a slide mass that parallel the direction of movement in calculating the 3D factor of safety (Stark and Eid 1998). Some of the software can incorporate the shear resistance along inclined sides of a slide mass, such as the scarp, but not along vertical sides such as the flanks or parallel sides of the slide mass. As a result a method for incorporating the shear resistance along the vertical sides of a slide mass is presented herein. A parametric study investigates the importance of incorporating 3D effects by providing a comparison of 2D and 3D analyses for various geometries and shear strength conditions. The objective of performing this parametric study was to present the results of 2D and 3D slope stability analyses in a manner that can be used by engineers to determine if a 3D slope stability analysis should be conducted for a particular situation. A field case history is used to illustrate the use of the parametric study results and the importance of conducting a 3D analysis in practice.

Parametric Slope Model

A translational failure mode was selected for use in the parametric study for the following reasons (Stark and Eid 1998): (1) Slopes failing in translational mode usually involve either a significantly higher or lower mobilized shear strength along the back scarp and sides of the slide mass than that along the base, e.g., the upstream slope failure in Waco dam (Beene 1967; Wright and Duncan 1972) and the slope failure in Kettleman Hills hazardous waste repository (Seed et al 1990; Byrne et al. 1992; and Stark and Poeppel 1994), respectively. These situations can result in a significant difference between the 2D and 3D factors of safety. This difference is less pronounced in slopes failing in rotational mode because they usually involve homogenous materials. (2) A translational failure can occur in relatively flat slopes because of the weak nature of the underlying material(s). The flatter the slope, the greater the difference between 2D and 3D factors of safety (Chen and Chameau 1983; Leshchinsky et al. 1985). (3) A translational failure often involves a long and nearly horizontal sliding plane through a weak soil layer [e.g., Maymont slide (Krahn et al. 1979), Gardiner Dam movement (Jasper and

Peters 1979), and Portuguese Bend slide (Ehlig 1992)] or geosynthetic interface, e.g., Kettleman Hills repository. The presence of a well-defined weak layer or interface provides some certainty in the shear strength input data. (4) A translational failure often involves a drained shearing condition. This facilitates estimation of the mobilized shear strength of the materials involved because shear-induced pore-water pressures do not have to be estimated.

Figs. 1 and 2 show the 3D geometry for the slope model used in the parametric study. Three slope inclinations, 1H:1V, 3H:1V, 5H:1V, were investigated. For each slope inclination, heights of 10 and 100 m were analyzed. For the slope height (H) of 10 and 100 m, width (W) to height ratios (W/H) of 1, 1.5, 2, 4, 6, 8, and 10 were analyzed. Thus, for a slope height of 10 m, slope widths of 10, 15, 20, 40, 60, 80, and 100 m were analyzed and for an H of 100 m, widths of 100, 150, 200, 400, 600, 800, and 1,000 m were analyzed. The slope length, L, is dependent on the slope inclination and slope height. The calculated slope lengths resulted in length to height ratios (L/H) of 1.03, 3.30, and 5.88 for slope inclinations of 1H:1V, 3H:1V, 5H:1V, respectively, for both slope heights of 10 and 100 m.

The sides parallel to the direction of movement, not the scarp, of the slide mass in the slope model were assumed to be vertical because the effective normal stress that acts on a vertical surface is only related to the lateral earth pressure and a vertical surface yields the minimum shear surface area. Therefore, in translational failures, vertical sides provide the minimum amount of 3D-shear resistance or end effect.

It can be seen that the slope model in Fig. 2(a) is essentially a rectangle. Of course, actual slide masses are more rounded at the head of the slide mass as well as having other rounded or curved areas. A rounded slide mass was not used because of the difficulties in varying slope length, width, and height with curved ends of the slide mass and obtaining consistent ratios of W/H and L/H. In summary, a rectangular slide mass was used to facilitate the parametric study and provide an insight to the importance of 3D stability analyses in practice.

Two materials, upper and lower, were incorporated in the parametric study as indicated in Fig. 2(b). The lower material was assumed to slope at 3 percent in the direction of sliding to simulate a natural bedding plane or landfill liner system. The saturated unit weights of the upper and lower materials were assumed to be 17 and 18 kN/m^3, respectively. Linear shear strength envelopes passing through the origin with friction angles of 30° and 8° were initially assumed for the upper and lower materials, respectively. The ratio of the friction angle for the upper and lower materials is varied subsequently to investigate the importance of the shear strength difference on the 3D factor of safety.

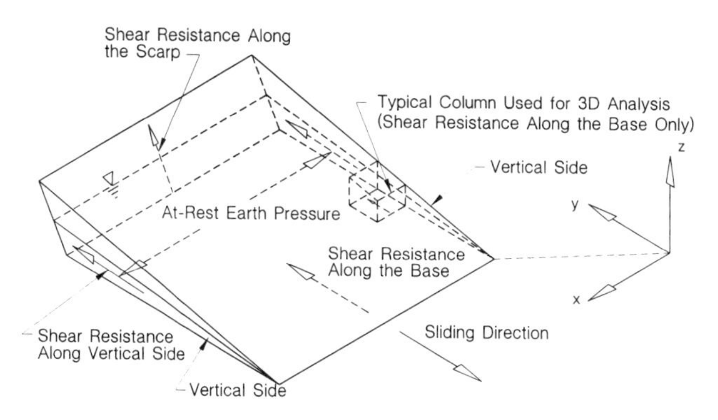

Figure 1. 3D View of Slope Model

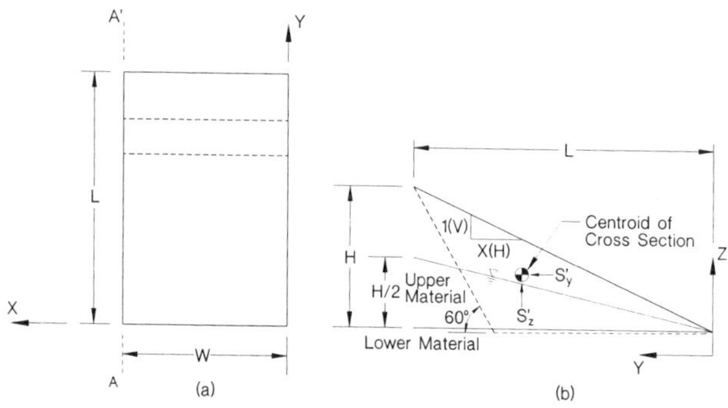

Figure 2. Plan View and Representative Cross Section for Slope Model: (a) Plan View; (b) Cross Section A-A'

In the parametric model, the groundwater level was placed at a height of H/2 as measured at a distance of L from the toe and linearly decreasing to a height of zero at the toe. The scarp was assumed to be inclined at $45° + \phi_{up}/2$ from the horizontal to simulate an active earth pressure condition where ϕ_{up} is the friction angle of the upper material. This inclination results in a minimum lateral earth pressure condition and a minimum shear resistance along the back scarp during failure. Based on a $\phi_{up} = 30°$, the back scarp of the slope model was initially assumed to be inclined at 60° from the horizontal. The bottom of the failure surface extends 0.2 m into the lower material and parallels the upper surface of the lower material until it daylights at the slope toe. The 0.2 m depth was randomly selected to ensure that the slope stability software modeled a translational failure mode with the base of the failure surface within the weak lower layer.

3D Slope Stability Software

In a study to investigate the performance of commercially available 3D slope stability software, Stark and Eid (1998) conclude that CLARA 2.31 (Hungr 1988) facilitated the input process for the slope geometry and pore-water pressure conditions, utilizes Janbu's (1954) simplified method for 2D and 3D analysis which is suitable for a translational failure mode, and can accommodate externally applied loads which can be used to simulate the resistance acting on the vertical sides. As a result, the microcomputer program CLARA 2.31 was used for the parametric study described herein due to the limitations of other 3D software and its capability of performing a 2D analysis from a 3D data file.

Specific information on CLARA 2.31 can be obtained from Hungr (1988, 1989). However, program options utilized for this study are discussed herein. CLARA 2.31 divides the slide mass into vertical columns that are the 3D equivalent of the vertical slices used in a conventional 2D analysis. Geometry data, which is input into CLARA 2.31 through a series of 2D cross sections, for vertical columns between the input 2D cross sections are generated by orthogonal interpolation in this study. The percentage of available array memory, which was used to automatically set the column length and width, used for this parametric study was set at 85 percent for all of the various slope model runs to provide uniform precision between the analyses. The Janbu's simplified method was utilized for this study because the method is suitable for a translational failure mode. The factor of safety obtained from Janbu's simplified method is based on horizontal and vertical force equilibrium. Moment equilibrium is not satisfied. Janbu's simplified method assumes that the resultant intercolumn forces are horizontal and an empirical correction factor is used to account for the interslice vertical force (Janbu 1954). The factor of safety provided by CLARA 2.31 does not include the Janbu correction factor extended to 3D for the effect of the intercolumn force distribution and the results were not adjusted manually to account for this correction.

Stark and Eid (1998) determined that commercially available software does not consider the shear resistance along the parallel sides of a slide mass in calculating the 3D factor of safety. To include this side resistance, an external horizontal and vertical side force equivalent to the shear resistance due to the at-rest earth pressure acting on the vertical sides was added at the centroid of the two parallel sides (see Fig. 2 (b)). From Fig. 3, the at-rest earth pressure acting on the vertical side of the slide mass, σ'_x, is estimated by multiplying the coefficient of earth pressure at rest, K_o, by the average vertical effective stress over the depth of the sliding mass, σ'_z. The coefficient of earth pressure at rest is determined from $K_o = 1-\sin\phi'_{up}$ where ϕ'_{up} is the friction angle of the upper material corresponding to the average effective normal stress on the vertical sides of the slide mass. The shear resistance due to the at-rest earth pressure acting on the vertical sides of the slide mass, S', is estimated by multiplying σ'_x by the tangent of ϕ_{up}'. In determining S', the shear resistance due to the small thickness and area of lower material between the upper material and the base of the failure surface was not included to simplify the determination of the cross section centroid (see Fig. 2 (b)). Therefore, only the side resistance of the upper layer was included. Additionally, it was assumed that S' acted parallel to the base of the failure surface, at a slope of 3 percent down slope.

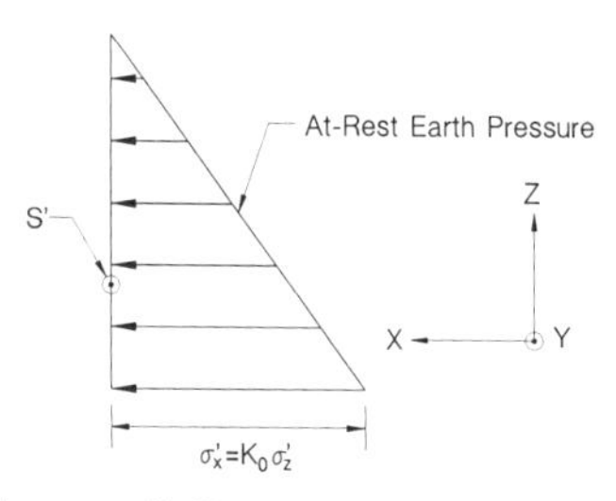

Figure 3. Shear Resistance, S', Due to the At-Rest Earth Pressure Acting on the Vertical Side of the Slope model

External loads are specified in CLARA 2.31 by its horizontal component (S_y') and vertical component (S_z') as depicted in Fig. 2(b) and by its point of application, X, Y, and Z coordinates. The vertical component of an external load is added to the total weight of the column directly below the vertical force if the vertical force is located within the plan area of the slide mass but is not included in the column total weight if the vertical force is located outside of the sliding mass plan area. The external forces applied to the parallel sides of the slide mass are considered to be within the sliding mass plan area. The horizontal component of external loads are included in the horizontal force equilibrium equation, whether or not the point of application is within or outside the plan area of the slide mass.

A 2D analysis was also performed for each geometry using CLARA 2.31 to provide a direct comparison with the 3D factor of safety. Since the slope model used in this parametric study is not rounded at the head scarp, it exhibits uniform cross sections across the slope that yield the same 2D factor of safety. When performing a 2D analysis using CLARA 2.31 from a 3D input file, CLARA 2.31 extracts the 2D cross section of interest from the mesh cross section nearest to the desired section. CLARA 2.31 reverts to a 2D problem by suppressing all input parameters in the third dimension and making all lateral column widths a value of unity. No external loads are utilized in the 2D analyses. It should be noted that the factor of safety obtained from a 3D analysis performed without considering the shearing resistance along the vertical sides of the slide mass (without external loads) yields a similar factor of safety as the corresponding 2D analysis. This confirms that the slope stability program does not consider the shear resistance along the parallel sides of a slide mass in calculating the 3D factor of safety.

Effect of Shear Resistance Along Vertical Sides

Fig. 4 presents a relationship between the ratio of 3D/2D factors of safety (3D/2D FS) and W/H for the three slope inclinations considered in the parametric study. The two different slope heights of 10 and 100 m were also used but there was little, if any, difference between the factors of safety. For example, the 3D/2D FS ratio versus W/H results at H=10 and 100 m for the 1H:1V slope were the same. For the 3H:1V slope, the 3D/2D FS ratio versus W/H results at H=10 and 100 m were nearly the same with differences in 3D/2D FS ratios not exceeding 0.05. For the 5H:1V slope, the 3D/2D FS ratio versus W/H results at H=10 and 100 m differed by less than 0.06 for W/H values greater than 1.5. At a W/H ratio of 1, the 3D/2D FS ratio difference was 0.19. The slight differences in the 3D/2D FS ratio versus W/H results obtained at H = 10 and 100 m for the three slope inclinations is probably caused by the affects of CLARA 2.31 moving each input cross section so the x-coordinate coincides with the nearest row of column center points and the interpolation between these input cross sections which influences determination of 3D column parameters. In summary, slope heights of 10 and 100 m did not significantly affect the relationship between the ratio of 3D/2D factors of safety and W/H for the slope inclinations considered. This observation is in agreement with the concept of geometric similarity.

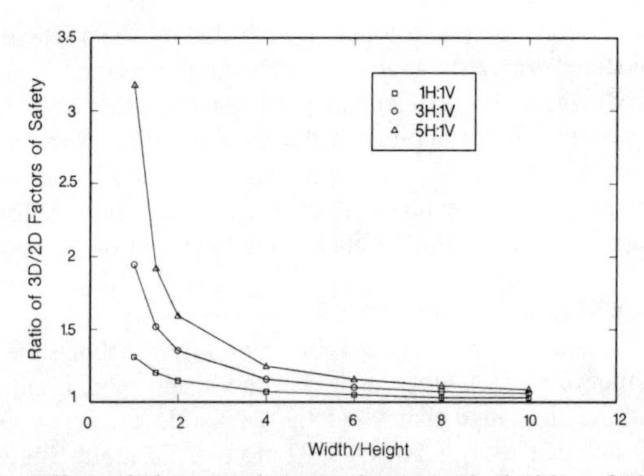

Figure 4. Effect of Shear Resistance along Vertical Sides of Slide Mass

The 3D factor of safety was greater than the 2D factor of safety for all of the W/H combinations used in the parametric study. It can be seen from Fig. 4 that the 3D/2D FS ratio increases with decreasing W/H ratios for a given slope inclination. The area of the vertical, parallel sides of the slide mass are the same for a given slope inclination and height for all width values. As the width decreases, the weight of the slide mass decreases and the shearing resistance along the parallel sides has a greater effect on the 3D stability. This is evidenced in Fig. 4 because the relationships increase rapidly at values of W/H less than 4. For W/H ratios less than 1.5, 3, and 5, the 3D factor of safety is at least 20 percent greater than the 2D factor of safety for the three slope inclinations. Therefore, the effect of including shearing resistance along the parallel sides of a slide mass increases as the slope width decreases.

Fig. 4 also shows that as the slope inclination decreases, for a given W/H ratio, the 3D/2D FS ratio increases. This increase in 3D/2D FS ratio for a given W/H ratio results from an increase in the area of the vertical sides of the slide mass due to the increase in L with decreasing slope inclination. Therefore, the importance of incorporating end effects in a slope stability analysis increases with decreasing slope inclination. Chen and Chameau (1983) and Leshchinsky et al. (1985) also indicate that the flatter the slope, the greater the difference between 2D and 3D factors of safety. Therefore, in translational failures, which can occur in relatively flat slopes because of the presence of underlying weak material(s), the back-calculated shear strengths may be too high if end effects on the sides of the slide mass are ignored.

Influence of Shear Strength

The results in Fig. 4 are based on a ratio of friction angles for the upper (30°) and lower (8°) materials, ϕ_{up}/ϕ_l, of 3.75. Additional ratios of ϕ_{up}/ϕ_l, e.g., 1, 1.5, 3, and 3.75, were used to investigate the influence of various friction angle ratios on the 3D/2D FS ratio values. To obtain these lower ratios of ϕ_{up}/ϕ_l, the friction angle of the lower material was increased. The friction angle of the upper material remained 30 degrees so the back scarp would remain inclined at 60° and thus simulate an active pressure condition as previously discussed. It was assumed that the value of unit weight of the lower material was 18 kN/m^3 and did not vary with ϕ_l. Figs. 5 through 7 show the results of these analyses for slope inclinations of 1H:1V, 3H:1V, and 5H:1V, respectively. Each relationship shown in Figs. 5 through 7 for a given W/H is the average obtained from the 3D/2D FS ratio versus ϕ_{up}/ϕ_l results obtained at H = 10 and 100 m. It can be seen from Fig. 7 that varying the ratio of ϕ_{up}/ϕ_l was most pronounced for the 5H:1V slope. The effect of varying ϕ_{up}/ϕ_l decreased for increasing W/H. For example, for a 5H:1V slope (Fig. 7) and a value of W/H of 10 and 1, the difference in the 3D/2D FS ratio ranged from 1.6 to 2.1, respectively, for the range of ϕ_{up}/ϕ_l ratios investigated.

Figs. 5 through 7 also show that for a given W/H ratio and back scarp angle, the influence of shear strength between the upper and lower layers increases with decreasing slope inclination. For a W/H of 1.0, back scarp angle of 60°, and a ϕ_{up}/ϕ_l of 3.75, the 3D/2D FS ratio increased from 1.31 to 3.18 for slopes of 1H:1V

(Fig. 5) to 5H:1V (Fig.7), respectively. For flatter slopes at a given W/H and back scarp angle, the L/H increases. Thus, the flatter slope yields a larger value of L and a larger shear stress is mobilized along the base of the sliding surface. In the previous section, it was shown that the influence of incorporating end effects in a slope stability analysis increases with decreasing slope inclination for a given W/H Ratio (see Fig. 4). The results of Figs. 5 through 7 indicate that this influence may become more substantial with larger differences in shear strength between upper and lower layers. Thus, in relatively flat slopes with large differences in in-situ material shear strengths, the back-calculated shear strengths may be too high or unconservative. This is especially true in translational failures that occur in relatively flat slopes where the underlying material(s) may be much weaker than the upper materials.

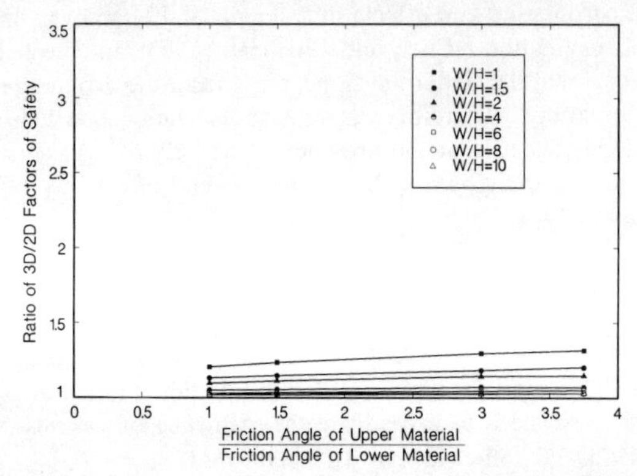

Figure 5. Influence of Shear Strength on Ratio of 3D/2D Factors of Safety for 1H:1V Slope

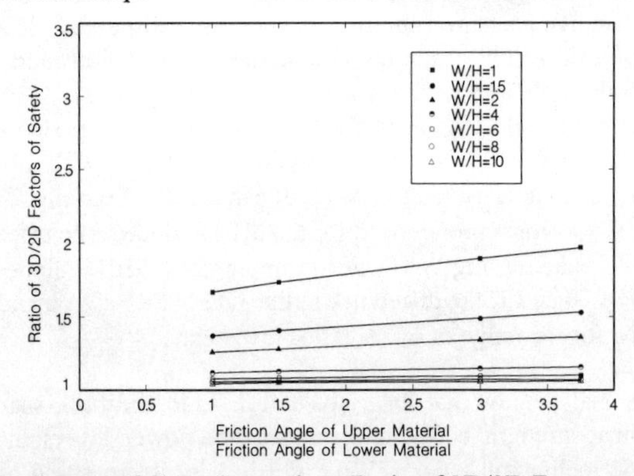

Figure 6. Influence of Shear Strength on Ratio of 3D/2D Factors of Safety for 3H:1V Slope

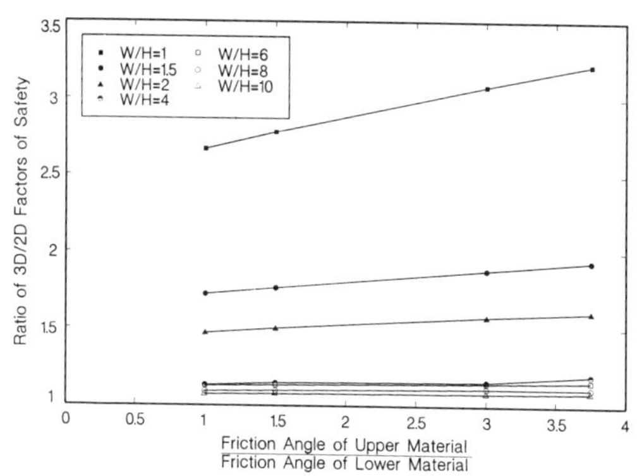

Figure 7. Influence of Shear Strength on Ratio of 3D/2D Factors of Safety for 5H:1V Slope

Figs. 8 and 9 illustrate the influence of ϕ_{up}/ϕ_l on the 2D and the 3D factors of safety, respectively, instead of using a FS ratio. These figures indicate that the effect of varying the friction angle between the upper and lower layers is more significant on the 3D factor of safety than the 2D FS value. The results of Figs. 8 and 9 indicate that the 3D/2D FS ratio differences obtained in Figs. 5 through 7 are due primarily to changes in the 3D factor of safety rather than to changes in the 2D factor of safety. Therefore, it is even more important that material shear strength parameters are adequately defined for a 3D analysis than a 2D analyses.

This parametric study was performed based on the assumption that materials along the vertical sides of the slide mass consist of cohesionless materials, i.e., cohesion, c=0. Previous studies have indicated that the 3D end effects are more pronounced for slopes of cohesive materials (Chen 1981; Lovell 1984; Leshchinsky and Baker 1986; Ugai 1988).

Field Case History

A field case history is presented to illustrate the use of the parametric study results and the importance of conducting a 3D analysis in practice. The 1979 Oceanside Manor landslide occurred in San Diego County, California and is described in detail by Stark and Eid (1998). The landslide occurred along a bluff approximately 20 m high in a residential area. The length of the scarp is approximately 130 m and the slide encompassed approximately 122,000 m^3 of soil. A plan view and representative cross section of the landslide prior to failure are presented by Stark and Eid (1998). The slope is underlain by the Santiago Formation, which is composed of a claystone and a sandstone. The sandstone is fine to medium grained and overlies the gray claystone. The remolded claystone classifies as a clay or silty clay of high plasticity, CH-MH, according to the Unified Soil Classification System. The liquid limit, plasticity index, and clay-size fraction are 89, 45, and 57 percent, respectively (Stark and Eid 1992).

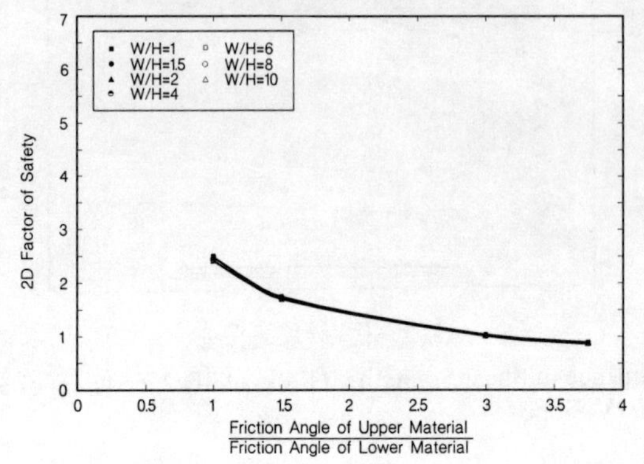

Figure 8. Influence of Shear Strength on 2D Factor of Safety for 5H:1V Slope

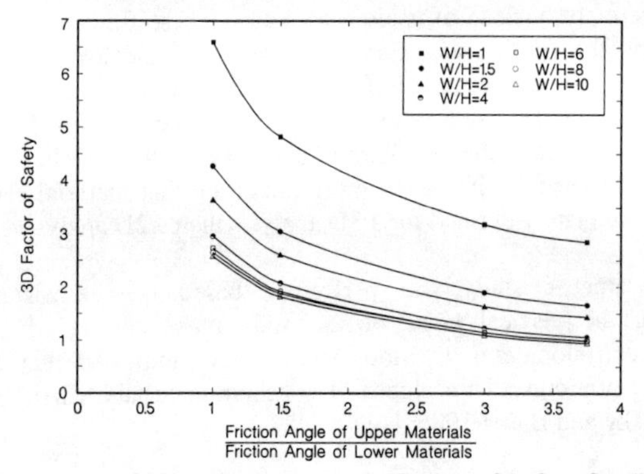

Figure 9. Influence of Shear Strength on 3D Factor of Safety for 5H:1V Slope

Field investigations showed that the claystone is commonly fissured, displaying numerous slickensided surfaces. The site has undergone at least three episodes of landsliding prior to this slide. Therefore, the claystone has undergone substantial shear displacement and has probably reached a residual strength condition along the base of the sliding surface. In addition, the largest portion of the sliding surface in the claystone is approximately horizontal through the Santiago Formation. This indicates that sliding occurred along a weak claystone seam or layer. As a result, Stark and Eid (1998) assumed that residual and fully softened shear strengths were mobilized during failure along the base and the scarp in the Santiago Formation, respectively. The slide surface was located using slope inclinometers and extensive borings and trenches. The groundwater levels were monitored using piezometers and water levels in borings and trenches shortly after movement started to occur.

Stark and Eid (1998) performed 3D and 2D slope stability analysis. The parallel sides of the sliding mass were assumed to be nearly vertical. In addition, the back scarp was taken to be inclined 60 degrees from the horizontal to simulate an active earth pressure condition. Moist unit weights for the Santiago claystone and the compacted fill were measured to be 19.6 kN/m^3. Based on residual and fully softened shear strengths measured on representative samples of the Santiago claystone using a ring shear test procedure, a residual friction angle of 7.5 degrees and a fully softened friction angle of 25 degrees was used for the claystone in the slope stability analyses. The fully softened friction angle of the claystone along the back scarp was used for calculating the mobilized shear strength of the sliding mass. The cohesion and friction angle of the compacted fill were measured using direct shear tests to be zero and 26 degrees, respectively.

An average 2D factor of safety of 0.92 was reported based on the analysis of 44 different cross sections and a 3D factor of safety of 1.02 was calculated. The 2D and 3D slope stability results were conducted using Janbu's simplified method and the microcomputer program CLARA 2.31. The factors of safety values are uncorrected for the effect of the interslice or intercolumn force distribution.

This case history is used to demonstrate the use of the results of the parametric study (Figs. 5 through 7). The slope had an average slope inclination of 3.5H:1V prior to failure. The W/H ratio is 130 m/20 m or 6.5. The ϕ_{up}/ϕ_l ratio (25 degrees divided by 7.5 degrees) is 3.3. Using Fig. 6, a W/H of 6.5, and $\phi_{up}/\phi_l = 3.3$, a 3D/2D ratio of 1.09 is obtained for a slope of 3H:1V. Similarly, from Fig. 7, a 3D/2D FS ratio of 1.14 is obtained for a slope of 5H:1V. For the landslide slope inclination of 3.5H:1V, a 3D/2D FS ratio of 1.1 can be interpolated from Figs. 6 and 7. Based on the average 2D factor of safety of 0.92 from 44 different cross-sections reported by Stark and Eid (1998), the 3D factor of safety can be estimated to be 1.01 using the 3D/2D FS ratio of 1.1. This estimated 3D factor of safety is in agreement with the 3D factor of safety value of 1.02 calculated by Stark and Eid (1998).

It should be noted that the slope inclinations in Figs. 4 through 9 represent an average inclination across the landslide. Additionally, the 2D value represented by the 3D/2D FS ratios in Figs. 4 through 8 is an average 2D factor of safety value across the landslide in the direction of movement. The ratio between the 3D factor of safety and the minimum 2D factor of safety may be slightly larger in practice due to the simplified model. For the Oceanside Manor case history the ratio between the 3D factor of safety and minimum 2D factor of safety is 1.6 (Stark and Eid 1998).

In summary, Figs. 5 through 7 can be used to determine the importance or necessity of performing a 3D slope stability analysis for a translational failure mode in practice. However, these figures should not be used as a substitute for performing an actual 3D slope stability analysis with site specific geometry, pore-water pressure condition, and material properties.

Conclusions

Commercially available 3D slope stability software do not consider the shear resistance along the two vertical sides of the slide mass that parallel the direction of movement in calculating the 3D factor of safety. A method for incorporating the shear resistance along the two sides of a slide mass is presented and consists of placing a horizontal and vertical external force equivalent to the shear resistance due to the at-rest earth pressure acting at the centroid of the sides parallel to the direction of movement of the slide mass. A parametric study was conducted to investigate the importance of incorporating 3D end effects for various slope geometries and shear strengths. The following conclusions are based on the 2D and 3D slope stability analyses performed in the parametric study:

(1) For a given slope inclination, the ratio of 3D/2D factors of safety increases with decreasing W/H ratios. As the width decreases, the weight of the slide mass decreases and the shearing resistance along the parallel sides has a greater effect on the 3D stability. Therefore, the effect of including the shear resistance along the parallel sides of a slide mass can be significant for slopes that have a W/H ratio of less than 4.

(2) As the slope inclination decreases for a given W/H ratio, the 3D/2D factors of safety ratio increases. This increase in the factor of safety ratio results from an increase in the area of the parallel sides of the slide mass caused by the increase in slope length with decreasing slope inclination. Therefore, the influence of incorporating end effects in a slope stability analysis increases with decreasing slope inclination. In translational failures involving slope inclinations less than 3H:1V, the back-calculated shear strengths may be too high if 3D end effects on the sides of the slide mass are ignored. For a given W/H ratio and back scarp angle, the impact of the shear strength in the upper and lower layers increases with decreasing slope inclination.

(3) The difference in shear strength between the upper and lower layers has a larger effect on the 3D factor of safety than on the 2D factor of safety. This effect of the shear strength difference increases with decreasing W/H ratios. Therefore, it is critical that material shear strengths be adequately defined for a 3D analysis.

(4) Figs. 5 through 7 presented herein can be used to determine the importance of performing a 3D slope stability analysis for a translational failure mode. The use of Figs. 5 through 7 is illustrated using a field case history. However, these figures should not be used as a substitute for performing an actual 3D slope stability analysis with site specific geometry, pore-water pressure condition, and material properties.

Commercially available 3-D slope stability software has inherent limitations that affect the calculated factor of safety for a translational failure mode (Stark and Eid 1998). These limitations include ignoring the shear resistance along the sides parallel to the direction of movement of the sliding mass, modeling a nonlinear failure envelope with a linear failure envelope, and using a 3-D slope stability method that ignores some of the internal shear forces. Commercially available 3D numerical modeling software that utilize numerical methods such as the finite element method provide a powerful alternative to the limit equilibrium approach for investigating slope stability problems because numerical methods may alleviate these limitations. The use of these numerical methods is the subject of subsequent research.

Acknowledgments

The first author acknowledges the support provided by the State of Illinois Minority Graduate Incentive Program. This study was performed as a part of National Science Foundation Grant CMS-98-02615. The support of this agency is gratefully acknowledged. The second author also acknowledges the support provided by a University Scholar Award. The authors acknowledge the support provided by Dr. Oldrich Hungr of O. Hungr Geotechnical Research Inc. The contents and views in this paper are the authors' and do not necessarily reflect those of any of the contributors or represented organizations.

References

Beene, R. R. W. (1967), "Waco Dam Slide," *Journal of Soil Mechanics and Foundations Division, ASCE*, 93(4): 35-44.

Byrne, R .J., Kendall, J., and Brown, S. (1992), "Cause and Mechanism of Failure, Kettleman Hills Landfill B-19, Unit IA," *Proceedings of Stability and Performance of Slopes and Embankments-II*, ASCE, Vol. 2, pp. 1188-1215.

Chen, R.H. (1981), *Three-Dimensional Slope Stability Analysis*, Report JHRP-81-17, Purdue University, West Lafayette, IN, USA.

Chen, R.H. and Chameau, J.-L. (1983), "Three-Dimensional Limit Equilibrium Analysis of Slopes," *Geotechnique*, 32(1): 31-40.

Duncan, J.M. (1992), "State-of-the-Art: Static Stability and Deformation Analysis," *Proceedings of Stability and Performance of Slopes and Embankments-II*, ASCE, Vol. 1, pp. 222-266.

Ehlig, P. L. (1992), "Evolution, Mechanics and Mitigation of the Portuguese Bend Landslide, Palos Verdes Peninsula, California," *Engineering Geology Practice in Southern California*, B. W. Pipkin and R. J. Proctor, eds., Star Publishing Co., Belmont, Calif., pp. 531-553.

Hungr, O. (1988), *User's Manual CLARA: Slope Stability Analysis in Two or Three Dimensions for IBM Compatible Microcomputers*, O. Hungr Geotechnical Research, Inc., West Vancouver, B.C.

Hungr, O., Salgado, F.M., and Byrne, P.M. (1989). "Evaluation of a three-dimensional method of slope stability analysis," *Canadian Geotechnical Journal*, 26: 679-686.

Janbu, N. (1954), *Stability Analysis of Slopes with Dimensionless Parameters*, Harvard University, Soil Mechanics Series, No. 46. Harvard University, Cambridge, Mass.

Jasper, J. L., and Peters, N. (1979), "Foundation Performance of Gardiner Dam," *Canadian Geotechnical Journal*, 16: 758-788.

Krahn, J., Johnson, R. F., Fredlund, D. G., and Clifton, A. W. (1979), "A Highway Cut Failure in Cretaceous Sediments at Maymont, Saskatchewan," *Canadian Geotechnical Journal*, 16:703-715.

Leshchinsky, D., Baker, R. and Silver, M. L. (1985), "Three Dimensional Analysis of Slope Stability," *International Journal for Numerical and Analytical Methods in Geomechanics*, 9(2): 199-223.

Leshchinsky, D. and Baker, R. (1986), "Three-Dimensional Slope Stability: End Effects," *Soils and Foundations*, 26(4): 98-110.

Lovell, C.W. (1984), "Three Dimensional Analysis of Landslides," *Proceedings of the 4th International Symposium on Landslides*, Toronto, pp. 451-455.

Seed, R.B., Mitchell, J.K. and Seed, H.B. (1990), "Kettleman Hills Waste Landfill Slope Failure. II: Stability Analysis," *Journal of Geotechnical Engineering*, ASCE, 116(4): 669-689.

Stark, T.D. and Eid, H. T. (1992), "Comparison of Field and Laboratory Residual Shear Strengths," *Proceedings of Stability and Performance of Slopes and Embankments-II*, ASCE, Vol. 1, pp. 876-889.

Stark, T.D. and Eid, H. T. (1998), "Performance of Three-Dimensional Slope Stability Methods in Practice," *Journal of Geotechnical and Geoenvironmental Engineering*, ASCE, 124(11): 1049-1060.

Stark, T.D. and Poeppel, A.R. (1994), "Landfill Liner Interface Strengths from Torsional Ring Shear Tests," *Journal of Geotechnical Engineering*, ASCE, 120(3): 597-615.

Ugai, K. (1988), "Three-Dimensional Slope Stability Analysis by Slice Methods," *Proceedings of the 6th International Conference on Numerical Methods in Geomechanics*, Innsbruck, pp. 1369-1374.

Wright, S. G., and Duncan, J. M. (1972), "Analyses of Waco Dam Slide," *Journal of Soil Mechanics and Foundations Division*, ASCE, 98(9): 869-877.

A Numerical Technique for Two-Dimensional Slope Stability Problems

Mohsen Beikae [1]

Abstract

All limit equilibrium methods of slope stability analysis 1) calculate a factor of safety which is assumed to be the same at all points along the potential slip surface, 2) use only force and moment equilibrium and employ varying assumptions to make the problem statically determinate, 3) assume that the potential sliding mass is a rigid body, 4) need a direction of movement of a potential sliding mass for three-dimensional cases, and 5) calculate a yield acceleration to be used in the Newmark method. To facilitate the slope stability analysis and to avoid some of the above limitations, a numerical technique has been developed and a computer program written to carry out the analysis technique. A single analysis determines local factors of safety and a pattern of induced deformations of a potential sliding mass due to gravity, hydrostatic forces, and base motions. Results of analyses for two example slopes are given to demonstrate the comparison between this technique and the conventional methods.

Introduction

A commonly accepted practice in slope design is to use limit equilibrium methods of slope stability analysis. As noted by Duncan and Wright (1980), all limit equilibrium methods have four characteristics in common: 1) a factor of safety is placed on shear strength parameters, 2) the strength parameters are independent of stress-strain behavior, 3) some or all of the equations of equilibrium are used to determine the factor of safety, and 4) forces involved in equilibrium methods are statically indeterminate. Two other common characteristics, which may be added to the above list, are: 1) the potential sliding mass is assumed to be a rigid body and 2) the direction of least resistance to sliding for a three-dimensional problem is not in general obvious, therefore, a critical direction is assumed.

The limit equilibrium methods calculate a factor of safety which, by definition, is assumed to be the same at all points along the potential slip surface. This is reasonable only at failure, when all the slices are on the verge of failure; that

[1]Engineer, Metropolitan Water District of Southern California, 700 N. Alameda Street, Los Angeles, California 90012

is, when the factor of safety equals unity for each slice. In reality, the local factors of safety will vary somewhat along the slip surface, for some slices it might be one and high value for others. In the case of brittle materials, even small hydrostatic loads can reduce the local factors of safety to less than unity and a progressive failure mechanism may be triggered. This happens, for instance, in overconsolidated clay that can exhibit residual shear strength under drained loading and in loose, saturated sands under undrained loading. Therefore, the methods cannot explicitly model the mechanism of progressive failure (Pyke, 1991).

The methods assume that the slope materials are rigid-plastic and the strength parameters are independent of stress-strain behavior. The methods use both rotational and/or translational equilibrium and ignore strain compatibility. As a result, the number of equilibrium equations is less than the number of unknowns. Thus, all the methods need to employ assumptions to make the problem statically determinate, by making up the balance between the number of equilibrium equations and the number of unknowns in the problem. The most critical of these assumptions typically deals with the side forces. For example, the ordinary method of slice assumes that the resultant of side forces acting parallel to the base of the slice; Bishop's Simplified Method assumes that the vertical component of the interslice forces is zero; Spencer's method assumes that the interslice forces are parallel (Spencer, 1967); Morgenstern and Price Method assumes that the ratio of side forces are given (Morgenstern and Price, 1965)

The methods imply that the potential sliding mass is a rigid body, which is uncommon in reality. Thus for failure surfaces which are concave everywhere and convex at the toe the factor of safety calculated by the methods may be overestimated. As such, the limit equilibrium solutions are not necessarily kinematically admissible (Pyke, 1991).

The limit equilibrium methods are also used to calculate the horizontal yield acceleration for a specified failure surface. The yield coefficient is a horizontal acceleration which, when applied to the slices, would reduce the factor of safety to unity. If the history of acceleration applied horizontally to the potential failure mass can be determined, then the movement, which occur during those portions of the history of loading in which the applied acceleration exceeds the yield acceleration, can be calculated. Therefore, the yield acceleration is used in the Newmark method to estimate the seismically induced deformation of slopes due to earthquake motions. This is done, however, by running multiple computer analyses including seepage, slope stability, static and dynamic finite element, and the Newmark double integration (Beikae, 1996a). In this study, the Newmark double integration is called the simplified Newmark method.

To facilitate the slope stability analysis and to avoid some of the above limitations, a numerical technique has been developed and a computer program written to carry out the analysis technique. A single analysis, discussed below, provides local factors of safety and a pattern of induced deformations of a potential sliding mass due to gravity, hydrostatic forces, and base motions. In addition, results of analyses for two example slopes are given to demonstrate the comparison between this technique and the conventional methods.

Methodology

A computer program DSLOPE was developed to carry out the slope stability analysis. DSLOPE uses the Lagrangian formulation of momentum equations, representing Newton's second law of motion. The Lagrangian formulation inherently takes into account the mass conservation law and allows soil slices with fixed masses to translate, compress, expand, and distort in space. The full dynamic equations of motion are used, even with the modeling systems that are essentially static. This enables DSLOPE to follow a physically unstable system without numerical problems. The equations of motion are cast into a discrete algebraic form, which is amenable to numerical solution. The algebraic equations are then solved at the center of gravity of each slice. The key mathematical basis and the numerical implementation of the program, which are discussed elsewhere by Cundall (1974) and Beikae (1996b), are briefly presented in the following sections.

Numerical Solution Process - In this technique, the specified potential sliding mass is divided into an array of slices, represented by grid points, and the basal slip surface by an assemblage of line elements connected at nodes, as shown in Figure 1. The mass of a slice is allocated to a grid point located at its center of gravity. Each grid or node represents two degrees of freedom, one horizontally and one vertically. Each grid point is connected to its adjacent grid points and basal node via "spring" and "dashpot" elements. The grid spacing and node spacing determine the ability to resolve fine details of deformations. In general, the smaller the grid spacing and the node spacing, the better is the accuracy of results.

The analysis consists of two parts: 1) Turn-on gravity analysis to compute local factors of safety, initial interslice and base forces due to gravity and hydrostatic forces and 2) dynamic analysis to compute seismic responses and displacements due to the simultaneous effects of gravity, hydrostatic forces, and earthquake motions. In Part 1, the drained strength parameters are used for all materials. However, the shear strengths of those materials that are expected to behave in undrained conditions during earthquake shaking are used in Part 2.

During Part 1, the basal surface (nodes) is fixed and the gravity and hydrostatic loads are turned-on gradually. Under this condition, the slices in the potential sliding block start moving and interacting laterally with their neighboring slices and vertically with the basal surface. If the global factor of safety is greater than or equal to one, the potential sliding block will quickly become stable. Otherwise, the block keeps moving along a specified slip surface until it becomes eventually stable. At the end of Part 1, the program computes local factors of safety. This is done by dividing the shear strength with that of shear force at the base of each slice. The technique can also model the progressive failure of slopes made of brittle materials and show that a progressive failure mechanism is being triggered due to gravity and hydrostatic forces.

If it is required, Part 2 is started automatically. In this part, the entire model subject to horizontal and vertical components of an input motion. As a result, the basal surface starts moving and inducing basal contact forces on the slices, and the slices generate side forces between themselves.

At each time step, during both Part 1 and Part 2, the incremental x and z deformations for a given contact are computed from the incremental x and z deformations at the contact of two adjacent grid points (n) and (i). These incremental deformations at the contact are then resolved into shear and normal deformations. The new incremental shear and normal forces are calculated from the force-deformation relationships, discussed below, and then added to the old forces. These forces are subject to the lower and upper bound conditions shown in Figure 1. These forces are then resolved into equivalent x and z forces and finally added to the other x and z forces acting on each grid point. Thus, forces built up on each slice are calculated. Then, Newton's second law is invoked to compute the slice accelerations. The accelerations are then numerically integrated to get velocities and integrated once again to get deformations. With this new set of deformations the calculation cycle is repeated.

Each slice moves through time in a series of time steps, discussed below. Essentially, the new position of a slice is its old position plus the distance it travels during a given time step. If no forces acted on a slice, the distance it traveled would be a function of its velocity at the previous position, because distance equals speed multiplied by time. During a time step, however, the forces exerted by other slices cause the slice to accelerate which in turn changes its velocity. If the forces are constant during the time step, Newton's law dictates that the change in velocity is proportional to the force, so the updated velocity can be calculated. The updated velocity is then used to calculate the new position of slice. In this repetitive manner the array of slices moves on the basal surface and the analysis marches forward with time. As time proceeds, dynamic equilibrium of the sliding mass is developed naturally satisfying both force equilibrium and strain compatibility. This is carried out with no prior assumptions regarding the critical direction of movement to be considered, and no need to perform multiple computer analyses to get seismically induced deformation.

No iteration is involved in the calculation discussed above. The velocities and deformations of all the grid points are assumed fixed while the forces are being computed as explicit functions of grid point velocities and deformations. Similarly, the forces associated with all the grid points are assumed fixed while velocities and deformations are being calculated as explicit functions of forces. This method of performing finite difference calculations using explicit functions is called the "explicit" finite difference method.

DSLOPE program can be compared to some of the other tools already available in the market such as FLAC (ITASCA, 1995). DSLOPE is specifically tailored to carry out slope stability analysis in two-dimensional cases and can be extended with minimum effort to three-dimensional cases, where as FLAC can simulate the behavior of structures built of soil, rock or other materials for two-dimensional cases only. Both programs use interface elements, consisting of normal and shear springs, to model slip surfaces. Both codes can use several constitutive models to simulate highly nonlinear material properties, and both programs use the Lagrangian calculation scheme which is well suited for modeling large distortion.

Time Step - The explicit method assumes that, in one time step, a given grid point can not communicate with its neighbors. Suppose the shortest of such time steps is designated as the critical time step, and suppose a slope, as shown on Figure 1, can be visualized as consisting of a series of grid points connected by springs at contact points. In such a system, the critical time step is controlled by the fundamental period of a grid point connected to other grid points by spring elements. To achieve a stable solution, the time step used in the explicit method must be smaller than the critical time step, which is proportional to the square root of the mass of the grid point divided by the stiffness of the grid point.

Initial and Boundary Conditions - The computation processes are subject to appropriate initial and boundary conditions. The initial conditions normally consist of a specified initial geometry of grid points and nodal points having zero deformations, velocities, and accelerations at the beginning. The boundary conditions include zero accelerations, fixed accelerations, or acceleration histories at the grid points. In addition, hydrostatic loads such as hydrostatic forces can be specified at grid points.

Material Properties - Each slice and element has a material type, which is characterized by attributes that are relatively easily obtainable, geometry, total unit weight, drained and undrained shear strength, small strain shear modulus and bulk modulus. The shear strength for each material type may be specified by means of a curved Mohr-Coulomb envelope or Hoek and Brown (Hoek and Brown, 1988) failure criterion for fractured rock masses.

Force-Deformation Relationships - Spring elements need force-deformation relationships. The program makes it possible to specify any force-deformation relationship. However, for this study, a linear relationship, or a spring constant, is used as shown Figure 1. The value of the spring constant reflects the expected shear and compression wave velocities in the embankment and is subject to two counteracting considerations: 1) keeping the spring constant relatively low to prevent the time step from becoming too small (resulting in an analysis that requires too much computational time) and 2) keeping the spring constant relatively high to reflect the expected shear and compression wave velocities in the slope. For practical purposes, however, the analysis results are not sensitive to values of spring constants within a reasonable range.

The force-deformation relationships are also subject to the following conditions, as shown on Figure 1: 1) forces in shear springs are less than or equal to available shear strengths, 2) forces in horizontal normal springs are less than or equal to available active forces and greater than or equal to available passive forces, and 3) forces in normal springs at the base are less than or equal to zero. It is noted that in the force-deformation relationships, normal forces in tension and compression are considered positive and negative, respectively.

Damping - Certain irreversible processes that convert kinetic energy to heat should take place between grid points and nodes. This effect was approximately reflected in the analyses by allowing some damping in the system. The program includes two forms of viscous damping: local damping and global damping. Local damping operates on the relative velocities between two adjacent grid points and between a grid point and its adjacent basal node. Global damping operates on the absolute velocities of the grid point. Local dampings are represented by dashpots oriented in the shear and normal directions between two adjacent grid points and between a grid point and its adjacent node. Similarly, global dampings are modeled by dashpots connecting grid points to the inertial reference (Beikae, 1996b).

Example Problems

Two example slopes were used in this study. Example Nos. 1 and 2 are both two-dimensional plane strain slope problems with potential logspiral slip surfaces. Based on conventional methods of slope stability, closed form solutions for global factor of safety were given for two-dimensional slopes with a logspiral slip surface. Figure 2 shows the equations for the calculated factor of safety and yield acceleration coefficient for two- dimensional cases. The global factor of safety was derived by equating the resisting moment with that of the driving moment. The resisting moment consists of the moment of the mobilized shear strength, divided by the global factor of safety, along the slip surface about the center of the slip surface. The driving moment consists of the moment of the total weight of potential sliding block and horizontal inertia force about the same center. Once the equation for factor of safety is determined, yield acceleration coefficient can be calculated from the equation by setting the factor of safety to unity, as shown on Figure 2.

For seismic analysis, Caltech B-1 ground acceleration (Jennings et al., 1968), with a time step of 0.025 second, was used as an input motion, representing a Mw-7 1/2 earthquake with a peak ground acceleration of 0.5 g, shown in middle part of Figures 4 and 6. Table 1 summarizes the geometry of slope, material properties, results of the closed form limit equilibrium analyses and simplified Newmark method, as well as the results of DSLOPE static and seismic analyses.

Figure 3 shows the geometry and results of closed form solution and DSLOPE static analysis for Example No. 1. As Table 1 and Figure 3 indicate, the cohesion and friction angle were selected so that the *FS* was 1.0 and k_y was zero using equations shown on Figure 2. As such, all slices were on the verge of failure: and they mobilized 100 percent of their shear strengths along the logspiral slip surface. Based on the results, the local and global factors of safety are the same; and the interslice side forces calculated by DSLOPE analysis are between the corresponding active and passive forces. At the crest of the slope, the side forces are equal to those of the active forces. However, in the middle of the slope, the side forces become greater than those of active forces. Finally, at the toe of the slope, the side forces gradually approach the passive forces. The variation of side forces suggests that, the upper part of the slope is under tension and the middle and the lower part of the slope are under compression.

Table 1

	Material Properties			Closed Form Solution Results			DSLOPE Analysis Results	
Example No. [1]	γ (pcf) (kg/m^3)	C (psf) (kPa)	φ (•)	FS[2]	k_y[3]	d_{max}[4] (feet) (meters)	FS[5]	d_{max}[6] (feet) (meters)
1	120 1918	1400 67	10	1.0	0.0	8.0 2.4	1.0	12.5 3.8
2	120 1918	2100 100.5	10	1.5	0.05	3.3 1.0	1 to >10	2.0 0.6

Notes:

1) Logspiral slip surfaces were assumed for two examples
2) Global factor of safety was calculated based on the first equation shown on Figure 2
3) Yield coefficient was calculated based on the second equation shown on Figure 2
4) Seismically induced deformation calculated by simplified Newmark method
5) Local factors of safety
6) Maximum induced deformation at the toe of slope (point A)

Figure 4 shows the deformed geometry provided by DSLOPE seismic analysis and the simplified Newmark method for seismically induced deformation. The results of DSLOPE seismic analysis were plotted for the toe (point A) and the crest (point B) of the potential sliding slope. As the figure indicates, the seismic induced deformation at the toe of the slope (point A) calculated based on DSLOPE seismic analysis is about one and half times greater than that of the simplified Newmark method. Where as, the seismic induced deformation at the crest of the slope (point B) is less than that of the simplified Newmark method.

Figure 5 shows the geometry and results of closed form solution and DSLOPE static analysis for Example No. 2. As Table 1 and Figure 5 indicate, the cohesion and friction angle were selected so that the *FS* was 1.5 and k_y was 0.05 using equations shown on Figure 2. As such, all slices mobilized 67 percent of their shear strengths along the logspiral slip surface. Based on the figure, the global factor of safety is 1.5 everywhere, where as the local factor of safety, calculated by DSLOPE analysis, varies from one to more than 10. The plot in the middle of Figure 5 shows that about 75 % of slices at the upper part of the slope are on the verge of failure and mobilize 100 % of their shear strengths along the logspiral slip surface, and the remaining 25 percent located at the toe of the slope are not at failure and mobilize about 10% to 67% of their shear strengths. In addition, the interslice side forces are between the corresponding active and passive forces. Similar to the results of Example 1, at the crest of the slope the side forces are equal to those of the active forces; however, in the middle and at the toe of the slope the side forces become greater than those of active forces. The variation of side forces suggests that, the upper part of the slope is under tension and the middle and the lower part of the slope are under compression.

Figure 6 shows the deformed geometry provided by DSLOPE seismic analysis and the simplified Newmark method for seismically induced deformation. The results of DSLOPE seismic analysis were plotted for the toe (point A) and the crest (point B) of the potential sliding slope. As the figure indicates, the seismic induced deformation at the toe and crest of the slope (points A and B, respectively) calculated based on DSLOPE seismic analysis are smaller than that of the simplified Newmark method.

Summary and Conclusions

The conventional limit equilibrium method of slope stability was used to estimate the global factors of safety and yield accelerations for the example slopes. Simplified Newmark method was also used to compute seismically induced deformation of the slopes due to Caltech B-1 record. Results of DSLOPE static and seismic analyses for the slopes were compared with those of the conventional methods. For Example No. 1, which is on the verge of failure, the global and local factors of safety are equal to unity. For Example No. 2, the local factor of safety varied somewhat from one to more than 10, the closed form solution estimated a global factor of safety of 1.5. For Example No. 1, the maximum seismically induced deformation at the toe of the slope computed by DSLOPE is about one and half times than that calculated by simplified Newmark method. However, for Example No. 2, the maximum seismically induced deformation at the toe of the slope computed by DSLOPE is less than that calculated by simplified Newmark method. For the two examples, DSLOPE also provided the two-dimensional pattern of slope deformation, whereas, the conventional methods calculate only the amount of deformation. The discrepencey between the results of the DSLOPE seismic analysis and simplified Newmark method may be due to some underlying assumptions for the simplified Newmark method. These assumptions are as follows: 1) a horizontal planar slip surface, which in reality is non-horizontal and non-planar and 2) a yield acceleration, which is based on force/moment equilibrium ignoring both vertical acceleration of the sliding block and kinematically admissible slip surface.

Acknowledgement

Review of this paper by Mr. Mike Smith from URS Greiner Woodward Clyde and two anonymous reviewers is appreciated.

References

Beikae, M., M. Lubbers, J. Barneich & D, Osmun, 1996a, "Seismic Deformation Analyses of Eastside Reservoir Dams," Seismic Design and Performance of Dams, Sixteenth Annual USCOLD Lecture Series: Los Angeles, California, July 22-26, pp. 315-331

Beikae, M., 1996b, "A Seismic Displacement Analysis Technique for Embankment Dams," Seismic Design and Performance of Dams, Sixteenth Annual USCOLD Lecture Series: Los Angeles, California, July 22-26, pp. 91-109

Cundall, P.A., 1974, "A Computer Model for Rock-Mass Behavior Using Iterative Graphic for the Input and Output of Geometrical Data," Report MRD-2-74, prepared at the U. of Minnesota, Under Contract NO. DACW 45-74-c-006, for Missouri River Division, U.S. Army Corps of Engineers.

Duncan, J.M. and Wright, S.G., 1980, "The Accuracy of Equilibrium Methods of Slope Stability Analysis." Engineering Geology, vol. 16, Elsevier Scientific Publishing Company, Amsterdam, pp. 5-17.

Jennings, P.C., G.W. Housner, and M.C. Tsai, 1968, "Simulated Earthquake Motions," Report on research conducted under a grant from National Science Foundation Pasadena California, April.

Hoek, E. and Brown, E. T., 1988. "The Hoek-Brown Failure Criterion – a 1988 Update," in Rock Engineering for Underground Excavations, Proceedings 15th Canadian Rock Mechanics Symposium, University of Toronto, Toronto, Ontario, pp 31-38.

ITASCA Consulting Group, Inc., 1995. FLAC Version 3.3.

Morgenstern, N.R. and Price, V.E., 1965. "The Analysis of the Stability of General Slip Surfaces." Geotechnique, Vol. 15, No. 1, pp 79-93,

Pyke, R.M., 1991, "TSLOPE3 User's Guide." TAGA Engineering Systems & Software, Lafayette, California, 15 pages.

Spencer, E., 1967, "A Method of Analysis of the Stability of Embankments Assuming Parallel Interslice Forces." Geotechnique, vol. 17, No. 1, pp. 11-26.

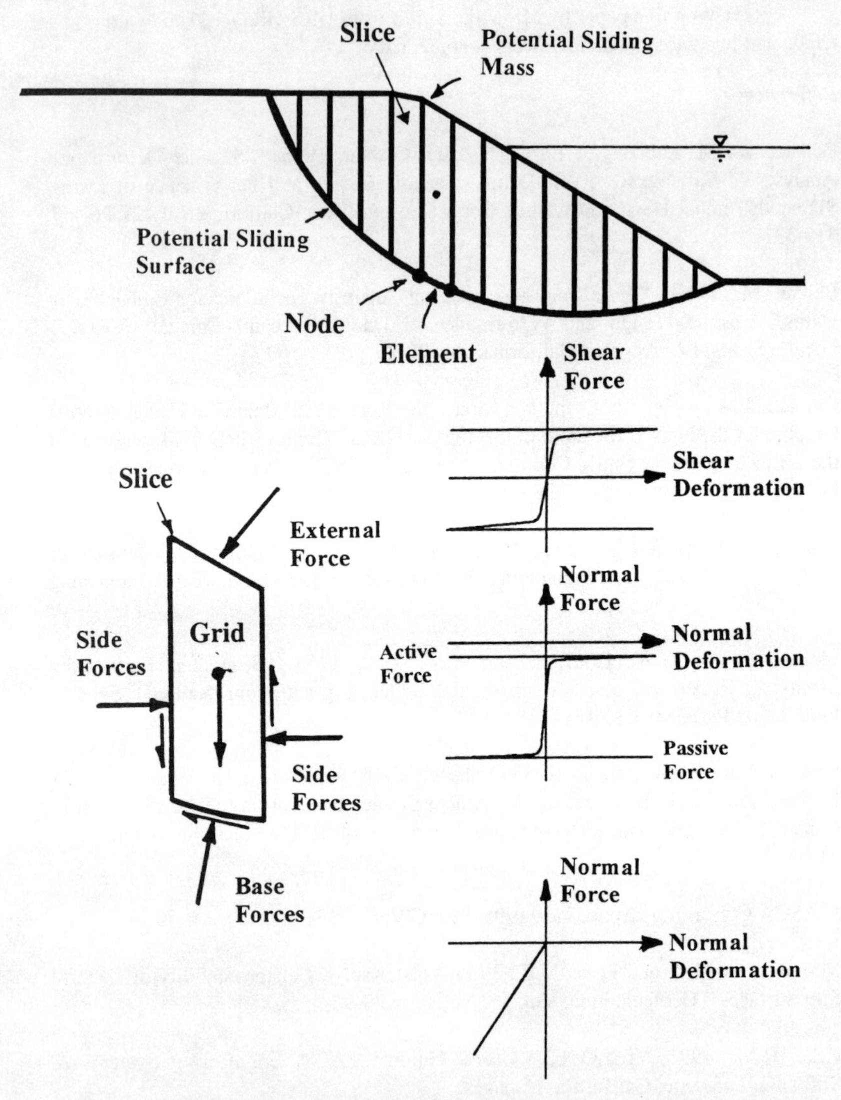

Figure 1. Two Dimensional Model of a Potential Sliding Block, Array of Slices, Contact and Body Forces, and Force Deformation Relationships

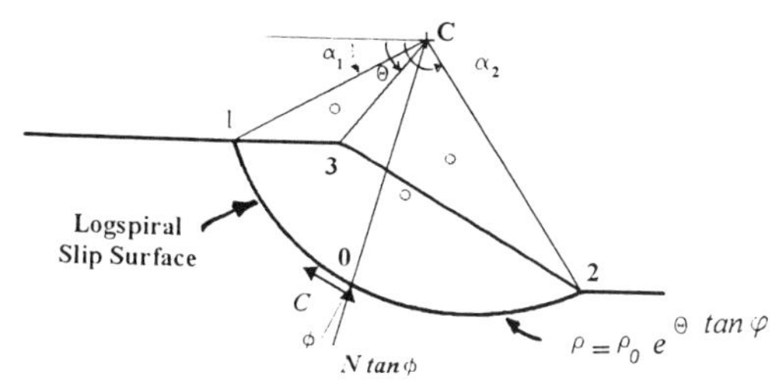

$$FS = 0.5\, C\, \rho_0^{\,2} \cos\varphi\, [\, \exp(2\alpha_2 \tan\varphi) - \exp(2\alpha_1 \tan\varphi)\,] / [\tan\varphi\,(W_X + A_h W_Y)]$$

$$k_y = \{0.5\, C\, \rho_0^{\,2} \cos\varphi\, [\, \exp(2\alpha_2 \tan\varphi) - \exp(2\alpha_1 \tan\varphi)\,] / \tan\varphi - W_X \} / W_Y$$

Where:

FS = Global factor of safety
α_1 = Slope angle between horizontal line and line C1 as shown above (radian)
α_2 = Slope angle between horizontal line and line C2 as shown above (radian)
ρ_0 = Radius as shown in the equation above (ft)
γ = Total unit weight of soil (pcf)
φ = Friction angle (degrees)
C = Cohesion (psf)
A_h = Horizontal acceleration coefficient
k_y = Horizontal yield acceleration coefficient
$W_X = W_G (X_C - X_G) - W_1 (X_C - X_{G1}) - W_2 (X_C - X_{G2})$
$W_Y = W_G (Y_C - Y_G) - W_1 (Y_C - Y_{G1}) - W_2 (Y_C - Y_{G2})$
W_G = Weight of segment C102 (p/ft)
W_1 = Weight of triangle C13 (p/ft)
W_2 = Weight of triangle C32 (p/ft)
X_C = X-coordinate of center of logspiral (ft) point C
Y_C = Y-coordinate of center of logspiral (ft) point C
X_G = X-coordinate of center of gravity of segment C102 (ft)
Y_G = Y-coordinate of center of gravity of segment C102 (ft)
X_{G1} = X-coordinate of center of gravity of triangle C13 (ft)
Y_{G1} = Y-coordinate of center of gravity of triangle C13 (ft)
X_{G2} = X-coordinate of center of gravity of triangle C32 (ft)
Y_{G2} = Y-coordinate of center of gravity of triangle C32 (ft)
X_1 = X-coordinate of point 1 as shown on the drawing (ft)
Y_1 = Y-coordinate of point 1 as shown on the drawing (ft)
X_2 = X-coordinate of point 2 as shown on the drawing (ft)
Y_2 = Y-coordinate of point 2 as shown on the drawing (ft)
X_3 = X-coordinate of point 3 as shown on the drawing (ft)
Y_3 = Y-coordinate of point 3 as shown on the drawing (ft)

Figure 2. Calculation of Factor of Safety and Yield Acceleration Coefficient for a Potential Sliding Block with a Logspiral Slip Surface

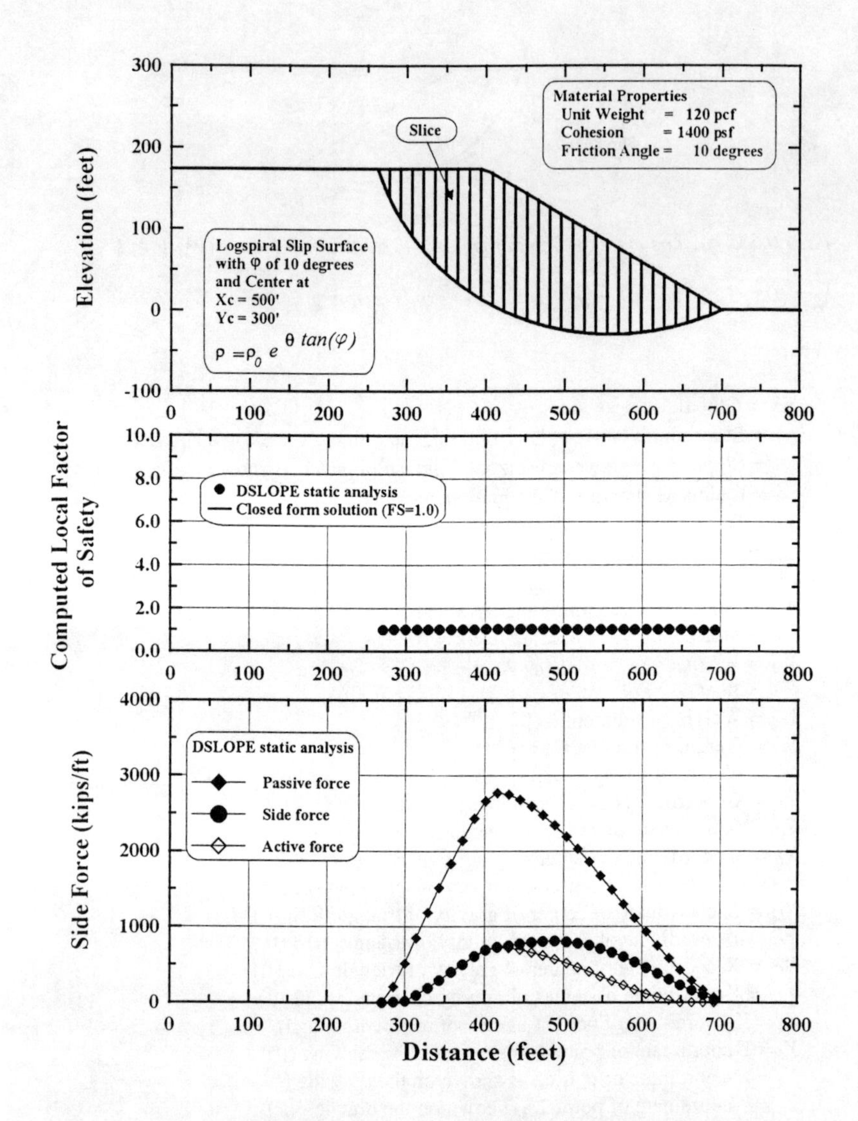

Figure 3. Results of DSLOPE Static Analysis and Closed Form Solution for Example No. 1

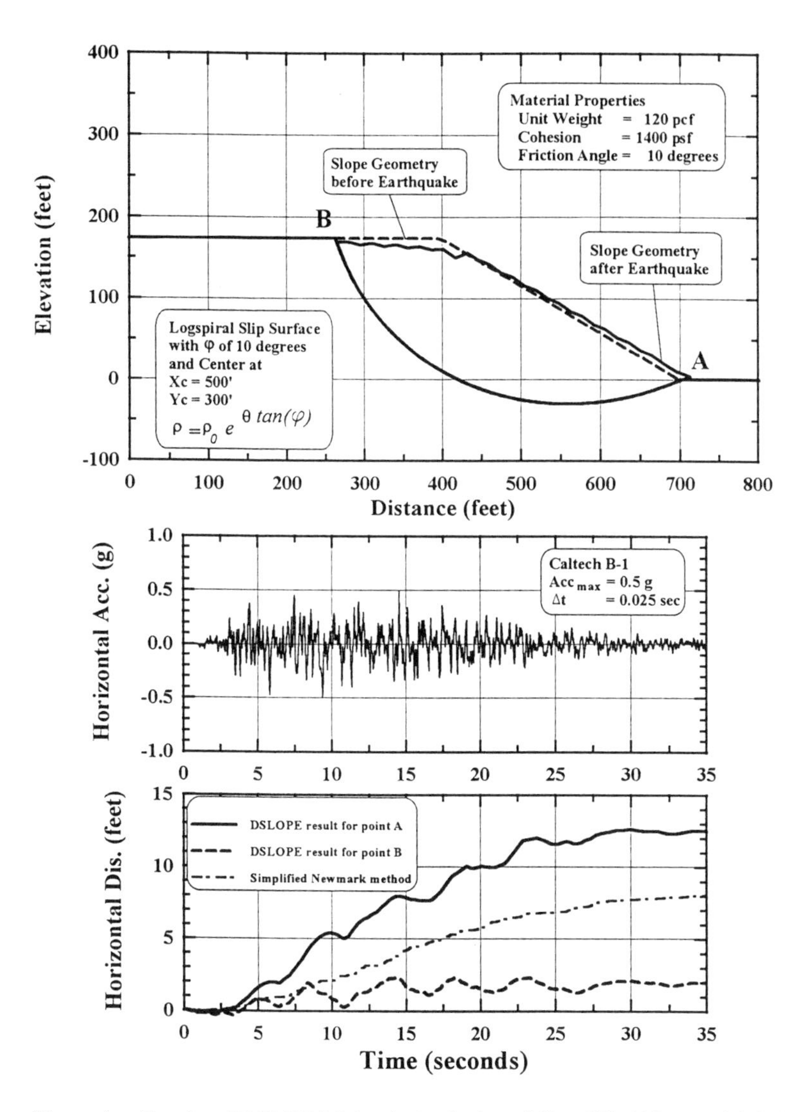

Figure 4. Results of DSLOPE Seismic Analysis and Simplified Newmark Method for Example No. 1

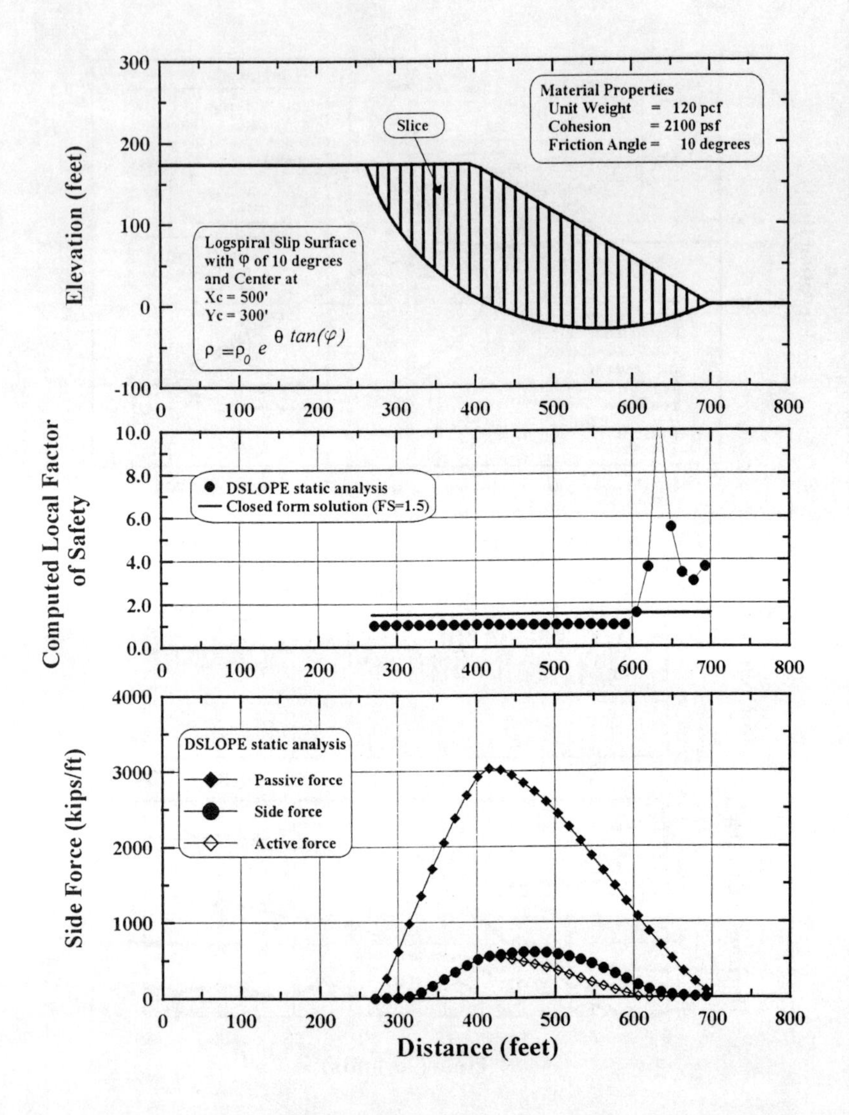

Figure 5. Results of DSLOPE Static Analysis and Closed Form Solution for Example No. 2

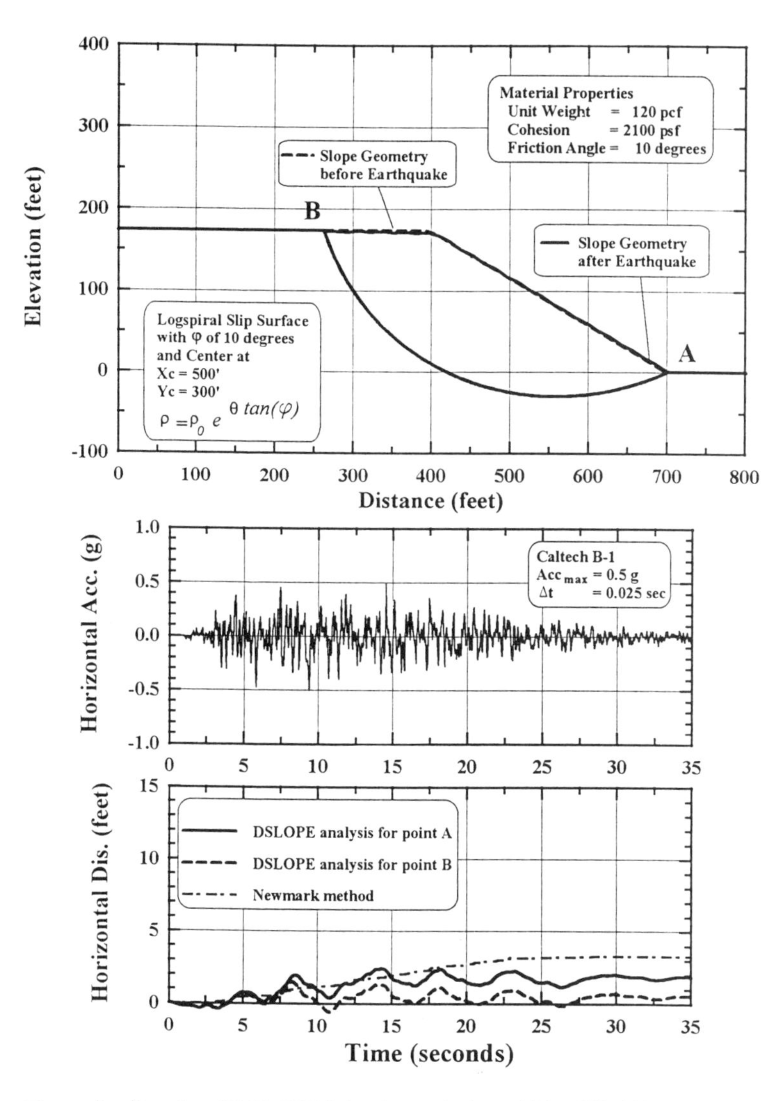

Figure 6. Results of DSLOPE Seismic Analysis and Simplified Newmark Method for Example No. 2.

Stability Assessment of an Old Domestic Waste Slope in Warsaw (Poland)

Abdelmalek Bouazza[1] and Michel Wojnarowicz[2]

Abstract

The slope stability analysis for MSW landfills is usually performed using conventional method of slices or translational wedge method considering potential failure surfaces at limit equilibrium. The safety factor is then determined by comparing the sum of the resisting forces to the driving forces, mobilised along the potential failure surfaces. Usually a safety factor of 1.5 is considered acceptable, other specific values can also be mandated by regulation. It is however uncertain whether or not conventional limit equilibrium methods are applicable to MSW because of their ability to undergo large strains without reaching failure. This paper presents a case study concerning the stability analysis of the Warsaw old town Gnojna Grora hill. The problem of its stability has its roots in the 17th century when Gnojna Grora was used as a municipal landfill. The paper describes its historical background, and the engineering analysis carried out to verify its stability. The paper shows that there is a high risk of failure of the old town hill if the hydraulic conditions change.

Introduction

In light of recent waste slope failures reported by Byrne et al. (1992), Mitchell (1996), Milanov et al. (1997), Pardo de Santayana & Veiga Pinto (1998), Schmucker & Hendron (1999), stability of domestic waste piles has started to receive more attention than in the past. In this respect, assessing the stability of waste fills has become a very important aspect of waste containment system analysis and design.

---

[1] Senior Lecturer, Department of Civil Engineering, P.O. Box 60 Monash University, Victoria 3800, Australia. Tel. 61-3-99054956, Fax: 61-3-9905 4944, email: malek.bouazza@eng.monash.edu.au

[2] Director, TERRASOL, Immeuble Helios, 72 Avenue Pasteur, Montreuil 91108 Cedex, France Ph: 33-1-49 88 24 42, Fax:33-1-49 88 06 66, email:Geokontruckja@terrasol.com

As with any stability study the selection of the most probable mode of failure and the accurate assessment of the necessary physical and mechanical properties and hydraulic conditions of the waste and the foundation soils are the most critical aspects. However, stability analyses for MSW landfills are more complex than those for classical earth structures as a result of the difficulties involved in evaluating the physical and mechanical properties of the waste, the interface interactions, and the variation of these parameters with depth. In addition, the variation of the waste properties with time may need to be considered in the analysis. As part of the stability analysis, the shape of the potential failure surface must be evaluated. Failure surfaces passing through the waste are generally circular. On the other hand if the stability along one of the interfaces (waste/liner, liner/foundation soil, etc.) is the most critical, the analysis may need to be performed considering a non-circular failure surface passing along the interface having the lowest strength (Bouazza & Donald, 1999, Brandl, 1999).

The purpose of this paper is to present a case study concerning the stability analysis of the Warsaw old town Gnojna Grora hill shown in Figure 1. The problem of its stability has its roots in the 17th century when Gnojna Grora was used as a municipal landfill

Figure 1 Gnojna Grora Hill

Site History

In order to carry out any type of geotechnical analysis on this particular site it was first necessary to conduct a historical research work to shed some light on the characteristics of the site. This search unearthed a wealth of interesting information, which showed that some form of waste management already existed in the Warsaw old town in the 15th century. Indeed, it was found that a landslide of part of the Vistula river bank, which occurred in the 15th century has led to the formation of a basin. This natural open basin became, with time, a uncontrolled municipal solid waste landfill where people tossed most of their garbage. The archeological work undertaken in the sector of the landfill shows as matter of fact vestiges of 13th and 14th centuries

The first historical data concerning the old city landfill goes back to 1620. The first project of refitting and rehabilitation of the landfill presented by the local city council date back to 18 June 1691. In 1722, a municipal law was promulgated to halt all landfill operations and to start the rehabilitation of the site. This led to the creation of a recreational park. A new municipal landfill was opened in the proximity of the old town in 1765. In 1766, a project was set up to partially recycle some of the domestic waste into commercial compost. However, in 1774 the king of Poland, Stanislas Poniatowski, ordered all landfill activities to stop and that the site be transformed into a landscape park. In the years after, the site was entirely developed and housed many thousands of people until its total destruction during the second world war. The reconstruction effort after the war restored the site to its original state. The current topographical and architectural character of this old landfill site stems from the various landscapes and architectural work completed in 1965. Structural disorders in the different buildings (some are shown in Figure 2) were noticed almost immediately after the completion of the reconstruction and renovation of the old town.

Figure 2. Some of the damages encountered in the old town of Warsaw

MSW Properties

One of the important tasks facing the geotechnical engineer involved in stability analysis is the quantification of relevant geotechnical properties of waste materials. Quantification of these properties is very difficult because: **1)** Municipal solid waste is inherently heterogeneous and variable among different geographic locations; **2)** There are no generally accepted sampling and testing procedures for waste materials; **3)** The properties of the waste materials change with time more drastically than those of soils. Furthermore, the complexity of the mechanical behaviour of domestic wastes makes the problem of landfill stability even more complicated to solve. The mechanical properties are dependent on the individual composition of the waste material and on the mechanical properties of its constituents. In addition, the mechanical parameters are time-dependent and related to the state of decomposition. In order to provide applicable parameters for stability or deformation analysis it might be useful to conduct appropriate tests consistent with the kinematics of potential failure problems for the specific cases. In any case, all so-called geotechnical parameters have to be implemented with engineering judgement.

Field investigations were undertaken to gather all the necessary information needed for the stability analysis. The investigations included the use of pressuremeters, a test boring program, and test pits. The purpose of these investigations was to determine the waste fill thickness in various locations and to characterise the nature of the existing waste fill and the underlying natural overburden materials. The study was conducted in two zones as shown in Figure 3. Zone 1, which included Gnojna Grora Hill and zone 2, which included the flatter side of the old town (sections 1 to 6).

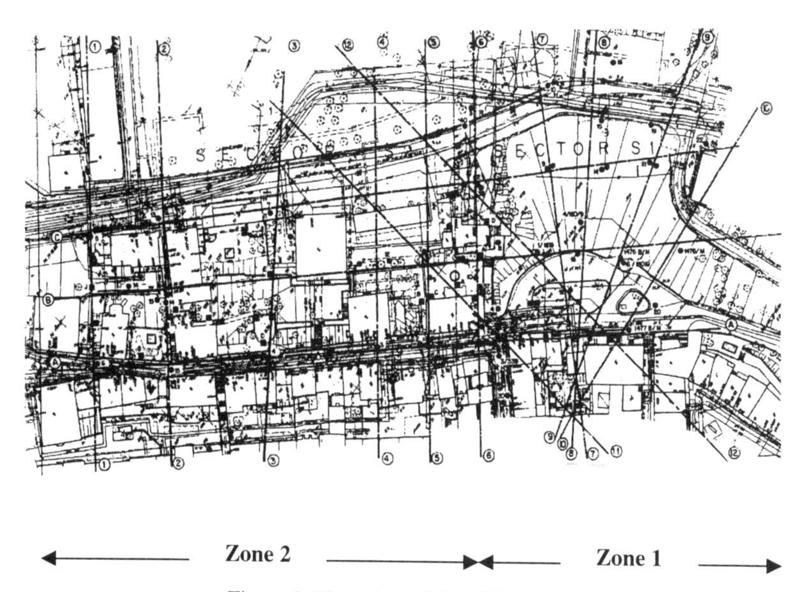

Figure 3. Plan view of the old town site

The waste depth varied between 5 m to 30 m, the deepest being in zone1. Only zone 1 will be considered in the present paper since it represented the most critical area. Four types of materials were encountered in zone 1, namely, a waste fill (WF) with strong concentration of demolition debris mixed with old domestic waste. Under this layer three types of wastes could be distinguished: Upper waste (UW) layer, relatively "young" in appearance, intermediate waste (IW) layer and lower waste (LW) layer. The pressuremeter test results and geotechnical properties of the wastes are given in Table 1and 2 respectively. Based on the piezometers records, the leachate/water table was found to vary between 3 to 5 m below ground surface depending on piezometers locations.The pressuremeter test results are given in Table 1 and compared to other values previously published (Gotteland et al., 1995). The pressure limit and pressuremeter modulus, obtained from the present tests, compares reasonably well with the results obtained for soils. This suggests that the waste material is relatively stiff with the lower bound values corresponding to the waste being underwater. The testing on the waste fill (WF) were inconclusive due to its high heterogeneity and therefore were not included in Table 1.

Table 1: Pressuremeter tests results in old domestic waste fill

	Soft Clay*	Marl*	Silt*	Road fill*	Young waste*	UW	IW	LW
Limit pressure, Pl (Mpa)**	0.05 to 0.3	0.6 to 4	0.2 to 1.5	0.4 to 1	0.5 to 1.4	0.9	0.6	0.5
Pressuremeter modulus, E_m (Mpa)	0.5 to 3	5 to 60	2 to 10	4 to 15	4 to 10	6	4	3.4

* (from Gotteland et al., 1995),

** the limit pressure corresponds to the limit of material rupture at the cavity wall

The geotechnical properties of materials UW, IW, and LW were assessed based on field and laboratory tests. The fact the materials were very homogeneous facilitated greatly their assessment. However, the properties of the waste fill (WF) material were difficult to estimate due to its extreme heterogeneity. Nominal parameters were assumed for its mechanical properties based on failure envelopes proposed by Kavazanjian et al (1995) and Manassero et al. (1996). The selected values of ϕ' and c' represented a baseline case for this study. The cohesion was also assigned values varying up to 15 kPa to account for the "Reinforced soil" like behaviour of wastes reported by Gabr & Valero (1995), Manassero et al. (1996), Mitchell (1996) and Brandl (1999). The value of the friction angle was also varied to account for the uncertainties in its estimation.

It should be stressed that the interpretation of shear tests on waste using soil mechanics models can be useful, at least at the present state of knowledge. However, one should also acknowledge the fact that there are significant deviations from conventional soils; the void ratio of waste is very high, which implies an unusually large volumetric compressibility. The waste particles are of very different natures, and some are weak and very deformable or crushable in themselves. There is a process of decomposition with time, which causes self-induced settlements and a variation of properties with time (Bouazza & Wojnarowicz, 1999). In this respect, the interpretation of the results of the tests on waste remains subject to many uncertainties.

Table 1: Geotechnical properties of domestic waste encountered in zone1

Type of Waste	Average Unit Weight, γ (kN/m^3)	Moisture content, w (%)	Effective friction angle, ϕ' (°)	Cohesion, c' (kPa)
WF	17	-	25	0
UW	14	28	30	12
IW	12	41	27	5 to 10
LW	12	80	24	5

Slope stability analysis

The slope stability analysis for MSW landfills is usually performed using conventional method of slices or translational wedge methods, considering potential failure surfaces at limit equilibrium. The safety factor is then determined by comparing the sum of the resisting forces to the driving forces mobilised along the potential failure surfaces. According to current practice, a safety factor of 1.5 is considered acceptable; other specific values can also be mandated by regulation. It is however uncertain whether or not conventional limit equilibrium methods are applicable to MSW because of their ability to undergo large strains without reaching failure. Since these large deformations are not acceptable for the performance of the collection and containment systems, an at-serviceability state approach may be more appropriate.

The stability of the Gnojna Grora Hill was analysed using the TALREN97 computer program developed by Terrasol. TALREN 97 is based on classical slope stability methods considering a failure surface at limit equilibrium. The Bishop's simplified method of slices for circular slips was chosen for this work. The following cases were considered:

1- a leachate/water table at about 3 to 5 m below ground surface as measured on site. In the present context, it will represent a normal leachate/water table level experienced by the landfill.
2- a leachate/water table at about 2.5 to 4.5 m below ground surface to account for a potential rise of leachate/water head. It will be referred here as high leachate/water level.

A sensitivity analysis using different ϕ' and c' values for the waste fill was also conducted to identify the worst case scenario. A typical cross section of the hill is shown in Figure 4.

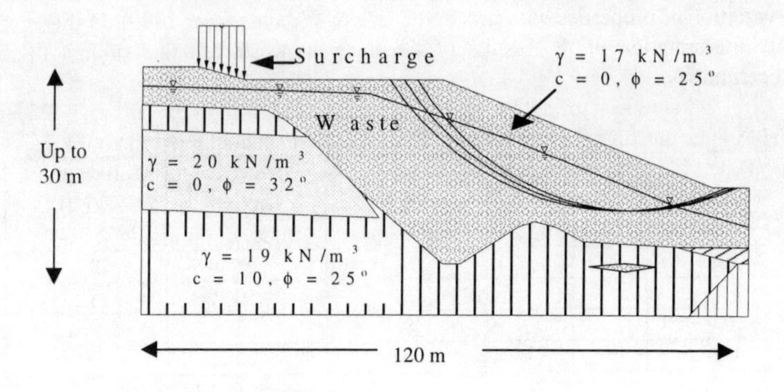

Figure 4. Cross section 7 of the old town hill

The analysis gives a minimum factor of safety (FS) corresponding to the various assumed values of ϕ' and c' for the waste fill and the leachate/water conditions. These results are summarized in Table 3.

Table 3: Summary of safety factors (FS).

Normal leachate/water level				High leachate/water level			
$\gamma = 17$ kN/m^3, ϕ'=25°		$\gamma = 17$ kN/m^3, c'=0 kPa		$\gamma = 17$ kN/m^3, ϕ'=25°		$\gamma = 17$ kN/m^3, c'=0 kPa	
c' (kPa)	FS	ϕ' (°)	FS	c' (kPa)	FS	ϕ' (°)	FS
0	1.26	25	1.26	0	1.14	25	1.14
3	1.34	28	1.43	3	1.22	28	1.30
6	1.42	31	1.62	6	1.30	31	1.46
9	1.50	34	1.82	9	1.39	34	1.64
12	1.58	37	2.00	12	1.47	37	1.84
15	1.66	40	2.12	15	1.55	40	1.96

The effect of cohesion and angle of shearing on the factor of safety can be readily seen in the above table. Obviously, the slightest reduction in the waste properties can lead to a critical situation where the factor of safety will fall in the marginal zone (i.e FS =1 to 1.2). Although it appears logical to include cohesion in the evaluation of any waste slope stability analysis, it is strongly suggested to not consider it as a design parameter since its value is unreliable. Brandl (1999) pointed out that it is likely that the cohesion values will decrease in the long term due to bio-chemical and physical degradation of the waste. Considering that the waste in the case of Gnojna Grora hill is about 300 years old, it is expected that the cohesion value will be about 0 kPa. The fact that the landfill is unlined and does not have any leachate collection or removal system is also a concern since no leachate has been removed from the site. Obviously, this situation has led to a build up with time of a leachate/water head of 20 m in certain areas acting on the base of the landfill and also to a saturation of most of the waste. Evidently, the repercussions of any changes on the actual leachate/water level conditions will be on the stability of the hill since the safety factor can drop dramatically to a dangerous level. This is illustrated in Table 3 where a hypothetical increase of 0.5 m in the leachate/water level lowered the factor of safety to 1.1 (for c' = 0 kPa and $\phi' = 25^{o}$).

A stability risk map of the area shown in Figure 5 has been produced on the basis of the stability analysis carried out in the present work. It identified the areas with a high risk of instabilities. A remedial work based on the installation of a drainage system and strengthening of the slope and existing building has been proposed, a monitoring system including settlement plates and inclinometers has been installed.

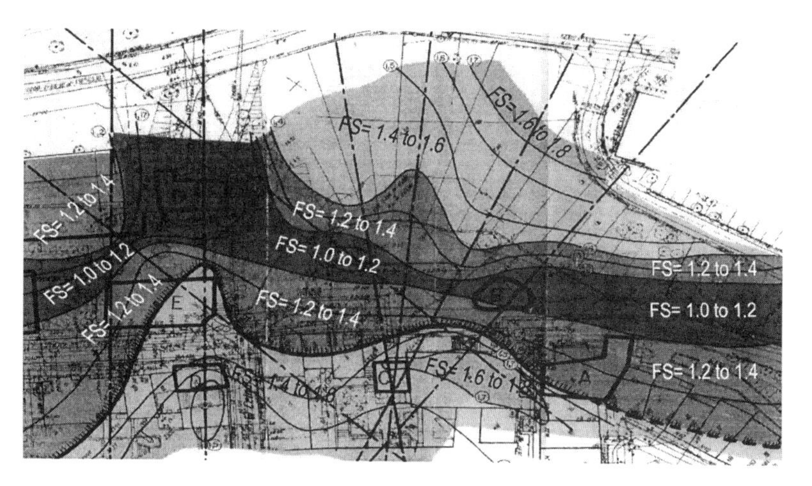

Figure 5. Stability risk map for the Gnojna Grora Hill

Conclusions

The stability analysis of municipal solid waste slope is a challenging task as the mechanical behaviour of waste is of a very complex nature. Unfortunately, the present state of knowledge is very limited, resulting in a renewed interest to have a better quantification of the geotechnical properties of wastes. Nevertheless, analysis of stability can be made using conventional stability analysis methods. For the present case, the stability analysis shows there is a high risk of failure of the old town Gnojna Grora hill if the hydraulic conditions change. A remedial work based on the installation of a drainage system and strengthening of the slope and existing building has been proposed. A monitoring system including settlement plates and inclinometers has also been installed to follow the movements of the hill and detect any sudden changes in its stability.

Acknowledgments

This paper has been prepared while the first author was on sabbatical leave at the University of Missouri-Columbia. The support provided by the Faculty of Engineering, Monash University and the Department of Civil & Environmental Engineering, College of Engineering and University Office of Research, University of Missouri-Columbia during his stay is gratefully acknowledged.

References

Bouazza, A. & Wojnarowicz, M. (1999). Geotechnical properties of municipal solid waste and their implications on slope stability of waste piles. Proc. XI Panamerican Conf. On Soil Mechs. & Geotech. Engng., Fos de Iguassu, Brazil, vol. 1, pp.489-495.

Bouazza, A. & Donald, I.B. (1999). Overall stability of MSW landfills: Use of an improved limit equilibrium analysis. Proc. Int. Conf. On Slope Stability, Shikoku, Japan, Vol. 2, pp.863-868.

Byrne, R.J., Kendall, J. & Brown, S. (1992). Cause and mechanism of failure, Kettleman hills landfill B-19, Unit 1A. Stability and Performance of Slopes and Embankments, ASCE. Geotech. Spec. Publ. No 31, pp. 1188-1520.

Brandl, H. (1999) Slope failure and spreading of waste deposits. Proc. 2nd Int. Conf. On Landslides, Slope Stability & The Safety of Infrastructures, Singapore, pp. 1-12.

Gabr, M.A. & Valero, S.N. (1995). Geotechnical properties of municipal solid waste. Geotechnical Testing Journal, Vol 18, No.2, pp. 241-251

Gotteland, P., Lemarechal, D. & Richard, P. (1995). Analysis and monitoring of the stability of a domestic waste landfill. Proceedings 5th Int. Landfill Symposium, Cagliari, Italy, pp. 777-787.

Kavazanjian, E.Jr., Matasovic, N., Bonaparte, R. and Schmertmann, G.R. (1995). Evaluation of MSW properties for seismic analysis. Geoenvironment 2000, Geotech. Spec. Publ. No 46, ASCE, Vol. 2, pp. 1126-1141.

Manassero, M., Van Impe, W.F. & Bouazza, A.(1996). Waste disposal and containment. Proc. 2nd Int. Congress Env. Geotech., Osaka, Japan, Vol.3, pp.1425-1474.

Milanov, V., Corade, J.M., Bruyat-Korda, F.& Falkenreck, G. (1997). Waste slope failure at the Rabastens landfill site. Proceedings 6th Int. Landfill Symposium, Cagliari, Italy, pp. 551-556.

Mitchell, J.K. (1996). Geotechnics of soil waste material interactions. Proc. 2nd Int. Congress Env. Geotech., Osaka, Japan, Vol.3, pp.1425-1474.

Pardo de Satayana, F. & Veiga Pinto, A.A. (1998). The Beirolas landfill eastern expansion landslide. Proc. 3rd Int. Congress on Env. Geotech., Lisbon , Portugal, Vol. 2, pp 905-910,

Scmucker, B.O. & Hendron, D.M. (1999). Forensic analysis of the 9th march 1996 landslide at the Rumpke sanitary landfill, Hamilton county, Ohio. Proc. 12th GRI Conference, Philadelphia, USA, pp. 269-295.

Important Factors to Consider in Properly Evaluating the Stability of Rock Slopes

Douglas D. Boyer, P.E., P.G., C.E.G.[1]
Keith A. Ferguson, P.E.[2]

Abstract

Evaluating sliding stability of rock slopes is probably one of the most overlooked and least understood factors in geotechnical engineering today. The literature has numerous case histories of rock sliding failures that have lead to catastrophic failure of critical structures, some of the most notable being St. Francis Dam, California and Malpasset Dam, France. These case histories serve as reminders that the inadequate or improper identification, characterization, and/or evaluation of the sliding of rock blocks formed by critically oriented discontinuities within a rock mass can lead to catastrophic loss of life and/or personal property.

Inherent in any characterization and evaluation of subsurface features, whether it is for dams, tunnels, slopes, foundations, etc., is the uncertainty in identifying the critical features present in the rock mass. Of utmost importance in rock slope stability evaluations are the characteristics of the discontinuities and surfaces of weakness in the rock mass. These characteristics include: orientation and attitude, infilling type, seepage conditions, roughness, among others. In addition to single discontinuities, the intersection of critical discontinuities forming potential 3-dimensional blocks must also be considered.

---

[1] Senior Engineering Geologist/Geotechnical Engineer, GEI Consultants, Inc., 6950 S. Potomac Street, Suite 200, Englewood, CO 80112, phone (303) 662-0100, fax (303) 662-8757, e-mail: dboyer@geiconsultants.com.

[2] Senior Technical Manager/National Director of Marketing, GEI Consultants, Inc., 6950 S. Potomac Street, Suite 200, Englewood, CO 80112, phone (303) 662-0100, fax (303) 662-8757, e-mail: kferguson@geiconsultants.com.

Introduction

It is not surprising that the evaluation of sliding stability of rock slopes and foundations is generally overlooked or improperly evaluated. The primary reason for this stems from the general limited understanding of three technical fields of expertise: 1) engineering geology, 2) geological or geotechnical engineering, and 3) rock mechanics. Knowledge of all three fields of expertise must be employed by the individual or team evaluating or reviewing the sliding stability of rock slopes to adequately characterize the threat of sliding.

Important factors in evaluating the sliding stability of rock slopes include: 1) identification and evaluation of potentially removable blocks within the rock mass, 2) evaluation of the shear strength of discontinuities in the rock mass, 3) evaluation of uplift pressures acting on the joint plane(s) within the rock mass, and 4) use of appropriate analysis methods.

It is the experience of the authors of this paper that the evaluation of sliding stability of rock slopes is generally either overlooked or improperly evaluated by engineers today. The authors have been involved in the review of a number of rock slope stability evaluations by others where critical elements in the evaluation process have been missed or not properly considered. Some of these critical elements have included:

1. Not properly considering or ignoring the geology and structural discontinuities in the rock mass.
2. Not properly evaluating the shear strength of the rock mass (intact or discontinuities).
3. Not properly identifying appropriate loading conditions, and particularly possible water or uplift pressures.
4. Not using proper or available analysis methods (two dimensional vs. three dimensional).

The premise that the evaluation of sliding stability of rock slopes is generally overlooked or improperly evaluated is further evidenced by the fact that, in a recent, well-known publication on landslides (Transportation Research Board, 1996), less than one-half of a page is devoted to a discussion of three-dimensional rock slope stability analyses in the 35 page chapter entitled "Rock Slope Stability Analysis". Evaluating the sliding stability of rock slopes is NOT the same as analyzing soil slopes. Typically, the stability of rock slopes involves the three-dimensional intersection of discontinuities and/or other planes of weakness in the rock mass that can only be properly evaluated using three-dimensional analytical techniques.

Approach

To properly evaluate the sliding stability of rock slopes, the investigator(s) must use an integrated approach which draws knowledge from three fields: engineering geology, geotechnical engineering, and rock mechanics. The lack or limited knowledge of any one of these fields often results in the improper characterization and/or evaluation of rock stability problems today. This may lead to the over-design of slope repair modifications through the use of overly conservative and often inappropriate assumptions or failure of the slope due to neglect of critical loading combinations or other site conditions.

Important factors to consider in evaluating the stability of rock slopes are briefly discussed below.

Geologic Data - The proper identification and collection of geological site information by qualified personnel (i.e., engineering geologists) is the first step in identifying and evaluating the sliding stability of any structure founded on rock. This not only includes the type and characteristics of the rock, but also, and sometimes more important, the discontinuities or planes of weakness in the rock. The following key geologic discontinuity factors are required to adequately identify potential sliding foundation blocks and to provide input in the evaluation of the shear strength of the discontinuities.

1. The location of the discontinuity relative to the structure or slope geometry in question.
2. The orientation (strike and dip) of the discontinuity.
3. The persistence or length of the observable discontinuity.
4. The roughness (both micro-roughness and macro-roughness) of the discontinuity surface.
5. The type and characteristics of any infilling within the discontinuity, if present.
6. The seepage characteristics of the discontinuity.

Two excellent references that provide procedures for measuring and describing the above factors include: 1) The U.S. Bureau of Reclamation Engineering Geology Field Manual (USBR, 1976), and 2) the International Society for Rock Mechanics Suggested Methods for the Quantitative Description of Discontinuities in Rock Masses (ISRM, 1978).

Geologic Data Interpretation - Once collected, the geologic data will require evaluation and interpretation to identify individual and "sets" of discontinuities for further investigation. For example, the use of structural geologic tools will be required to identify sets or groups of discontinuities with similar orientations. This usually entails the use of stereonet projections to identify groupings or sets of joints and the angular relation of multiple joint sets. This task can be aided by the use of commercially available computer software. Again, this task should be performed by qualified personnel familiar with the evaluation and interpretation of geologic data.

Identification of Potential Failure Modes - In general, the identification of potential failure modes of the rock mass is more complex than for soil slopes. In general, the failure modes of hard rock masses include: planar, wedge, and topple. A fourth failure mode, block, is a subset of the wedge failure mode. Existing commercially available computer software can assist in the identification of potential failure modes.

Estimation of Material Properties - Material properties will be required for any sliding stability evaluation. The most common required material properties for input to the stability analyses include the unit weight of the rock mass and the shear strength of the identified discontinuities. Additional material properties may include the shear strength of the intact rock mass. The investigator(s) must select a credible lab capable and experienced in performing joint direct shear testing.

Shear Strength Determination of Discontinuities - Depending of the scope of the stability evaluation, site-specific measurements of the shear strength of the rock mass discontinuities may not be available. A number of references are available in which to guide (or mis-guide) the investigator(s). (A partial reference list is provided at the end of this paper). The estimation of the shear strength of rock mass discontinuities is one of the most important and difficult factors in evaluating the sliding stability. Far too often investigators rely solely on the literature to estimate the shear strength of discontinuities. Although conservative estimates of shear strength can be helpful in performing preliminary stability analyses, this practice often results in inadequate factors of safety for sliding and unnecessary costs to mitigate unfounded stability problems.

In general, there are four main components of the shear strength of the discontinuity that must be considered:

1. Basic or Fundamental Friction - This is the basic friction angle of the rock independent of irregularities on the discontinuity surface. The basic friction angle is a function of the texture and mineralogy of the rock.
2. Micro-roughness -Micro-roughness is the small-scale roughness (on the order of inches) attributed to small irregularities or asperities on the discontinuity surface which require the rock to dilate to shear under low normal stress conditions.
3. Macro-roughness - Macro-roughness is similar to micro-roughness, but differs in scale. Macro-roughness is typically on the order of feet. Macro-roughness is the larger-scale effect of irregularities on the discontinuity surface, similar to waviness or changes in the orientation over distance of the discontinuity surface. Again, in the field, the block

must dilate to shear under low normal stress conditions. Macro-roughness is a field measurement and cannot be determined in the laboratory.

4. Infilling - Infilling characteristics of the discontinuity, if present, may reduce or increase the shear strength of the discontinuity depending on the type and characteristics of the infilling material.

Many investigators ignore the beneficial aspects of the micro- and macro-roughness on evaluating the available shear strength of the discontinuities. Micro-roughness can be estimated in the lab if both natural discontinuity surfaces and saw-cut surfaces are tested using direct shear apparatus. However, care must be employed in the use of micro-roughness and macro-roughness values. For example, at high normal stress conditions, small- and/or large-scale asperities on discontinuity surfaces may be sheared off or through resulting in no long term benefit of these values. Also, both micro-roughness and macro-roughness may have directional qualities that must be considered when evaluating the potential sliding direction of the rock block.

Laboratory direct shear strength testing of appropriate samples of discontinuities obtained from a project site can be invaluable sources of information to estimate the overall shear strength of a discontinuity surface. However, one must realize that testing the shear strength of a small sample of rock in the laboratory should only be used as a guide to evaluate the actual shear strength of the entire joint surface. Other factors that must be considered include:

1. The macro-roughness or waviness variables of the joint surface.
2. The representativeness of the laboratory sample. Many times only the "best" samples can be selected for testing since poorer quality samples many not have been obtainable or suitable for testing.
3. The continuity of the joint surface and whether or not intact portions of the rock are present along the principal plane of the joint surface.
4. The appropriateness of the normal stresses used in the laboratory testing as compared to the estimated normal stresses on the discontinuity surface(s) in the field.

Loading Conditions - The loading conditions evaluated for the sliding stability are subject to the site and project conditions and configurations, but generally consist of static and earthquake loading. However, other loading conditions may exist that may not be readily apparent. As an example, under certain circumstances, ice may form at discharge areas of critical joints that may reduce or eliminate the drainage of discontinuities in a slope. The subsequent build-up of water pressures behind the slope face may be sufficient to cause failure of the slope.

In evaluating the loading conditions, the investigator(s) must identify all the forces acting on the block, including gravitational forces, and all external forces, including uplift, and seismic loads, if necessary. Individual forces should be determined by magnitude and direction (trend and plunge) and summed to obtain a single resultant load to evaluate the factor of safety.

Analysis Methods - The investigator has many possible analysis methods in which to perform the stability analyses. In general, these tools are different from the analysis methods and computer programs available for evaluating the stability of soil slopes.

Some of the available rock slope stability computers programs/analysis methods include:

Program/Method	**Available through**	**Type of Surfaces**
ROCKPACK II	Rockware	Planar and wedge
Keyblock	PanTechnica	Planar, wedge, and block
Generalized Wedge Method	University of Toronto	Planar and irregular
U.S. Army Corps of Engineers Method	U.S. Army Corps of Engineers	Planar and irregular
--------	University of Tennessee	Planar, wedge, and irregular
Three-dimensional DDA	still in research phase	Planar, wedge, and block

Each of these analysis methods and computer programs have their advantages and limitations as to the appropriateness of the application. Some are limited to geometry-specific applications, while other are more general and more thorough in their ability to evaluate complex geometric block configurations. It should be noted that the above list does not represent every available program or method available to evaluate rock slopes, but rather a sampling of programs and methods known and/or used by the authors for rock slope stability. It should also be noted that the authors do not specifically endorse nor warrant the results obtained from any of the above computer programs or suppliers.

Interpretation of the Results - Once the "moment of truth" has arrived (i.e., the numerical value of the factor of safety has been calculated) the work is over. Right? WRONG! It is still up to the investigator to interpret the appropriateness of the results and compare the results with either generally accepted or recommended minimum factors of safety or otherwise determine if the stability results indicate satisfactory or unsatisfactory results. Uncertainties in the data, conservatism in the approach and evaluations, the risk of failure, among other factors, must be considered in evaluating the results of the analyses and the appropriate "minimum acceptable factor of safety".

Case Histories

The following case histories have been selected to illustrate some of the important factors discussed in the sections above.

New Elmer Thomas Dam, Oklahoma - New Elmer Thomas Dam is a 113-foot-high roller-compacted concrete dam located in southwestern Oklahoma. The dam, which was completed in 1993, is owned and operated by the U.S. Department of the Interior, U.S. Fish and Wildlife Service. The dam is founded on moderately jointed, medium-grained granite. During the foundation excavation in 1992, the foundation bedrock was mapped in detail, and observations of exposed abutment slopes indicated that there were potential foundation sliding stability concerns in the right abutment and central portion of the foundation.

After completion of the foundation excavation, detailed geologic mapping of the bedrock discontinuities was performed. More than 400 discontinuities were mapped and described in the foundation area. In general, the discontinuities were described as slightly rough to smooth, very closely to widely spaced, low to very high persistence, and very tight to tight aperture. The shear strength of the joint surfaces (basic friction and micro-roughness components) were determined from laboratory direct shear strength testing performed during design from samples collected from the final design exploration program.

Static and seismic loading conditions were evaluated with a horizontal ground acceleration component of 0.80g used for the seismic loading condition.

Two critical foundation blocks were identified based on the detailed geologic mapping and subsequent evaluation of the data. These blocks were formed by the intersection of three discontinuity surfaces. The plan view of the right abutment block is shown on Figure 1. A three-dimensional stereonet projection modal analysis method was used to evaluate factor of safety of the foundation blocks (Goodman, 1976). This analysis method is capable of identifying all possible modes of failure of a particular tetrahedral block. The actual potential mode of failure for a particular block is dependant on the location of the resultant block force on the projection. The frictional resistance against sliding of the planes are represented by friction circles on the plot. The factor of safety is then determined graphically based on the plotted location of the resultant force relative to the friction circle on the stereonet projection. The modal analysis of the right abutment block is shown on Figure 2.

Florence Lake Dam, California - Florence Lake Dam is an existing 150-foot-high multiple arch buttress dam owned and operated by Southern California Edison. The dam is founded on granitic rock with exfoliation, cooling, and tectonic jointing. Recent inspections of the dam identified the need to evaluate the static and seismic stability of the foundation.

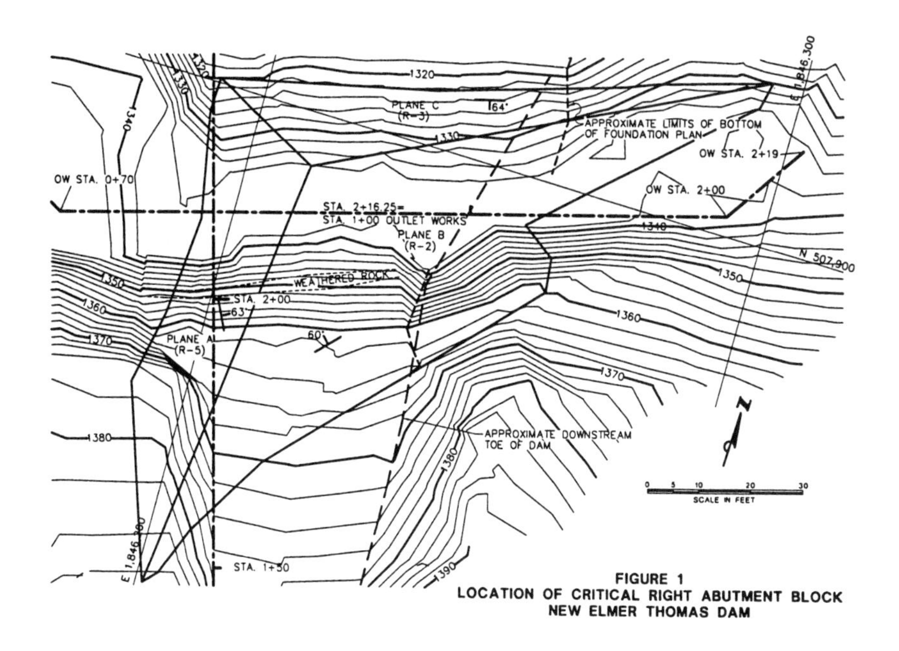

FIGURE 1
LOCATION OF CRITICAL RIGHT ABUTMENT BLOCK
NEW ELMER THOMAS DAM

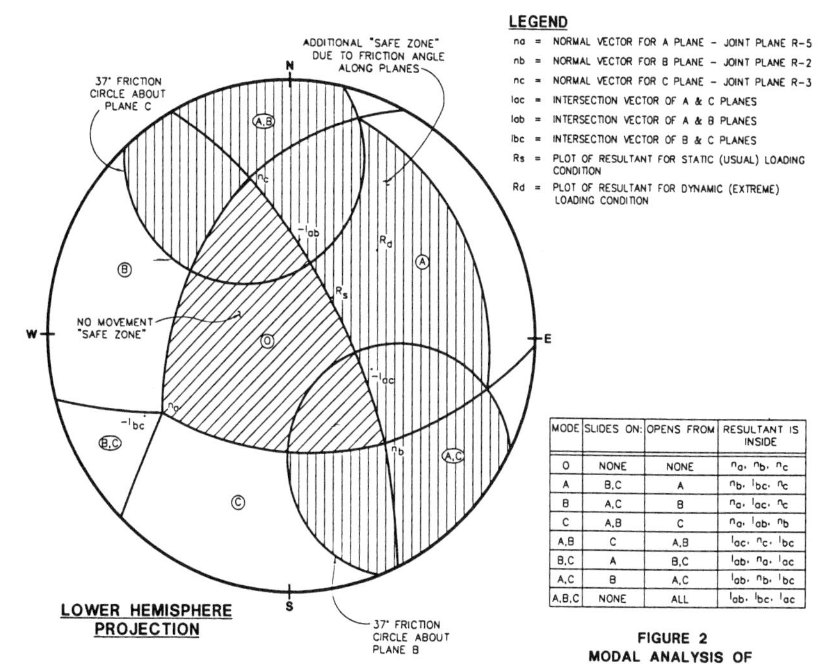

MODE	SLIDES ON:	OPENS FROM	RESULTANT IS INSIDE
O	NONE	NONE	n_a, n_b, n_c
A	B,C	A	n_b, I_{bc}, n_c
B	A,C	B	n_a, I_{ac}, n_c
C	A,B	C	n_a, I_{ab}, n_b
A,B	C	A,B	I_{ac}, n_c, I_{bc}
B,C	A	B,C	I_{ab}, n_a, I_{ac}
A,C	B	A,C	I_{ab}, n_b, I_{bc}
A,B,C	NONE	ALL	I_{ab}, I_{bc}, I_{ac}

FIGURE 2
MODAL ANALYSIS OF
CRITICAL RIGHT ABUTMENT BLOCK

Field mapping and review of available construction information were used to locate and identify discontinuity surfaces in the foundation rock in the vicinity of the dam. Geologic evaluations identified three potential three-dimensional blocks in the foundation below the dam, as shown on Figure 3. Both planar and block failure modes were identified. Three-dimensional preliminary stability analyses were performed using the same graphical procedure as described for the previous case history. The results of the preliminary stability analyses using conservative shear strength assumptions and loading conditions indicated that more rigorous evaluations should be performed.

Site investigations were performed in the areas of the identified three-dimensional blocks which included drilling of borings to intercept identified discontinuities and additional geologic mapping to identify the macro-roughness (waviness) of the discontinuities. Samples of the discontinuity surfaces were obtained for laboratory testing and piezometers were installed to measure water pressures at the discontinuities. Direct shear strength testing was performed on selected samples of the discontinuities to measure the shear strength of the samples (basic and micro-roughness components). Based on the data collected from the geologic mapping of discontinuities and the results of the shear strength testing performed on the discontinuities, estimates of the discontinuity shear strength were performed. This information was used in the supplemental stability evaluations to estimate the usual, extreme, and post-earthquake stability of the three identified foundation blocks. The extreme factors of safety were calculated for each 0.01 second of the earthquake time history. The plan view and modal analysis of one of the blocks are shown on Figures 4 and 5 respectively.

Milltown Hill Dam, Oregon - Milltown Hill Dam is a proposed new 190-foot-high roller-compacted concrete dam located in southwestern Oregon. Explorations performed for final design of the dam identified the presence of weak, downstream dipping claystone interbeds sandwiched between sub-vertical jointed basalt flows in the foundation of the dam. Sliding along continuous claystone interbeds within the foundation of the dam is considered to be the most critical mode of potential instability in the foundation. Since the claystone interbeds dip downstream, a passive wedge of intact basalt will contribute to the overall sliding resistance.

Stability analyses were performed at 13 cross section locations along the dam profile. A typical stability section is shown on Figure 6. Two-dimensional limit equilibrium sliding stability analyses were performed using the U.S. Army Corps of Engineers method (USCOE, 1981). Three-dimensional stability analyses were not possible due to the very limited information available on the geologic discontinuities from the dam site and sparse bedrock outcrops in the area.

The strength of the claystone interbeds was determined by laboratory joint direct shear strength testing (basic and micro-roughness). The shear strength of the intact rock mass portion of the failure surface was estimated using the Hoek-Brown failure criterion (Hoek and Brown, 1997). The Hoek-Brown failure criterion is one of the few techniques available for estimating the rock mass strength from geological and limited

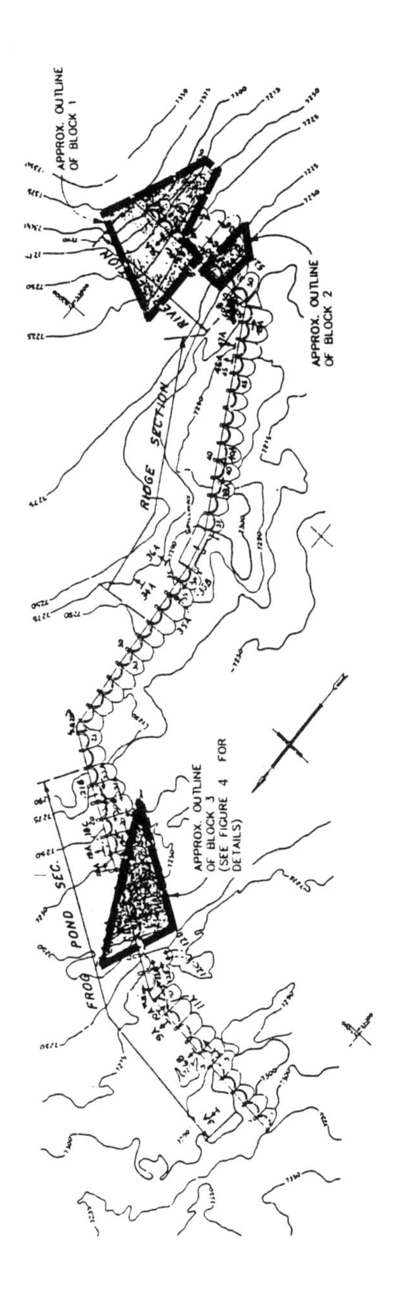

(NOT TO SCALE)

FIGURE 3
LOCATION OF CRITICAL FOUNDATION BLOCKS
FLORENCE LAKE DAM

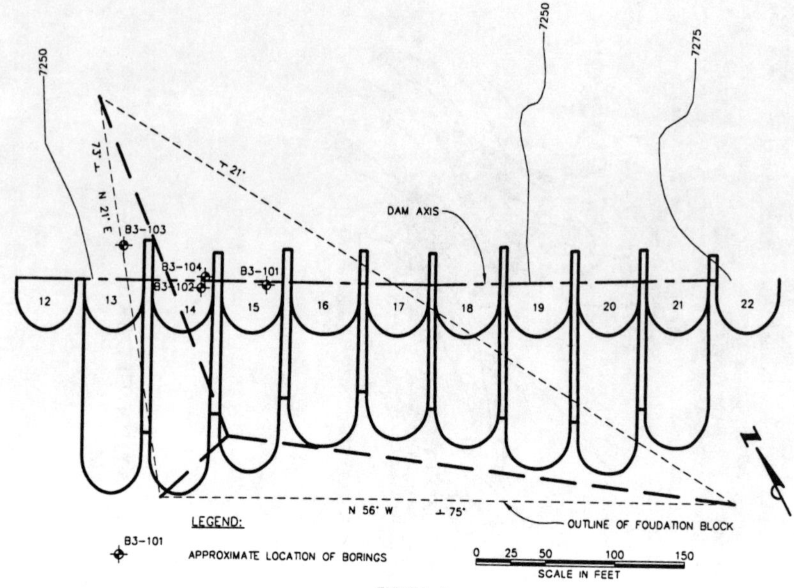

FIGURE 4
LOCATION OF CRITICAL FOUNDATION BLOCK
FLORENCE LAKE DAM

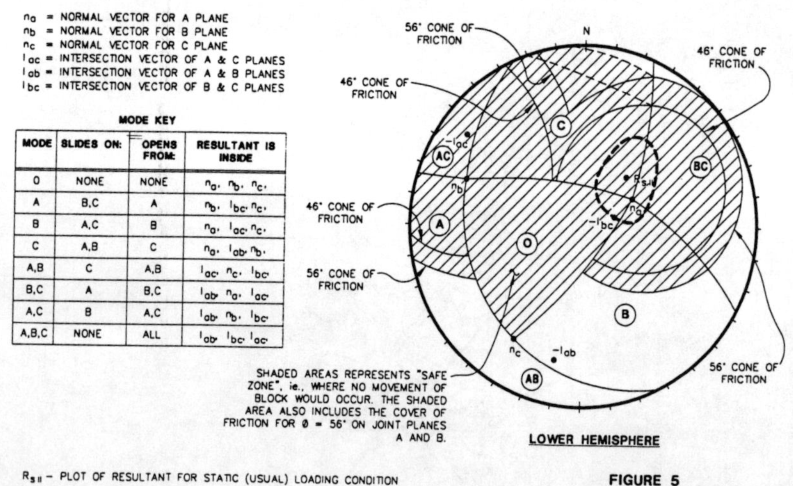

MODE KEY

MODE	SLIDES ON:	OPENS FROM:	RESULTANT IS INSIDE
O	NONE	NONE	n_a, n_b, n_c
A	B,C	A	n_b, I_{bc}, n_c
B	A,C	B	n_a, I_{ac}, n_c
C	A,B	C	n_a, I_{ab}, n_b
A,B	C	A,B	I_{ac}, n_c, I_{bc}
B,C	A	B,C	I_{ab}, n_a, I_{ac}
A,C	B	A,C	I_{ab}, n_b, I_{bc}
A,B,C	NONE	ALL	I_{ab}, I_{bc}, I_{ac}

FIGURE 5
MODAL DIAGRAM OF CRITICAL FOUNDATION BLOCK
FLORENCE LAKE DAM

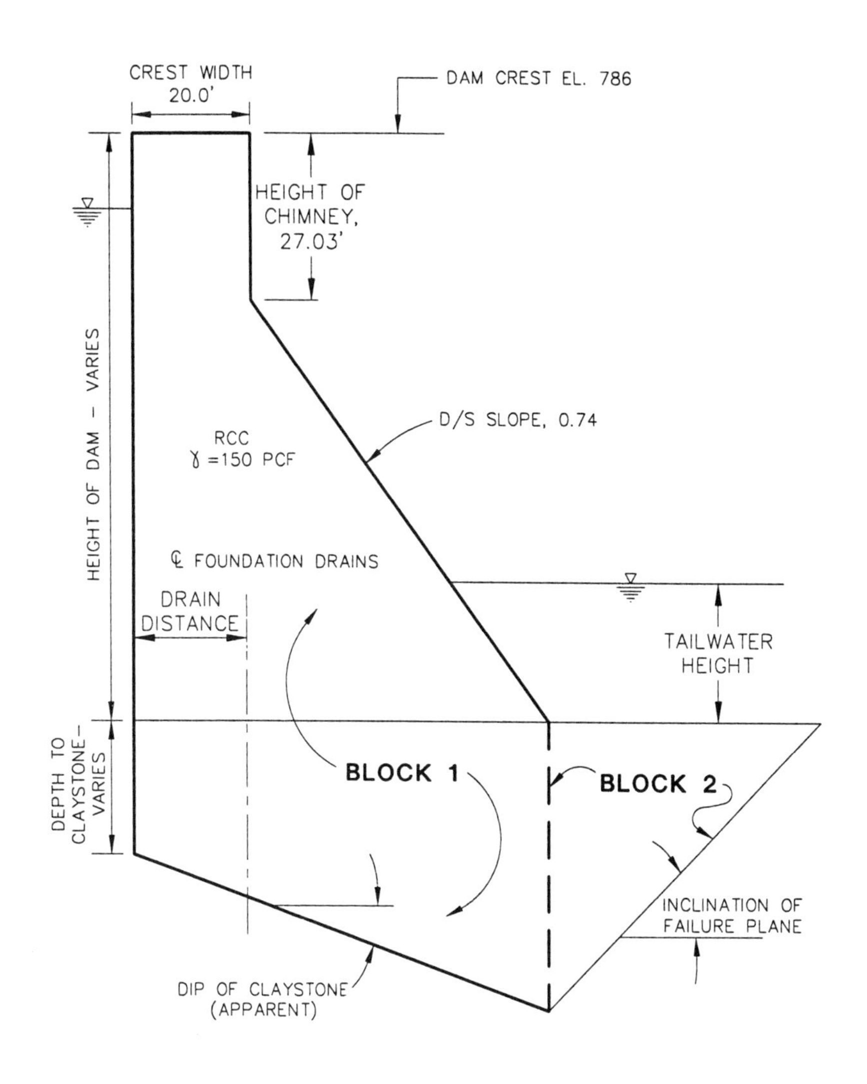

FIGURE 6
TYPICAL STABILITY CROSS SECTION
MILLTOWN HILL DAM

laboratory test data. Although rarely used in traditional geotechnical applications, the criterion has been widely used in rock mechanics analyses. This empirical approach incorporates the use of the Hoek-Brown failure criteria in conjunction with the rock mass rating (RMR) system (Bieniawski, 1988).

Forces acting on the dam and foundation block, including uplift pressures were estimated for each cross section location and the factor of safety was calculated using the iterative Corps of Engineers procedure (USCOE, 1981) in a spreadsheet model.

Conclusions

It is the experience of the authors of this paper that the evaluation of sliding stability of rock slopes is generally either overlooked or improperly evaluated by engineers today. The primary reason for this stems from the general limited knowledge and understanding of three critical fields of expertise: 1) engineering geology, 2) geological or geotechnical engineering, and 3) rock mechanics.

Important factors in evaluating the sliding stability of rock foundations include: 1) identification and evaluation of potentially removable blocks within the foundation (failure modes), 2) evaluation of shear strength of discontinuities in the rock mass, 3) evaluation of uplift pressures acting on discontinuities within the foundation, and 4) use of appropriate analysis methods.

References

Bieniawski, Z.T., "The Rock Mass Rating (RMR) System (Geomechanics Classification) in Engineering Practice, in Rock Classification Systems for Engineering Purposes, Louis Kirkaldie, editor, ASTM STP 984, 1988.

Goodman, R.E., Methods of Geological Engineering in Discontinuous Rocks, West Publishing, St. Paul, MN, 1976.

Hoek., E. and E.T. Brown, "Practical Estimates of Rock Mass Strength", Int. Journal of Rock Mechanics and Mining Sciences & Geomechanics, Vol. 34, pp. 1165-1186, 1997.

International Society of Rock Mechanics, "Suggested Methods for the Quantitative Description of Discontinuities in Rock Masses", Pergamon Press, Ltd., 1978.

Transportation Research Board, National Research Council, Landslides, Investigation and Mitigation, Special Report 247, National Academy Press, Washington, D.C., 1996.

U.S. Army Corps of Engineers, "Sliding Stability for Concrete Structures," ETL-1110-2-256, Washington, D.C., June 1981.

U.S. Bureau of Reclamation, Engineering Geology Field Manual, Denver, Colorado, 1976.

Additional References

The following references provide additional discussion and are excellent references for investigators performing rock slope stability analyses.

Barton, N.R., "The Shear Strength of Rock and Rock Joints", Int. Journal of Rock Mechanics and Mining Sciences & Geomechanics, Vol. 13, pp. 255-279, 1976.

Bieniawski, Z.T. and Orr, C.M., "Rapid Site Appraisal for Dam Foundations by the Geomechanics Classification", ICOLD, Mexico, 1976.

Goodman, R.E., Introduction to Rock Mechanics, John Wiley and Sons, New York, 1980

Goodman, R.E. and G-h. Shi, Block Theory and Its Application to Rock Engineering, Prentice-Hall, Inc., 1985.

Hoek, E., and J.W. Bray, Rock Slope Engineering, Institution of Mining and Metallurgy, London, 1994.

U.S. Federal Highway Administration, Rock Slope Engineering; Planning, Design, Construction and Maintenance of Rock Slopes for Highways and Railways, Parts A through E, Washington, D.C., 1976-1981.

Bayesian Calibration of Slope Failure Probability

Raymond W.M. Cheung[1], Wilson H. Tang[2]

ABSTRACT

A limitation of reliability approach to slope safety has been that reliability evaluation yields only notional probability of failure. It would be desirable if the assessed notional probability can be indeed calibrated to the real probability of failure. In this way, landslide risk could be managed with confidence and choices among alternative options could be made more defensible. This paper proposes a calibration procedure based on the Bayesian approach. It will first integrate the analytically assessed failure probabilities of a large set of soil slopes in Hong Kong with the corresponding observed performances of these soil slopes. Both failed and unfailed slopes were used in the calibration process. Through this exercise, the reliability assessed for a given slope in Hong Kong with similar soil material to those used in the calibration study can be calibrated to a realistic reliability measure.

INTRODUCTION

Reliability of a slope is conventionally assessed through deterministic factor of safety approach. In this approach the factor of safety (FS) of a slope is determined from an acceptable slope stability analysis method coupled with an assessed pore water pressure distribution and soil strength parameters that are estimated from the soil tests conducted. The value of FS has to meet a required minimum FS which has been prescribed by prior experience and professional judgment. However, it has been shown that slopes with same factor of safety may exhibit different probability of failure. This discrepancy is not unexpected as the failure probability would be influenced by the uncertainty in the ground condition, extent of site exploration and laboratory testing efforts, and the choice of soil parameters. Hence, as an alternative to the conventional approach, probabilistic approach is developed which attempts to systematically treat the uncertainties in the evaluation of the reliability of a slope. A probabilistic model is formulated for the stability of the slope on the basis of a conventional slope stability computation model, supplemented with a statistical

[1] Geotechnical Engineer, Geotechnical Engineering Office, Government of HKSAR, Hong Kong

[2] Head & Professor of Civil Engineering, Hong Kong University of Science & Technology, Hong Kong

characterization of the pertinent geotechnical parameters. By using the First Order Reliability method, the reliability index, β as first defined by Hasofer and Lind (1974) for the slope is then determined. Nonetheless, the assumed probabilistic model is only an abstraction of the real world and is subject to unavoidable model error, which is difficult to evaluate. Researchers have attempted to estimate the model error from a detailed analysis of failed cases where they believed that accurate measurements of the soil properties were made at the site. The limitation here is that there are not many failure sites where the detailed in situ soil properties are available and completely known to provide the information required for the assessment of the model error (e.g. Gilbert et al 1998). Other researchers have tried to compare the model prediction to a detailed finite element analysis (or sometimes a sophisticated laboratory test) in order to assess the error of a given geotechnical model. Those exercises undoubtedly furnish an assessment of the crudeness of the model relative to a full-scale finite element model (or a well conducted laboratory-simulated model). However, it could be still far from duplicating those environmental factors at the site. Centrifuge modeling could alleviate some of the problems by simulating the in-situ behavior in full scale, and by effectively repeating many site conditions to provide an overall assessment of the model error. Nevertheless, the construction and installation error at the site would not be inferred without actual field tests and observed performance data. Therefore, the values obtained from a formal reliability evaluation above are referred to as notional probabilities; the calculated notional probability of failure may not reflect the actual failure frequency of a given slope. Practicing engineers often claim that these calculated notional probabilities of failure cannot be used for decision making.

On the other hand, one could estimate the probability of failure by observing the failure statistics of a large number of similar slopes. This approach is impractical as slopes are generally different between sites, it is difficult to identify a sufficiently large number of similar slopes that have failed. If we lump all slopes into one population for statistical analysis, the estimated probability value would fail to distinguish a well from a poorly designed slope or between slopes with different levels of uncertainty. The statistical approach may be satisfactory in providing an estimate for the entire class of slopes or at best a crude estimate for the given slope; however the probability value is not generally applicable for a specific site. Moreover, it is desirable to study if certain design options or slope improvement work could bring the slope to an acceptable reliability level.

By recognizing the limitations of the current reliability analysis and the approach based on historical record of failed slopes, a method is proposed herein that retains the systematic evaluation of the formal reliability procedure and whose calculated notional probability of failure can be then calibrated to yield the realistic failure frequency. The proposed calibration is based on the performance record of 80 cut slopes in completely decomposed volcanic (CDV) soil; 30 of which failed and the rest remained non-failed.

PERFORMANCE FUNCTION AND RELIABILITY INDEX

The measure of stability performance of a slope may be described by the performance function, G(**x**) where **x** is the vector of variables including the geometrical, geotechnical and environmental ones. A convenient way is to define in terms of the difference between the total sliding resistance along a potential slip surface and the driving force caused by the soil weight and other imposed loads. The performance function can be derived based on moment equilibrium or force equilibrium. Failure of the slope is defined by the event G(**x**) < 0 and vice versa. The probability of slope failure is given by P(G(**x**) < 0). For a given set of statistics of the random variables **x**, the reliability index, β as first proposed by Hasofer and Lind (1974), for the slope can be expressed in the following matrix form (Ditlevsen 1981):

$$\beta = \min_{x \in F} \sqrt{(\mathbf{x}-\mathbf{m})^T \, \mathbf{C}^{-1} \, (\mathbf{x}-\mathbf{m})} \tag{1}$$

where **x** = vector of input parameters, **m** = vector of mean values, **C** = covariance matrix, *F* = failure domain. This reliability index would correspond to a failure probability of Φ(−β) where Φ( ) is the cumulative normal distribution function as shown in Curve A in Figure 5.

In this study, the reliability calculation of a slope is performed conveniently by using the spreadsheet method (Low and Tang 1997; Low et al 1998) that implements the minimization of (1) for determining the reliability index. The numerical stability model is based on the extended generalized method of slice (Chen & Morgenstern, 1983). Details of the uncertainty analysis are described in the later section in the context of cut slopes in volcanic soils in Hong Kong.

PROPOSED RELIABILITY-BASED METHOD

As described earlier, the probability of failure associated with the reliability index β represents only a notional probability. One would question: What would be the actual failure probability associated with the slope whose calculated reliability index is β? In other words, what is P(*F* | β) where *F* denotes the failure event? The following presents a procedure for determining the actual probability. Suppose substantial amount of failed and non-failed sites are available. The method calls for first performing reliability analysis of each of the slopes in these two groups of sites thus yielding the corresponding reliability index, β. The probability distribution of β for the set of failed sites and that for the set of non-failed sites are then determined accordingly. For a new site, if its reliability index has been assessed as β, its real probability of failure may be given by the following application of Bayes Theorem (see e.g. Ang and Tang 1975; Juang et al 1999) namely:

$$\mathrm{P}(F \mid \beta) = \frac{\mathrm{P}(\beta \mid F)\,\mathrm{P}(F)}{\mathrm{P}(\beta \mid F)\,\mathrm{P}(F) + \mathrm{P}(\beta \mid NF)\,\mathrm{P}(NF)} \tag{2}$$

where $\mathrm{P}(F \mid \beta)$ = probability of slope failure for a given value of β; $\mathrm{P}(\beta \mid F)$ = probability of β given that the slope did fail; $\mathrm{P}(\beta \mid NF)$ = probability of β given that the slope did not fail; $\mathrm{P}(F)$ = prior probability of slope failure and $\mathrm{P}(NF)$ = prior probability of no slope failure. The conditional probabilities $\mathrm{P}(\beta \mid F)$ and $\mathrm{P}(\beta \mid NF)$ can be further evaluated from the respective conditional distributions of β. Thus Eq. (2) becomes:

$$\mathrm{P}(F \mid \beta) = \frac{f_F(\beta)\,\mathrm{P}(F)}{f_F(\beta)\,\mathrm{P}(F) + f_{NF}(\beta)\,\mathrm{P}(NF)} \tag{3}$$

where $f_F(\beta)$ = probability density function of β for slopes with past failure record and $f_{NF}(\beta)$ = probability density function of β for slopes without past failure record.

Hence, from Eq (3), the actual probability of slope instability for a given reliability index can be determined provided that the prior probabilities are known. Reasonable estimates of the failure probability could be based on the global estimate for the entire class of slopes.

APPLICATION TO CDV SOIL CUT SLOPES IN HONG KONG

To demonstrate the above procedure, consider the performance of cut slopes in CDV soil, which is a major soil type in Hong Kong. Figure 1 shows the distribution of rock types in Hong Kong. The soil shear strength parameters and groundwater condition are considered as random variables in the present slope instability analysis.

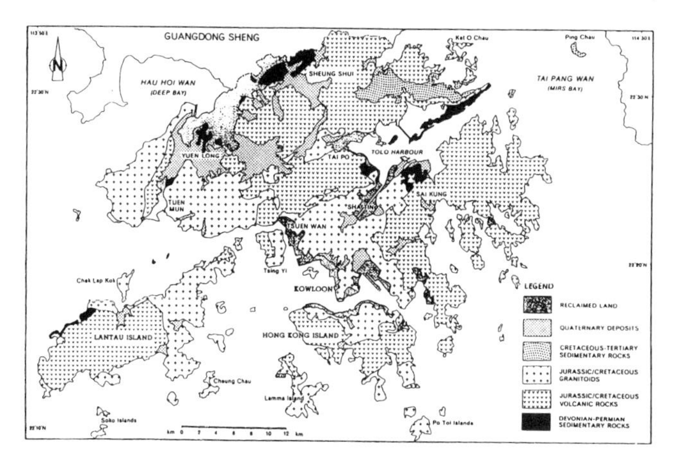

Figure 1 - Geology of Hong Kong (extracted from Geoguide 1, 1993)

In this study, shear strength parameters are assumed to follow normal distribution. Much has been published in literature for the range of soil property and coefficient of variation (*c.o.v.*) (e.g. Lumb 1966; Geoguide 1 1993). It is common to use the scatter exhibited by the test values of soil specimen to estimate the statistics of the soil property for evaluating the failure probability. However, this procedure does not reproduce the true picture of what governs the in-situ performance of a slope. The pertinent soil property controlling its performance often involves a much larger volume of soil than that in a typical test specimen. Thus, to evaluate correctly the slope performance, the probabilistic description of the spatial average property over the potential slip surface needs to be determined from the specimen property. The approach proposed by Tang (1984) is adopted for estimating the statistics of the average soil strength along the slip surface. The means and *c.o.v.* of the shear strength parameters of CDV soil are determined from 118 soil samples taken from 22 different sites based on the above procedure. The means of angle of friction, ϕ' and effective cohesion, c' are taken as 35.9^{o} and 8.1kPa with *c.o.v.*'s of 0.1 and 0.3 respectively. However, no strong correlation between the angle of friction and effective cohesion can be observed from the measurements and thus the correlation coefficient is assumed to be zero herewith. Groundwater table is another input random variables. In general, response of ground and perched water tables depends on a number of factors such as the rainfall intensity and duration, infiltration and evaporation rates of soil, and moisture content of soil. Various methods have been developed (e.g. Lumb, 1975) to evaluate the correlation between the groundwater response and these factors. However, further research is necessary to calibrate the proposed models and reality. In Hong Kong, a new cut slope is designed to sustain a rainfall of a 10-year return period. Hence, the probability of failure is calculated for a 10-year period for the present study. The ground and perched water tables are assumed to follow normal distribution with mean values of one-third of slope cutting height and 1.5m respectively, and a *c.o.v.* of 0.25 for a 10-year return period rainfall (see Figures 2a & 2b). These values are chosen to represent the current practice and prior experience in the industry. However, if site-specific ground water regime is of interest, it is recommended to install monitoring stations to establish the respective probability distribution.

Failure record for the past 13 years (1984 to 1996) in Hong Kong (GEO Report Nos. 1-6, 14, 20, 35, 43, 54, 59 & 70) has been used to yield a prior probability of failure of 0.7% per annum (or 0.07 over the 10-year period) for cut slopes in CDV soil.

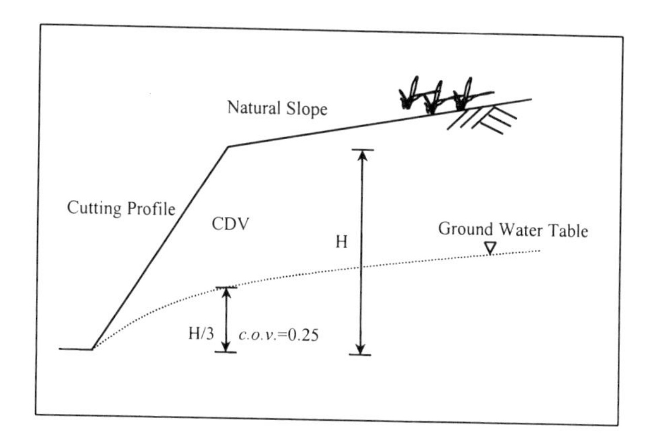

Figure 2a – Assumed Ground Water Table

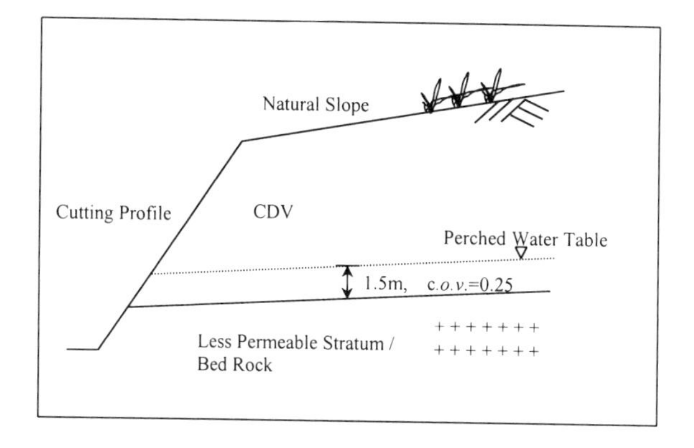

Figure 2b – Assumed Perched Water Table

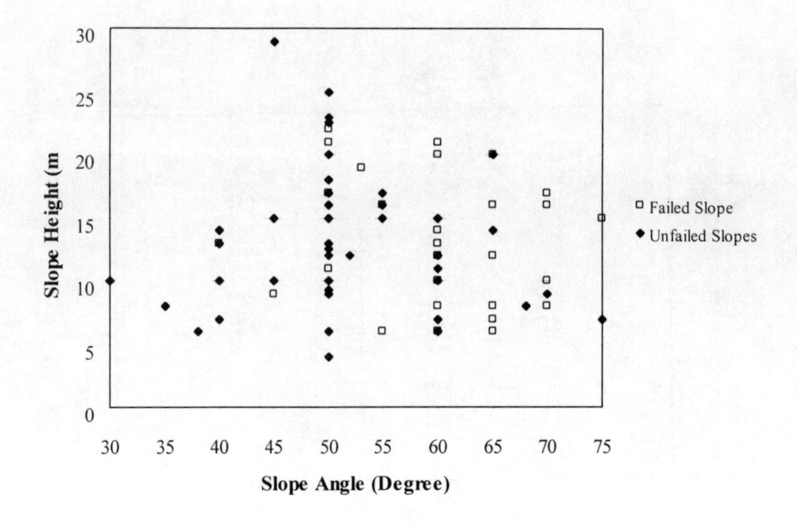

Figure 3 – Distribution of sampled slopes with respect to slope angle and height

Evaluation of the Probability of Failure

Reliability analysis has been carried out for a total of 80 cut slopes (30 failed and 50 unfailed) in Hong Kong CDV soil. Figure 3 shows the distribution of the slope height and angle of the sampled cut slopes. From observed statistics, it appears that cut slopes which have experienced past failure mainly lie on region of slope angle greater than 50° and height greater than 10m. The reliability indices obtained for these two groups (failed and unfailed) are observed to fit the normal distributions. The Kolmogorov-Smirnov (K-S) Goodness-of-fit test for probability distribution was not rejected at the 5% significance level for failed and non-failed groups of β. The results yield two probability distributions as shown in Figure 4. In principle, the density function of the non-failed group should increase monotonically with β. However, economic reasons have prevented slopes from being very conservative; hence, slopes with exceptionally high values of reliability index are expected to be few or non-existent thus preventing the density function curve from increasing. At the same time, the density function of the failed group should be in principle monotonically decreasing with β. However, in reality very few slopes would be associated with β values less than -2 as they would not have existed in the first place; hence the density function would begin with very low values for that range of reliability index. The shape of normal distributions fitted for the two groups are therefore reasonable.

Based on the prior probability of failure of 0.07 per ten-year, if a new cut slope whose reliability index has been assessed to be β, its real probability of failure is given by:

$$P(F \mid \beta) = \frac{0.07\, f_F(\beta)}{0.07\, f_F(\beta) + 0.93\, f_{NF}(\beta)} \tag{4}$$

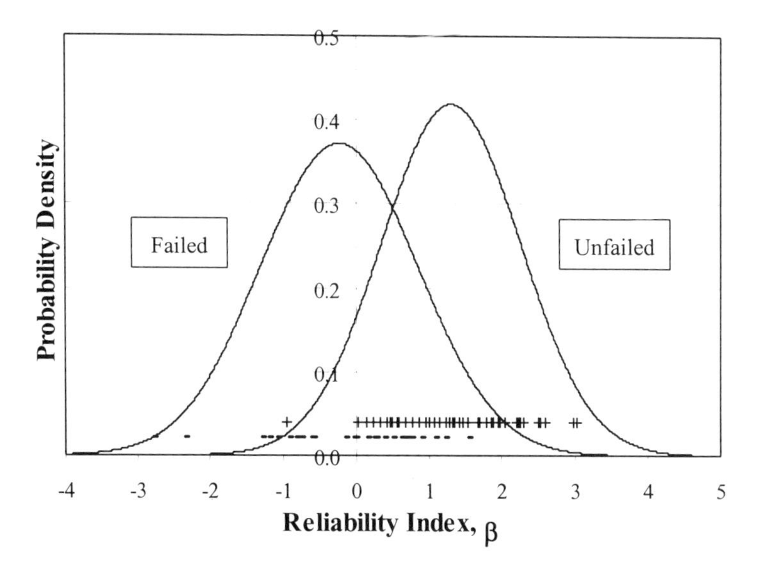

Figure 4 - Distributions of reliability indices, β for failed and non-failed slopes

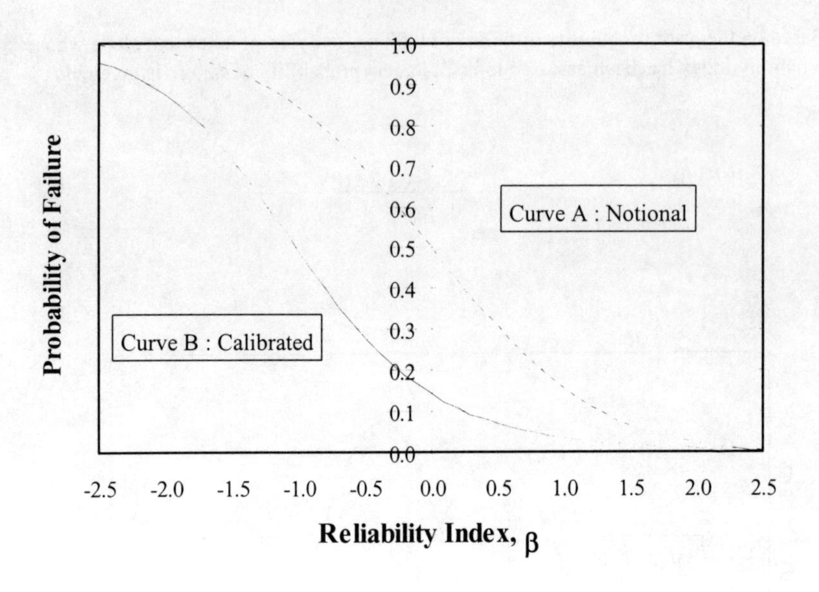

Figure 5 - Notional and calibrated probabilities of slope failure for a given reliability index, β

Figure 5 shows that the calibrated probability of slope failure is always less than the respective notional value. The significant reduction in the calibrated value is likely due to the conservatism of the geotechnical model adopted for the reliability evaluation; for example, neglecting the matric suction in soil mass could be a reason. In fact, the degree of separation between the two probability distributions in Figure 4 reflects the degree of discrimination between failed and non-failed cases by using the formal reliability evaluation. Consider the extreme case if the two probability distributions are perfectly overlapped, i.e. $f_F(\beta) = f_{NF}(\beta)$, the calibrated probability for any given reliability index, will be equal to the prior probability of failure. In other words, the probabilistic geotechnical model used in the reliability evaluation has not improved our estimate of the probability of failure. On the other hand, if the two probability distributions are much separated such that $f_F(\beta) >> f_{NF}(\beta)$ or $f_F(\beta) << f_{NF}(\beta)$, the calibrated probability of failure will tend either 0 or 1. In this case, the probabilistic geotechnical model is so discriminating that it almost becomes a deterministic model in the prediction of failure event.

Another observation from Figure 4 is that at the intersection of the two probability distributions, $\beta \approx 0.5$ and $f_F(\beta) = f_{NF}(\beta)$. The calibrated probability of failure will be the same as that estimated for the prior. In other words, for a new design cut slope in CDV soil, if the prescribed reliability index equals to 0.5, one would expect that the probability of failure is similar to the average failure frequency observed over the last 13 years. As stated before the historical global probability of slope failure is about 0.7% per annum. If this reliability level is considered within social acceptable limit, then a prescribed reliability index of 0.5 for the design of slopes over a 10-year period would be satisfactory. On the other hand, if further improvement in the safety level is desired, then a larger value of reliability index could be adopted. For example if the required minimum reliability index for a new slope design is 1.5, the expected probability of failure will be 0.18% per annum, which shows a significant improvement from the current safety level. Of course, the consequence of slope failure should also be considered to yield the acceptable risk level in overall risk management.

Sensitivity Analysis

The results as presented in previous sections are based on estimated prior probabilities from the 13-year historical failure record. It is of interest to study the influence of the prior probabilities on the calibrated failure probability. The change of calibrated failure probability with respect to the change of prior probability of failure can be expressed in the following equation:

$$\Delta \mathrm{P}(F \mid \beta) = \frac{f_F(\beta)\, f_{NF}(\beta)}{[f_F(\beta)\, \mathrm{P}(F) + f_{NF}(\beta)\, \mathrm{P}(NF)]^2}\, \Delta \mathrm{P}(F) \qquad (5)$$

where $\Delta \mathrm{P}(F \mid \beta)$ = change of calibrated probability of failure, $\Delta \mathrm{P}(F)$ = change of prior probability of failure

A sensitivity study has been performed in which the prior probability of failure was varied by ±50% from the previous estimated value. The corresponding calibrated probabilities of slope failure are shown as in Figures 6a and 6b. Observe that the percentage change in probability of failure for a positive reliability index is nearly the same as that in the prior probability of failure. However, the sensitivity is gradually reduced at negative region of β. Having recognized the sensitivity of the prior probabilities to the calibrated probability, especially in the positive region of β, the prior probabilities should be updated once additional information is available.

Recall Eq. (5) above, apart from the change in prior probability of failure with respect to its best estimate, the sensitivity of the calibrated probability is affected by the estimate itself. In the previous sections, the 13-year historical failure record has been used to provide the best estimate of the prior probabilities. Suppose there is no historical statistics available for such an estimation. Based on the principle of maximum entropy, it is assumed that the prior probability of slope failure is equal to the prior probability of no slope failure, i.e. P(*F*) = P(*NF*). The respective calibrated probability is shown in Figure 7. Note that in the absence of knowledge about the prior probabilities, the calibrated probability could be much greater than that inferred based on historical record before. The calibrated probability is even greater than the corresponding notional probability. Thus, even though the information of the 80 slopes has been integrated in the calibration, without knowledge of the prior probabilities, the failure probability could be much over-estimated. In other words, the prior probabilities are of paramount importance in the calibration such that it should be estimated at our best knowledge from historical record, if available, and/or professional judgement. Otherwise, an even more conservative result would be obtained. It is the reason that a collection of 13 years historical record has been used at the first place to provide a crude estimate for the performance of the entire slope class in the calibration.

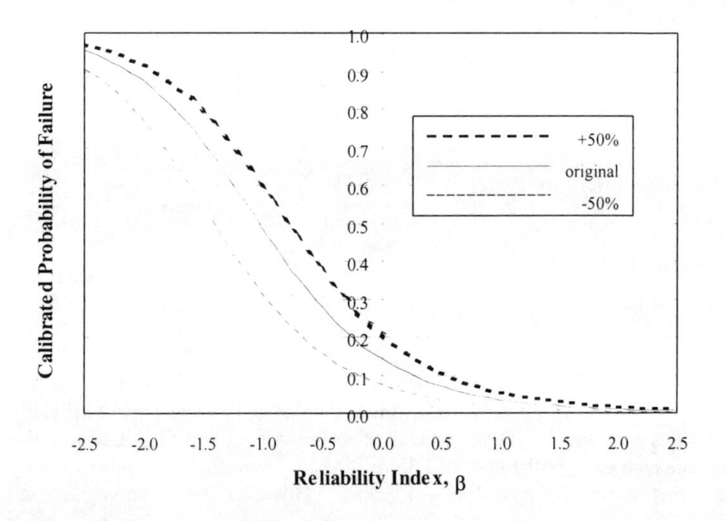

Figure 6a – Sensitivity of calibrated probability of slope failure to prior probability of failure

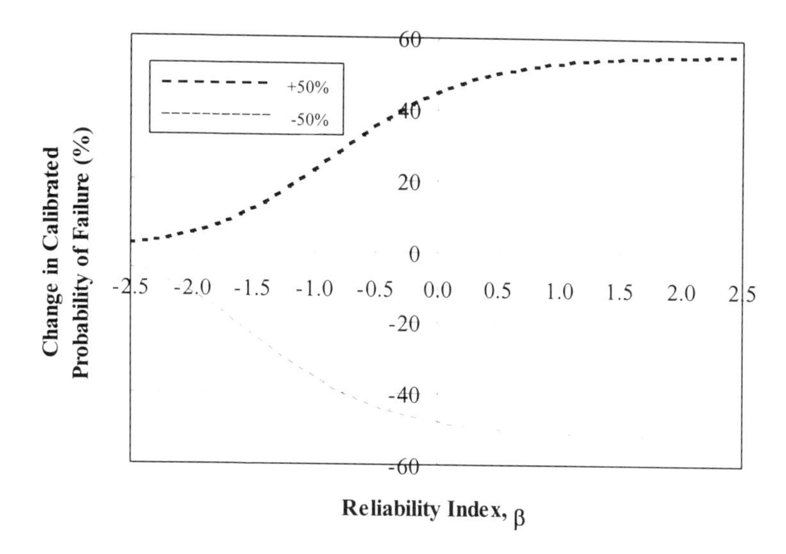

Figure 6b – Sensitivity of calibrated probability of slope failure to prior probability of failure

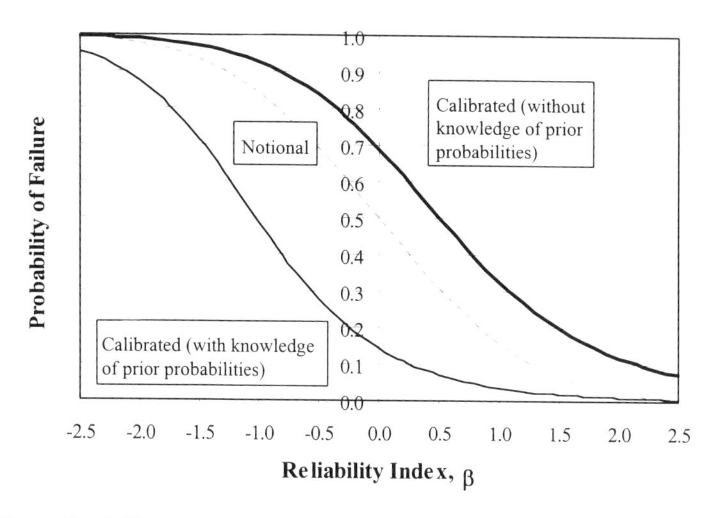

Figure 7 – Calibrated probability of slope failure with respective to knowledge of prior probability of failure

CONCLUSION

A Bayesian procedure has been presented to calibrate the calculated reliability index to a realistic probability of failure. The basis of the calibration is on an integration of analytical assessment with field performance in slope instability problem. The results also indicate that the prior probabilities are of paramount importance in the calibration process. With the proposed procedure, practising engineer could be more confident in performing reliability analysis and to decide among various design options in landslide risk mitigation.

REFERENCE

Ang, A.H-S. and Tang, W.H. (1975). *Probability Concepts in Engineering Planning and Design, Vol. 1 Basic Principles*, John Wiley & Sons.

Chen, Z. and Morgenstern, N.R. (1983). *Extensions to the generalized method of slices for stability analysis*. Can. Geotech. J., Ottawa, Canada, 20(1), 104-119

Ditlevsen, O. (1981). *Uncertainty Modeling with Applications to Multi-dimensional Civil Engineering Systems*. McGraw-Hill Book Co., Inc., New York

GEO (1993). *Geoguide 1 – Guide to Retaining Wall Design*. Geotechnical Engineering Office

Gilbert, R.B., Wright, S.G. and Liedtke, E. (1998). *Uncertainty in Back Analysis of Slopes : Kettleman Hills Case History*. J. Geotech. and Geoenvironmental Engrg., ASCE, vol.124 no.12, 1167-1176

GEO (1984 – 1996). *GEO Report Nos. 1 to 6, 14, 20, 35, 43, 54, 59 and 70 - Hong Kong Rainfall and Landslide*. Geotechnical Engineering Office.

Hasofer, A.M. and Lind, N. (1974). *An exact and invariant first-order reliability format*. J. Engrg. Mech., ASCE, 100(1), 111-121

Juang, C.H., Rosowsky, D.V. and Tang, W.H. (1999). *Reliability-based method for assessing liquefaction potential of soils*. J. Geotech. and Geoenvironmental Engrg., ASCE, vol.125 no.8, 684-689

Low, B.K. and Tang, W.H. (1997). *Efficient reliability evaluation using spreadsheet*. J. Engrg. Mech., ASCE, 123(7), 749-752

Low, B.K., Gilbert, R.B. and Wright, S.G. (1998). *Slope reliability analysis using generalized method of slices*. J. Geotech. and Geoenvironmental Engrg., ASCE, vol.124 no.4, 350-362

Lumb, P.(1966). *The variability of natural soils*. Can. Geotech. J., vol.3, 74-97

Lumb, P. (1975). *Slope failures in Hong Kong*. Quarterly J. of Engrg. Geology. 8:31-65

Tang, W.H. (1984). *Principle of probabilistic characterization of soil properties*. 74-89 in Probabilistic Characterization of Soil Properties: D.S. Bowles and Hon-kim Ko eds. New York, ASCE

ACKNOWLEDGMENT

The authors would like to acknowledge the support of the Research Grant Council of Hong Kong through grants Nos. HKUST 722/96E and HKUST 6039/97E. This paper is published with the permission of the Head of the Geotechnical Engineering Office and the Director of Civil Engineering, the Government of Hong Kong SAR.

A Model of Stability of Slopes

N. D. Cristescu

Abstract

A model of slope stability is presented using a non-homogeneous Bingham model. The constitutive parameters can be calibrated from in situ observations only (inclinometers). It is shown that the yield stress is varying nonlinearly with depth. An example is given. The case of possible slip at the bed rock is discussed.

Introduction

Gravitational creep flow of geomaterials as for instance slope movements, submarine landslides, snow avalanches, debris flow, volcanic lava flow, etc., has been studied by many authors. From the various aspects studied let as mention: deep-seated landslides at low strain rates (Petley & Allison [1997], Chiriotti [1999]), the study of the order of magnitude of the velocities of various land slides (Guzzetti [1998]), very large slope failures (Sousa & Voight, [1991]), snow slope stability during storms (Conway H., Wilbour C.[1999]), the influence of the water table depth on landslides velocity (Russo & Urciuoli [1999]), McDougall *et al.*[1999]), to mention just a few. Several models have been used to describe landslides (Vulliet [1999]), the most successful seems to be the viscoplastic constitutive equation (Desai *et al.* [1995], Samtani *et al.* [1996]), or Bingham models (Liu & Mei [1989], Sousa & Voight [1991]).

In the present paper the creep flow of a landslide is described starting from the idea that due to the gravitational forces the geomaterial compacts under its own weight. Thus, the mechanical properties of the layers vary with the depth. In particular the density and the yield stress are varying with depth. This is further analyzed to determine the depth where the sliding is incipient and the distribution with depth of the landslide velocity.

---

[1] Graduate Research Professor, Dept. of Aerospace Engineering, Mechanics & Engineering Science, University of Florida, P.O.Box 116250, Gainesville, Fl. 32611-6250.

Formulation of the problem

Let us consider a stratum of thickness h inclined with the angle θ to the horizontal (Fig.1). A geometrical deposited on this inclined slope will compact in

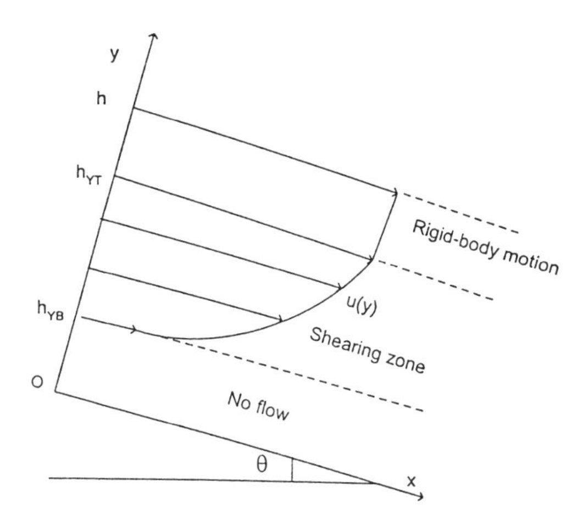

Fig.1 Creep flow of an inclined stratum

time under its own weight. If an elastic/viscoplastic constitutive law (Cristescu [1991]) is used one can show that the density varies nearly linearly with depth. For the analysis which follows this linear variation is accepted for simplicity of the presentation.

In the practical procedure to analyze a landslide any such variation, determined experimentally, can be used. Thus we assume

$$\rho = \rho_1 \; y + \rho_o = \frac{\rho_h - \rho_o}{h} \; y + \rho_o \tag{1}$$

where ρ_o is the density at the bottom $y = 0$ and ρ_h the density at the top $y = h$.

The constitutive equation which will be used is a nonhomogeneous Bingham one

$$D_{ij} = \begin{cases} 0 & \text{if} \quad II_{T'} \le k^2(\rho) \\ \dfrac{1}{2\,\eta(\rho)} \left\langle 1 - \dfrac{k(\rho)}{\sqrt{II_{T'}}} \right\rangle T'_{ij} & \text{if} \quad II_{T'} > k^2(\rho) \end{cases} \tag{2}$$

where D_{ij} is the strain rate tensor, T'_{ij} the stress deviator tensor, $II_{T'} = \frac{1}{2} T'_{ij} T'_{ij}$ the second invariant of the stress deviator, $k(\rho)$ is the yield stress depending on density and $\eta(\rho)$ is the viscosity coefficient (units in Poise).

If it is assumed that the particles move only along the slope, i.e., $v_x = u(y), v_y = v_z = 0$, from the equilibrium equation follow that the only nonzero stress components is

$$T_{xy} = -\,g \sin\theta \int_h^y \rho(y)\,dy \tag{3}$$

and for the linear variation (1) it follows

$$T_{xy} = g \sin\theta \left[(h-y) \left[\rho_o + \frac{(\rho_h - \rho_o)(h+y)}{2\,h} \right] \right] \tag{4}$$

if the surface $y = h$ is stress free.

It can be shown on physical arguments, that the variation of the yield stress k with depth cannot be linear. That has been shown in Cristescu and Cazacu [2000]. For convenience we summarize here the arguments. One starts from the observation that the shearing stress T_{xy} varies parabolically with depth (see (4)). This variation, apparently linear, is shown in Fig.2. If the thickness h of the stratum increases or if the slope θ is greater, then the slope of the $T_{xy}(y)$ curve as function of y will increase. At the same time the law of the variation with depth of the yield stress can be of the kind shown by dotted lines in Fig.2. At the ground surface the yield stress k_h is smaller, while at the bottom of stratum, it is bigger $k_o > k_h$. Let us assume that the variation of k with depth is linear (curve b). Thus if the stratum thickness h is steadily increased, the slope of $T_{xy}(y)$ steadily increases and the $T_{xy}(y)$ line will intersect $k(y)$ always at the bottom $y = 0$. Thus the incipient shear will take place always at the bottom. While that is a possible particular case, generally it is to be expected that the incipient shear would start at some other depth, mainly if h is very big. $k(y)$ may also change when h increases; however, if we assume that $k(y)$ varies linearly with depth the incipient

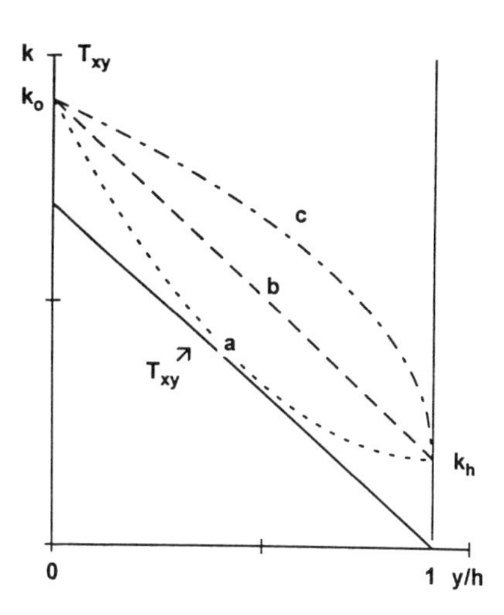

Fig.2. Three possible laws of variation of k(y) and variation of shearing stress $T_{xy}(y)$ with depth.

shear will take place always at the bottom. The same arguments hold against the choice of a law of type c (i.e., $n < 1$). That is why it was assumed that the law of variation of k(y) is of the form

$$k(y) = (k_o - k_h)\left(1 - \frac{y}{h}\right)^n + k_h \tag{5}$$

where $n > 1$ is a material constant and $k_o > k_n \geq 0$ (curve (a) in Fig.2). All these parameters may strongly depend on humidity. We will show later how this relationship can be determined from the data, straight from measurements performed with inclinometers, i.e. no laboratory tests are needed.

If additional geomaterial is added and the stratum becomes thicker (snow) or a density increase takes place (heavy rain, etc.), then T_{xy} given from (4) increase, i.e. the slope of T_{xy} shown in Fig.3 increase. If the line $T_{xy}(y)$ will touch the line k(y) at the depth h_{Y_i} it is at this depth that the sliding starts, since the inequality (2_2) which for our case becomes

$$T_{xy}(y) > k(y) \tag{6}$$

will there be satisfied. This depth can be determined precisely since both $T_{xy}(y)$ and $k(y)$ can be determined from the in situ data. In this case the upper layer $h_{Yi} \leq y \leq h$ slides as a rigid along the plane $y = h_{Yi}$ (plug flow).

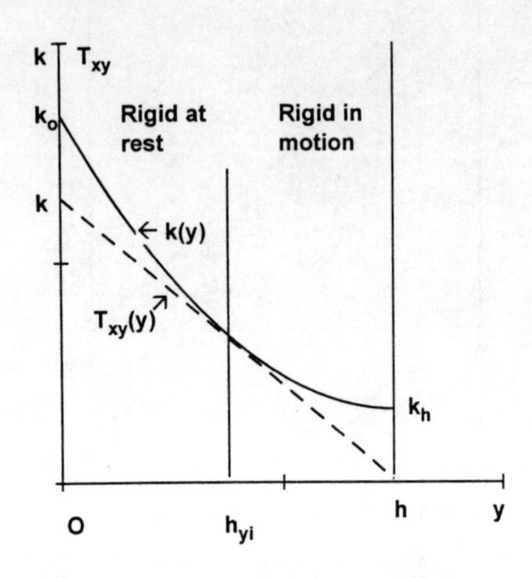

Fig.3 Incipient shear flow

Generally the flow condition (6) is satisfied in a certain interval $h_{YB} \leq y \leq h_{YT}$ which may be relatively small. The thickness of this interval follows from (4) and (5) with $T_{xy}(y) = k(y)$ and assuming $n = 2$ (just for an illustrative example)

$$\left.\begin{matrix} h_{YT} \\ h_{YB} \end{matrix}\right\} = \frac{\frac{\rho_o h g \sin\theta - 2(k_o - k_h)}{h} \pm \sqrt{\Delta}}{2\left[\frac{\rho_o - \rho_h}{2h} g \sin\theta - \frac{k_o - k_h}{h^2}\right]} \tag{7}$$

where

$$\Delta = \frac{1}{h^2}\left[\begin{matrix} h^2 \varrho_h g^2 \sin^2\theta + 2h(\varrho_o - \varrho_h)k_h g \sin\theta \\ + 4k_h(k_h - k_o) \end{matrix}\right] . \tag{8}$$

This case is shown in Fig.4: the top layer $h_{YT} \leq y \leq h$ moves as a rigid with the velocity $u(h_{YT})$ while the bottom layer $0 \leq y \leq h_{yB}$ is at rest. Depending on the mechanical properties of the geomaterial and on the preliminary compaction, the thickness of the upper stratum can be very small, close to zero. Also, if at the bottom sliding conditions exist at the bedrock, then if T_{xy} is steadily increasing, at a certain moment the bottom layer may start sliding also as a rigid.

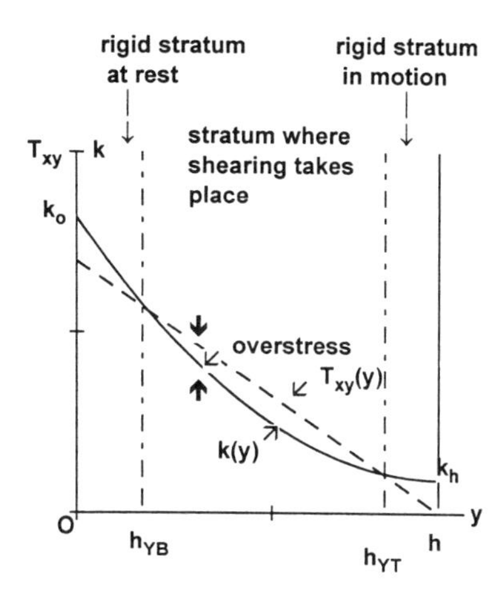

Fig.4 Determination of the thickness of the stratum where sliding takes place. The upper stratum slides as a rigid.

Velocity field

If the creep flow condition (6) is satisfied, one can integrate the constitutive equation in the interval $h_{YB} \leq y < h_{YT}$ to get the velocity distribution

$$u(y) = \frac{1}{\eta}\left\{\begin{array}{l} g\sin\theta\left[\frac{\rho_h - \rho_o}{2h}\left(h^2 y - \frac{1}{3}y^3\right) + \rho_o\left(hy - \frac{1}{2}y^2\right)\right] + \\ \frac{k_o - k_h}{(n+1)h^n}(h-y)^{n+1} - k_h y \end{array}\right\} + \mathbf{C} \qquad (9)$$

with

$$\mathbf{C} = -\frac{1}{\eta}\left\{ \begin{aligned} & g\sin\theta\left[\frac{\rho_h - \rho_o}{2h}\left(h^2 h_{YB} - \frac{1}{3}h_{YB}^3\right) + \rho_o\left(h\,h_{YB} - \frac{1}{2}h_{YB}^2\right)\right] \\ & + \frac{k_o - k_h}{(n+1)h^n}\left(h - h_{YB}\right)^{n+1} - k_h\,h_{YB} \end{aligned} \right\} \tag{10}$$

if we assume $u(h_{YB}) = 0$ for $y = h_{YB}$.

Formula (9) can be used to compare the in situ observed velocity distribution, as obtained with inclinometers, for instance, with the prediction of (9). In this way one can determine very easily the viscosity coefficient.

Example.

In order to apply the above theory we do not need any laboratory tests to determine the material properties. All the tests are to be made in situ without disturbing the specimens by carrying them in the laboratory. In order to apply the above theory, the only necessary data are the distribution of density with depth and the inclinometer readings which are giving the displacement in time at various depths. The base slope angle must also be known. The inclinometer reading of a landslide from Villarbeney in Switzerland as reported by Samtani et al. [1996] are used. The times of the recording are : 9 November 1978: 16 February 1979; 5 April 1979; 22 May 1979; 25 July 1979 and 29 October 1979.

The inclinometers readings are shown in Fig.5, together with the model prediction.

The procedure goes as follows. Since the authors are also giving the variation with depth of the unit weight and since the thickness h of the layer follows from Fig.5, one can obtain the variation of the shearing stress with depth (4). Further from Fig.5 follow h_{YT} and h_{YB} (see (7)). That is enough to determine (5) assuming for n a value $1 < n < 2$. The Fig.4 can be plotted for this particular case. The "overstress" $T_{xy}(y) - k(y)$ will determine the magnitude of strain rate and therefore the velocity u(y) given by (9). A single inclinometer reading is only needed in order to determine the viscosity coefficient η . For the borehole called E_2 we have from Fig.5:

$$\begin{aligned} & h_{YT}(=2\,m) \le y \le h\,(=8m) && \text{– top layer in rigid motion} \\ & h_{YB}(=0.5m) \le y \le h_{YT}(=2m) && \text{– slow creep shearing takes place} \\ & 0 \le y \le h_{YB}\,(=0.5m) && \text{– rigid at rest} \end{aligned} \tag{11}$$

with $k(h_{yB}) = 48$ kPa, $k(h_{yT}) = 38$ kPa. These values determine (5), with a value for n slightly bigger than one. A formula to exactly determine n is given by Cazacu and Cristescu [2000]. Comparing the displacement distribution with the data obtained after 100 days, one has obtained the viscosity coefficient

$$\eta(E_2) = 1.7 \times 10^{11} \text{ P.} \tag{12}$$

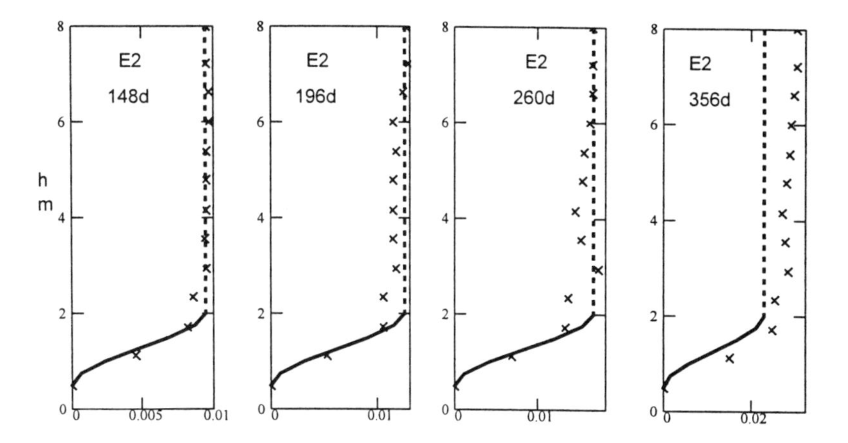

Fig.5 Displacement profiles for creep flow at borehole E_2 for time intervals shown; Stars are observed data, lines (full and dotted) are predicted profiles.

With this coefficient one can predict the displacement profile after any time interval. Fig.6 shows a comparison of the observed data and the predicted profiles after the time intervals shown: 148 d, 196 d, 260 d, and 356 d. A perfect matching was not expected since during one year interval the humidity is varying significantly and this variation was not recorded. However the prediction seems very good and can be improved if the annual humidity variation is recorded in long time intervals and thus the constitutive parameter can be considered to depend on humidity as well.

Slip with friction at the bottom

If at the bottom of the of the stratum a rigid rock surface exists, then the soil stratum may slide with friction along the bottom (Terlien, 1998).

For materials which are obeying a viscoplastic law and are sliding along a rigid surface, a friction law of the form

$$\tau = m\sqrt{II_{T''}} \tag{13}$$

was proposed (Cristescu, 1975); τ is the shearing friction stress at the interface, and $0 \le m \le 1$ is a "friction factor" with $m = 0$ means no friction, and $m = 1$ complete adherence to the bottom surface (no slip). For a Bingham model (13) can be written as

$$\tau = m\left(k + 2\,\eta\,\sqrt{II_{D'}}\right) \quad , \tag{14}$$

with $II_{D'}$ the second invariant of the strain rate deviator. Thus, the friction stress depends on the velocity as well. In the case of no shear, the ("static") friction shear stress between the bottom surface and the rigid layer of the material becomes

$$\tau = m\,k \tag{15}$$

where k is the yield stress of the material in the neighborhood of the rigid surface. Sliding takes place if τ is surpassing the value $m\,k$.

It can be shown that if at $y = 0$, a friction law (15) is assumed and $m < 1$, and if h is steadily increased and $\theta = \text{const.}$, then there is a minimum value of h for which a rigid stratum at the bottom slides, i.e., the shearing zone is never reaching the bottom $y = 0$ (unless $m = 1$).

The proof is straightforward (see Figure 6). Assume that at the bottom stratum $y = 0$, a friction law of the form (15) holds, and that, $0 < m < 1$. That means that the maximum shearing stress which the surface $y = 0$ can hold without slip is $m\,k_o$. Thus, when T_{xy} increases, due to the increase of h, $T_{xy}(0)$ cannot surpass this value without a slip to take place. Thus, when $T_{xy}(0)=m\,k_o$, a sliding along $y = 0$ will take place. Let us denote by h_s the value of h, corresponding to the reaching for the first time, at the bottom $y = 0$, of the equality

$$T_{xy}\big|_{y=0} = m\,k_o \,. \tag{16}$$

In the case when θ is fixed, and h is steadily increased, when h is reaching the value h_s obtained from (3) for $y = 0$ as

$$h_s = \frac{2\,m\,k_o}{(\rho_o + \rho_h)g\sin\theta} \tag{17}$$

plug sliding at bottom starts. The thickness of the rigid plug which is sliding along the bottom surface $y = 0$, is easy to be determined, since it is just h_{YBs}, the solution (7) in which $h = h_s$ (Fig.6).

If from tests (field observation) one can determine h_s, then from (17) one can determine m. Generally, if at $y = 0$ a friction law of the form (15) holds, and

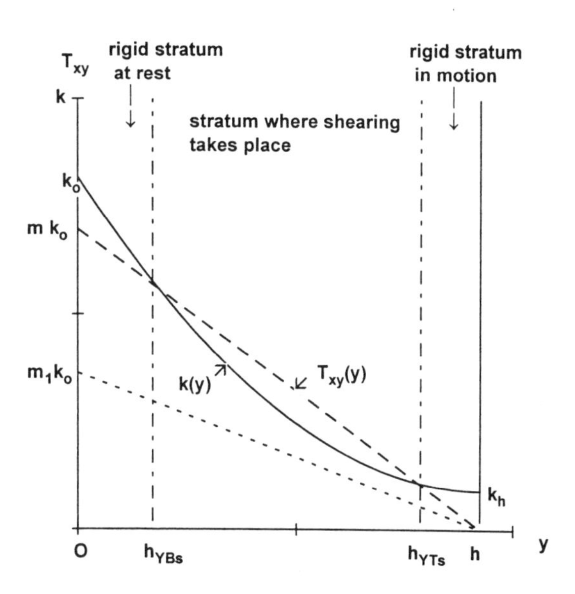

Fig.6 Procedure to determine the thickness of the bottom layer which is sliding along $y = 0$, for friction factor $0 < m < 1$. When at the bottom $T_{xy}(0) = m\,k_o$ the whole stratum slides as a rigid.

$m < 1$, then a rigid plug at the bottom which will slide without shearing will always exists (when h increases very much). This conclusion is true for any friction law which postulates that the shearing stress which the bottom surface can hold without sliding, is smaller than the shearing stress which can produce a shearing inside the body.

Similar arguments hold if somewhere in the stratum $0 \leq y \leq h$ a weaker plane $y = y_s$ exists along which the maximum possible shearing stress is much smaller than in the neighboring layers. In other words, that happens if a plane $y = y_s$ exists where the yield stress $k(y_s)$ is discontinuously smaller than in the strata above it (see for instance, the case mentioned by Alonso et al. (1993), where a thin marl layer has weaker shearing properties than all other materials in the moving stratum). Figure 1 shows the velocity profile when a sliding is taking place at the depth $y = h_{YB}$. If y_s can be determined from observations, the corresponding value for h follows from formulae similar to (7) and (17).

If at $y = 0$, $m \rightarrow 1$ it follows that the sliding velocity $u_s \rightarrow 0$ and the shearing flow is possible up to the bottom. The case $m \rightarrow 1$ corresponds to a very rough bottom surface.

In a similar way one can analyze hydrological triggered landsliding. However, in this case a decrease in $k(y)$ due to water infiltration and percolation should be considered. The friction coefficient is also decreased due to an increased moisture. T_{xy} is also changing due to an increase in soil density (see (4)).

A sudden sliding along the soil-bedrock interface can be triggered also by seismic vibration, which may diminish both the friction coefficient and the yield stress $k(y)$ (and, maybe, the viscosity coefficient as well).

Conclusions.

A theory of creep displacement of a landslide can be formulated using a nonhomogeneous Bingham model. The constitutive constants and functions can be calibrated from measurements in situ using inclinometers, determining the distribution of density with depth and of the displacement of the borehole after a certain time interval. The matching of the prediction with the data is reasonable. The method can be extended to the cases when the variation of the material properties with depth are not smooth, when there are discontinuity plane with a discontinuous smaller yield stress, when a variable in time humidity is influencing the sliding mainly due to the change the yield stress and of the viscosity coefficient (thus the speed of sliding may change during some long time intervals). This aspect as well as other examples are discussed by Cazacu and Cristescu [2000].

Generally if the humidity variation (rain, snow, etc.) is recorded both k and η become function on humidity as well, and these functions can be determined from in situ measurements.

The landslide mechanism described above is but a model and as such has its limitations.

References

Alonso, E.E., Gens A. and Lloret. A. "The landslide of Cortes de Pallas, Spain", Géotechnique, **43**, 507-521. 1993

Cazacu, O. and Cristescu, N.D. "Constitutive model and analysis of creep flow of natural slopes" Rivista Italiana Di Geotecnica, **34,** 2000, (submitted).

Chiriotti , E. "Un esempio di deformazione gravitativa profunda: monitoraggio e modellazione". Rivista Italiana Di Geotecnica, **33,** 1, 28-34, 1999.

Conway, H., Wilbour, C. "Evolution of snow slope stability during storms" Cold Regions Science and Technology, **30**: (1-3), 67-77, 1999

Cristescu, N.D. "Plastic flow through conical converging dies, using a Viscoplastic constitutive equation", Int. J. Mech. Sci., **17,** 425-433, 1975.

Cristescu, N.D. "Nonassociated elastic/viscoplastic constitutive equations for sand", Int. J.Plasticity, **7**, 41-64. 1991.

Cristescu,N.D. and Cazacu, O. "On creep flow of natural slopes" Pacific Rocks 2000; Rock Around The Rim. Seattle, Washington.

Desai, C.S., Samtani, N.C., Vulliet, L. "Constitutive Modeling and Analysis of Creeping Slopes" Journal of Geotechnical Engineering, **121**, 43-56, 1995.

Guzzetti, F. "Hydrological triggers of diffused landsliding" Environmental Geology, **35** (2-3), August 1998, Springer-Verlag.

Liu, K.F., Mei, C.C. "Slow spreading of a sheet of Bingham fluid on an inclined plane", J.Fluid. Mech., **207**, 505-520, 1989.

McDougall, J., Ng, CWW, Shi, Q. "Influence of rainfall intensity and duration on slope stability in unsaturated soils" by C.W.W.Ng & Q. Shi; Quarterly Journal of Engineering Geology, **32**: 303, Part 3 1999.

Petley, D.N., Allison, R.J. "The Mechanics of Deep-Seated Landslides" Earth Surface Processes and Landforms, **22,** 747-758, 1997, John Wiley and Sons, Ltd.

Russo, C., Urciuoli. G. "Influenza delle variazioni di pressioni neutre sugli spostamenti di frane Lente", Rivista Italiana Di Geotecnica, **33**, no.1, 47-55. 1999.

Samtani, N.C., Desai, C.S., Vulliet, L. "An Interface Model to Describe Viscoplastic Behavior" Int. J. for Numerical and Analytical Methods in Geomechanics, **20**, 231-252, 1996, John Wiley and Sons, Ltd.

Sousa, J., Voight, B. "Continuum simulation of flow failures" Geotechnique, **41**, 515-538, 1991.

Terlien, M.T.J. "The determination of statistical and deterministic hydrological Landslide-triggering thresholds" Environmental Geology, **35**, 124-130, 1998.

Vulliet, L. "Modelling creeping slopes" , Rivista Italiana Di Geotecnica", **33.** 1, 71-76, 1999.

Geotechnical Stability Analysis by Strength Reduction

Ethan Dawson,[1] Farid Motamed,[1] A. M. ASCE,
Saddanathapillai Nesarajah,[1] A. M. ASCE,
and Wolfgang Roth,[1] M. ASCE

Abstract

The factor of safety of a slope can be computed with a finite element or finite difference code by reducing the soil shear strength in stages until the slope fails. The resulting factor of safety is the ratio of the soil's actual shear strength to the reduced shear strength at failure. This "shear-strength reduction technique" can also be applied to geotechnical stability problems such as cantilever walls, braced excavations and retaining walls. The strength reduction technique is demonstrated for several geotechnical stability problems, illustrating some of the advantages and limitations of the technique.

Introduction

The method of slices, the most commonly used numerical method for geotechnical stability analysis, assumes that failure occurs by sliding along a slip surface. But for many stability problems, failure involves deforming wedges of soil rather than slip surfaces. Common examples are rotational failure of cantilever walls and retaining walls. The method of slices was not designed with this type of problem in mind, and can be awkward to use. Instead, engineers must resort to hand calculations, relying on various active and passive earth pressure assumptions.

An alternative tool for geotechnical stability analysis is the FEM-based strength reduction technique. This technique has been applied mostly for slope stability analysis, where it has been shown to produce results comparable to the method of slices and other limit-equilibrium methods. However, the technique can also be applied to earth retaining structures

In this paper, the strength reduction technique is demonstrated first for a traditional slope stability problem where failure occurs along a slip surface. The technique is then applied to analyze various earth-retaining structures including a

[1] Dames & Moore, 911 Wilshire Blvd, Suite 700, Los Angeles, CA 90017.

cantilever wall, a braced excavation and a gravity retaining wall. These examples illustrate failure mechanisms involving deforming wedges, and also illustrate some of the advantages and limitations of the strength reduction technique.

The Strength Reduction Technique

For slopes, the factor of safety *F* is traditionally defined as the ratio of the actual soil shear strength to the minimum shear strength required to prevent failure (Bishop, 1955). Duncan (1996) describes *F* as the factor by which the soil shear strength must be divided to bring a slope to the verge of failure. Hence, an obvious way of computing *F* with a finite element or finite difference program is simply to reduce the soil shear strength until collapse occurs. The resulting factor of safety is the ratio of the soil's actual shear strength to the reduced shear strength at failure. This "shear strength reduction technique" was used as early as 1975 by Zienkiewicz *et al.* (1975), and has since been applied by Naylor (1982), Donald & Giam (1988), Matsui & San (1992), Ugai (1989), Ugai & Leshchinsky (1995) and many others.

The shear strength reduction technique has a number of advantages over the method of slices for slope stability analysis. Most importantly, the critical failure surface is found automatically and failure mechanisms involving deforming wedges can be analyzed. Deforming wedges are important for cantilever walls and retaining walls, as well as for 3D slope stability (Dawson & Roth, 1999). Application of the technique has been limited in the past due to the long computer run times required. However, with the increasing speed of desktop computers, the technique is becoming an effective alternative to the method of slices, and is being used increasingly in engineering practice. Strength reduction is now a built-in feature of commercial geotechnical FEM codes such as PLAXIS (PLAXIS BV, 1998).

FLAC (Fast Lagrangian Analysis of Continua)

The strength reduction technique will be demonstrated using the explicit-finite-difference code, *FLAC* (Itasca Consulting Group, 1995). While, strictly speaking, FLAC is a finite difference code, it is not very different from an element-by-element FEM code. The set of algebraic equations solved by *FLAC* is identical to that solved with the finite element method. This set of equations is solved using dynamic relaxation (Otter, Cassell & Hobbs 1966), an explicit, time-marching procedure, in which the full dynamic equations of motion are integrated step by step. Static solutions are obtained by including damping terms that gradually remove kinetic energy from the system.

The convergence criterion for *FLAC* is the nodal unbalanced force, the sum of forces acting on a node from its neighboring elements. If a node is in equilibrium, these forces should sum to zero. For this study, the unbalanced force of each node is normalized by the gravitational body force acting on that node. A simulation is considered to have converged when the normalized unbalanced force of every node in the mesh was less than 10^{-3}.

All computations presented in this paper are performed in small-strain mode (the coordinates of the nodes are not updated according to the computed nodal displacements). Soil is modeled as a linear elastic-perfectly plastic material with a Mohr-Coulomb yield condition and an associated flow rule. An associated flow rule allows us to compare results to limit analysis solutions and also allows us to ignore the effects of elastic constants, the initial stress and the stress path. These have no effect on the collapse load for an associated material (Chen, 1975). However, for more realistic soil models, with non-associated flow rules, these factors cannot be ignored.

Stability of a Homogeneous Embankment

The strength reduction technique is demonstrated first for computing the factor of safety of a homogeneous embankment. Results are compared to a limit analysis, upper bound solution derived by Chen (1975), assuming a log spiral failure surface. Consider an embankment of height H = 10 m, sloping at angle β = 45° with friction angle ϕ = 20°, unit weight γ = 20 kN/m^3, cohesion c = 12.38 kPa, and with no pore pressure. With these soil properties the slope has a factor of safety of exactly 1.0, according to the solution of Chen (1975).

The model mesh is shown in Figure 1. Horizontal displacements are fixed for nodes along the left and right boundaries while both horizontal and vertical displacements are fixed along the bottom boundary. Simulations are run for a series of trial factors of safety F^{trial} with c and ϕ adjusted according to the equations:

$$c^{trial} = \frac{1}{F^{trial}} c \qquad (1)$$

$$\phi^{trial} = \arctan\left(\frac{1}{F^{trial}} \tan \phi \right) \qquad (2)$$

The value of F^{trial} at which collapse occurs is found by bracketing and bisection. First, upper and lower brackets are established. The initial lower bracket is any F^{trial} for which a simulation converges. The initial upper bracket is any F^{trial} for which the simulation does not converge. Next, a point midway between the upper and lower brackets is tested. If the simulation converges, the lower bracket is replaced by this new value. If the simulation does not converge, the upper bracket is replaced. The process is repeated until the difference between upper and lower brackets is less than a specified tolerance.

Using this procedure, the factor of safety computed for the embankment is 1.026, a few percent higher than the limit analysis solution. The velocity field at collapse is shown in Figure 2, along with the critical log spiral failure surface from the limit analysis solution.

A more detailed comparison between strength-reduction and limit-analysis solutions for homogenous embankments can be found in Dawson et al. (1999). In that study, computations were performed for a wide range of slope angles, soil friction angles and pore pressure coefficients. It was concluded that strength-reduction results converged to limit-analysis solutions as the numerical mesh was refined. But for coarser meshes, strength reduction results were generally a few percent higher than upper bound limit analysis solutions.

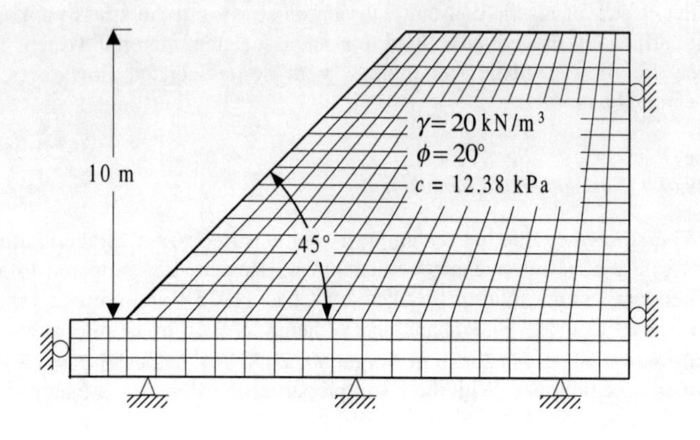

Figure 1. Numerical mesh for homogeneous embankment

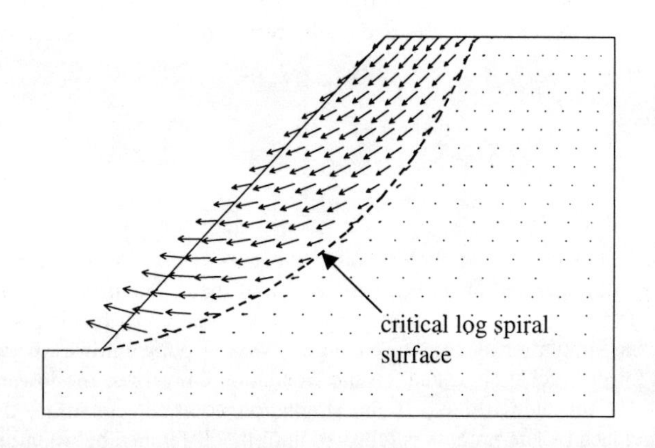

Figure 2. Velocity field at collapse, along with critical log-spiral surface

Cantilever Wall

For many geotechnical problems plastic failure occurs in deforming wedges of soil. Consider rotational failure of a cantilever wall. The wall prevents localization of deformation into a slip plane and, instead, distributes deformation throughout passive and active wedges. This failure mode is demonstrated by comparing FLAC simulations to static and kinematic limit analysis solutions. For simplicity, we consider only the case of a cohesionless soil and we assume that the wall-soil friction angle is zero.

A limit analysis static solution is any stress field that satisfies the equilibrium conditions, satisfies the stress boundary conditions, and nowhere violates the yield condition. For the cantilever wall, a static solution provides an upper bound to the depth of embedment D required for a given wall length L (see Figure 3). A limit analysis kinematic solution is any velocity field that satisfies the plastic flow rule and satisfies the kinematic boundary conditions. For the cantilever wall, a kinematic solution provides a lower bound to the depth of embedment required for a given wall length.

Static Solution

A static solution for the cantilever wall can be derived using the Rankine active and passive states shown in Figure 3. The wall is assumed to rotate about a point near its lower tip. Behind the wall the soil is in an active state above the point of rotation, and is in a passive state below the point of rotation. In front of the wall the soil is in a passive state above the point of rotation, and is in an active state below. For a given friction angle, the depth of embedment required for stability can be computed using the equilibrium equations for horizontal force and for moment. The solution procedure is described in geotechnical text books such as that of Das (1990). This solution is plotted in Figure 4.

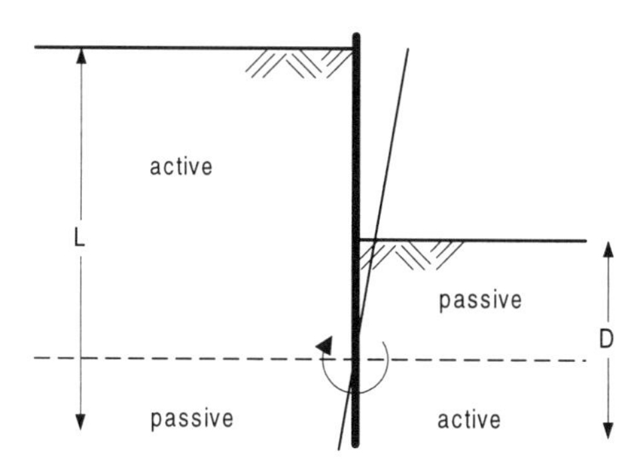

Figure 3. Stress field for limit analysis static solution for cantilever wall

Kinematic Solution

A kinematic solution is available from Derski et al. (1989). The failure mechanism assumed consists of rotation about the lower tip of the wall with homogeneously shearing soil wedges behind and in front of the wall, as illustrated in Figure 5. Since the material has no cohesion and obeys an associated flow rule,

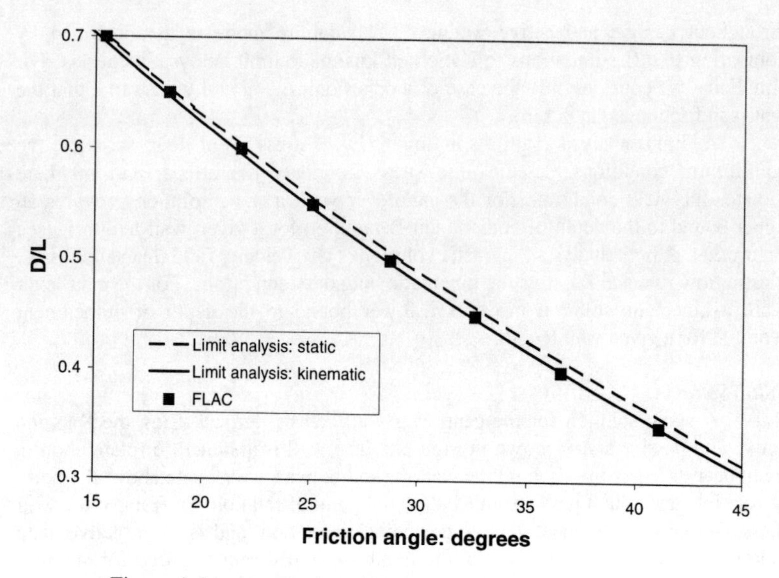

Figure 4. Limit analysis and FLAC solutions for cantilever wall

there is no energy dissipation due to plastic flow. Only the rate of work done by gravity forces need be considered. The solution is obtained by equating the rate of work done by the block behind the wall with the rate of work done by the block in front of the wall. The solution can be expressed in the form

$$\frac{D}{L} = \sqrt[3]{K_a / K_p} \tag{3}$$

where K_a and K_p are the active and passive earth pressure coefficients. This solution is plotted in Figure 4, along with the previous static solution.

FLAC Solution

The cantilever wall was simulated with FLAC using linear elastic beam elements connected to the soil mesh through interface elements upon which slip can take place. The interfaces were frictionless to match the limit analysis solutions. FLAC simulations were performed for walls with several different D/L ratios using an element size of 1% of the wall length. The friction angle at which each wall failed was found using the bracketing and bisection procedure described above. The FLAC results, plotted in Figure 4, closely match the limit analysis solutions. A numerical mesh with exaggerated deformation for D/L = 0.6 is shown in Figure 6. Note that for clarity, the mesh plotted is coarser than that used to produce the data shown in Figure 4. Contours of horizontal velocity have been plotted to demonstrate that deformation is distributed uniformly throughout the active and passive wedges. Strain is not localized into a shear band.

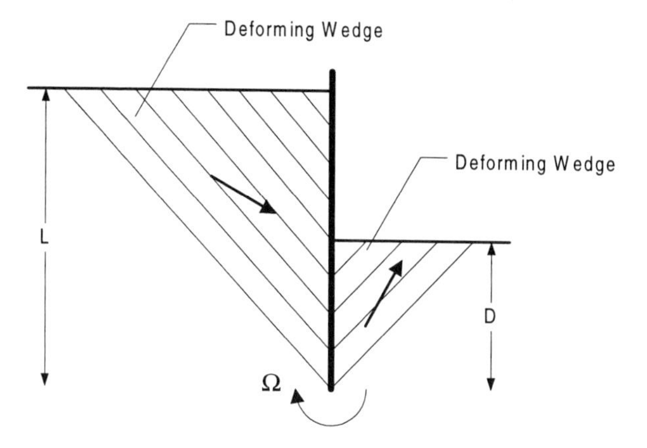

Figure. 5. Kinematic solution for cantilever wall

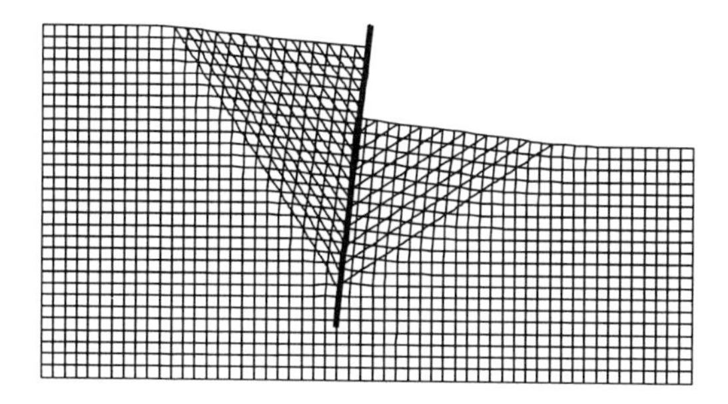

Figure 6. Deformed mesh and contours of horizontal velocity for cantilever wall

Braced Excavation

For a wall braced with a single row of struts at the top, the stability concern is "toe" or "kick-out" failure in which the bottom of the wall rotates into the excavation as illustrated in Figures 7 and 8.

Static Solution

Again, we consider only the case of a cohesionless soil with a wall-soil friction angle of zero. A static solution can be obtained by assuming active earth pressure behind the wall and passive earth pressure in front. The length of wall embedment required can then be computed from the balance of moment about the top of the wall. Moment equilibrium requires that

$$\left(\frac{D}{L}\right)^3 - 3\left(\frac{D}{L}\right)^2 + 2\frac{K_a}{K_p} = 0 \quad (4)$$

where K_a and K_p are the active and passive earth pressure coefficients. This static solution is plotted in Figure 9.

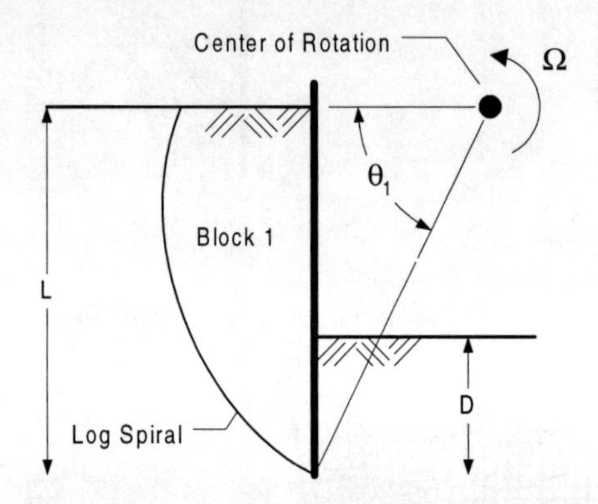

Figure 7. Rotational failure mechanism for braced excavation: Block 1

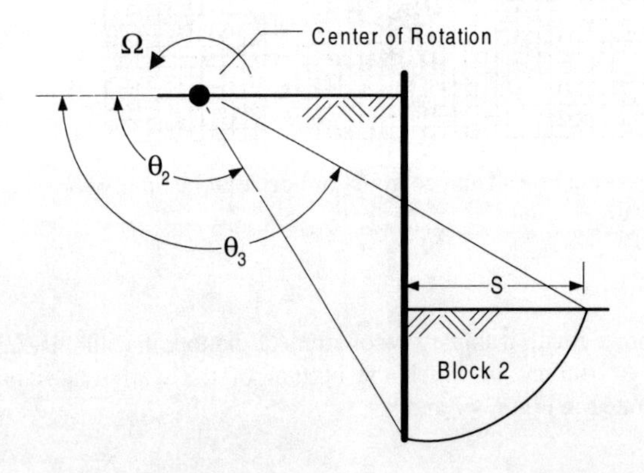

Figure 8. Rotational failure mechanism for braced excavation: Block 2

Kinematic Solution

A kinematic solution was derived for a failure mechanism consisting of two rigid blocks bounded by log-spiral failure surfaces as illustrated in Figures 7 and 8. It is assumed that horizontal displacement is prevented at the point of strut attachment. This constraint is satisfied as long as the centers of rotation of the blocks are at the same elevation as the strut (see Figures 7 and 8). Note that because the centers of rotation are not coincident with the point of strut attachment, slip occurs between the rotating blocks and the wall. The kinematic solution is derived by equating the rate of gravitational work done by Block 1 to the rate of work done against gravity by Block 2. The rate of work for each block was computed following the approach used by Chen (1975) in deriving the critical height of a vertical cut.

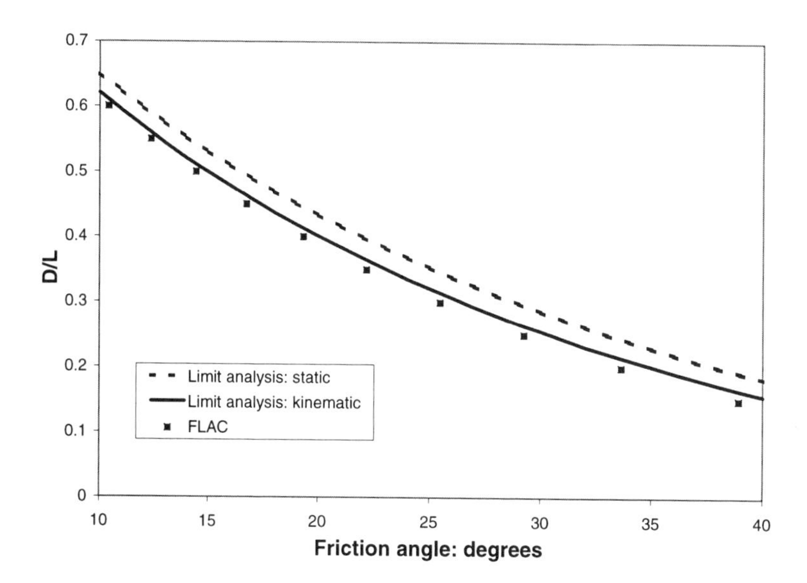

Figure 9. Limit analysis and FLAC solutions for braced excavation

The rate of work done by gravity forces for Block 1 is

$$\dot{w}_1 = \gamma L^3 \Omega \left[\frac{(3\tan\phi\cos\theta_1 + \sin\theta_1)\exp(3\theta_1\tan\phi) - 3\tan\phi}{3(1+9\tan^2\phi)\sin^3\theta_1\exp(3\theta_1\tan\phi)} - \frac{1}{3\tan^2\theta_1} \right] \quad (5)$$

where Ω is the angular velocity about the center of rotation and the angle θ_l is as defined in Figure 7.

The rate of work done by gravity forces for Block 2 (negative because the block is moving upward) is

$$\dot{w}_2 = \gamma L^3 \Omega (f_1 - f_2 - f_3) \tag{6}$$

where

$$f_1 = \frac{(3\tan\phi\cos\theta_3 + \sin\theta_3)\exp(3(\theta_3 - \theta_2)\tan\phi) - 3\tan\phi\cos\theta_2 - \sin\theta_2}{3(1 + 9\tan^2\phi)\sin^3\theta_2} \tag{7}$$

$$f_2 = -\frac{D}{L}\frac{\cos^2\theta_2}{3\sin^2\theta_2} \tag{8}$$

$$f_3 = \frac{S}{L}\frac{\sin\theta_3 \exp[2(\theta_3 - \theta_2)\tan\phi]}{6\sin^2\theta_2}\left(2\cos\theta_3 + \frac{S}{L}\frac{\sin\theta_2}{\exp[(\theta_3 - \theta_2)\tan\phi]}\right) \tag{9}$$

and where

$$\frac{S}{L} = \frac{1}{\sin\theta_2}(\cos\theta_2 - \cos\theta_3 \exp[(\theta_3 - \theta_2)\tan\phi]) \tag{10}$$

The distance S and the angles θ_2 and θ_3 are as defined in Figure 8.

For a given friction angle ϕ, a greatest lower bound for the ratio D/L was found by adjusting D/L so that $\dot{w}_2$, numerically minimized with respect to the angles θ_2 and θ_3, was equal but opposite $\dot{w}_1$, numerically maximized with respect to the θ_1. This solution is plotted in Figure 9, along with the previous static solution.

FLAC Solution

As in the cantilever case, FLAC simulations were performed for walls with several different D/L ratios using an element size of 1% of the wall length. The FLAC results, plotted in Figure 9, are a few percent below the kinematic, lower bound solution. This discrepancy illustrates one difficulty encountered with the strength reduction technique. Shear bands in FLAC are approximately three elements wide. Thus, the centerline of the shear band, which is the numerical approximation of a slip surface, is located not at the base of the wall, but at a distance of approximately 1.5 element widths below the base. Because of this, the effective length of the wall is slightly greater than intended. The numerical results are thus sensitive to the size of the elements at the wall base.

This point is illustrated in Figure 10 which shows the FLAC velocity field at failure along with the critical log spiral surfaces for D/L = 0.3. The critical friction angle for the kinematic solution for this case is 26.5°. Note that for clarity, a coarser mesh has been used for this figure, than was used to produce the data plotted in Figure 9.

Stability of a Gravity Retaining Wall

The strength reduction technique is particularly useful for problems in which several different failure modes are possible. The technique finds the critical mode automatically. Consider the gravity retaining wall shown in Figure 11. The wall is 3-m high, 1-m wide and is embedded 1-m. The soil has a unit weight of 18 kN/m^3, while the wall material has a unit weight of 24 kN/m^3. The friction angle of the soil is 30° and the friction angle between the wall and soil is 20°. The conventional rotational factor of safety computed from the balance of moments about the toe is 1.40, using the Coulomb approach to compute earth pressures. Note that the Coulomb approach takes into account the wall friction. The factor of safety for base sliding using the Coulomb approach is 1.95. These factors of safety are listed in Table 1.

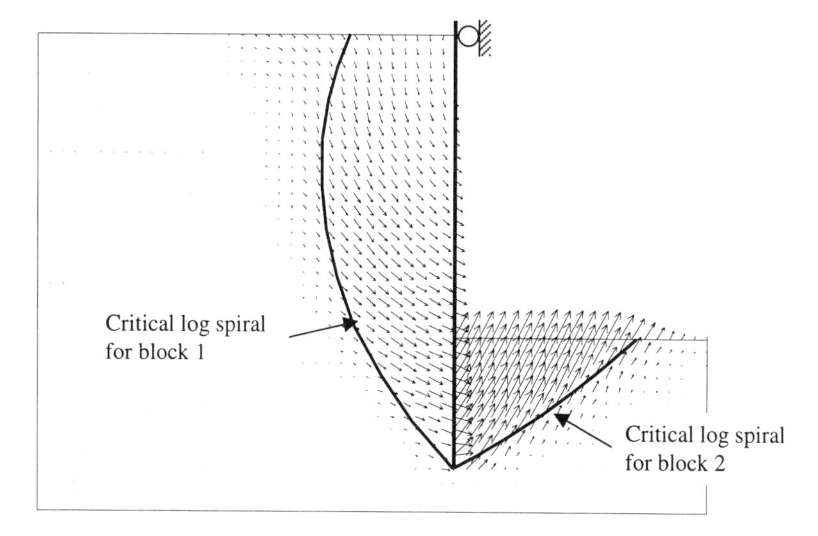

Figure 10. Velocity field at collapse along with critical log-spiral surfaces for the limit analysis solution for D/L = 0.3.

The strength reduction factor of safety is computed by reducing both the soil friction angle and the wall-soil friction angle. The wall fails by rotation as shown in Figure 12, which shows the numerical mesh with greatly exaggerated displacements. The factor of safety of is 1.13, significantly lower than the Coulomb rotational factor of safety (1.40).

There are two reasons for this difference. First, the two factors are defined in different ways. The strength-reduction factor is a ratio of shear strengths, while the Coulomb factor is a ratio of moments about the toe. Second, the Coulomb analysis assumes a particular failure mode (rotation about the toe). The strength reduction analysis finds a more critical mode: rotation about a point slightly in from the toe combined with bearing capacity failure (see Figure 12). This suggests that traditional hand calculations that assume simplified failure modes can be unconservative in some situations. Perhaps this is why the relatively high factor of safety of 2 is usually specified for rotational failure of retaining walls.

Table 1. Retaining wall factors of safety: Coulomb-approach versus strength reduction.

Coulomb Approach		Strength Reduction
Rotation	Sliding	
1.40	1.95	1.13

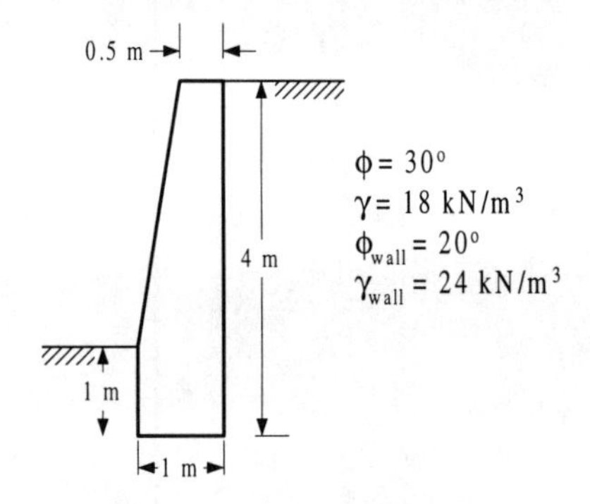

Figure 11. Gravity retaining wall

Strength reduction is not the only technique for computing factors of safety with a finite element or finite difference code. Conventional factors of safety for sliding and rotation can be computed by imposing a failure mode on the wall and measuring the resulting earth pressures acting on the wall. To compute the sliding factor of safety, the wall is moved horizontally away from the backfill until active pressures develop behind the wall and passive pressures develop in front of the toe. Figure 13 shows the stresses acting on the wall after this procedure has been applied. Defining the resisting force as the horizontal component of the forces acting on the toe and the base, and defining the driving force as the horizontal component of the forces acting on the back of the wall, the sliding factor of safety is 1.97. This is almost identical to the sliding factor of safety computed using the Coulomb approach (see Table 2).

A factor of safety with respect to rotation can be computed in a similar fashion. The wall is rotated about the toe until active pressure develops behind the wall and passive pressure develops in front of the toe. The factor of safety is computed from the resisting and driving moments about the toe. This procedure yields a factor of safety of 1.41, very close to the rotational factor of safety computed using the Coulomb approach.

Table 2. Retaining wall factors of safety: Coulomb-approach versus FLAC with imposed failure modes.

	Rotation	Sliding
Coulomb Approach	1.40	1.95
FLAC with imposed failure mode	1.41	1.97

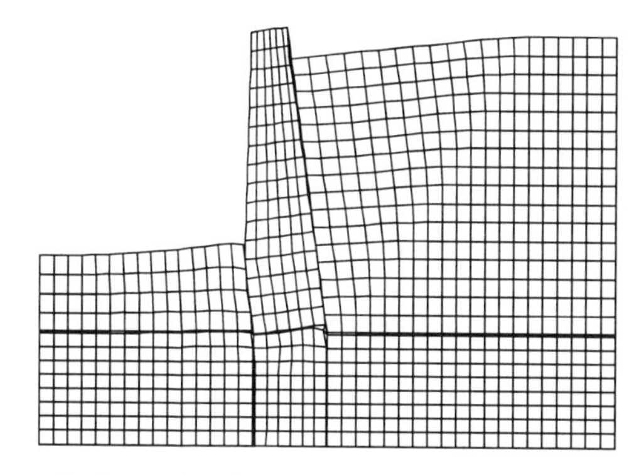

Figure 12. Strength reduction failure mode for gravity retaining wall

Conclusion

Finite element and finite difference programs can be used for complicated, realistic geotechnical analyses, but they are also powerful tools for simple limit-equilibrium-type analyses. The strength reduction technique is a convenient method for performing these analyses. If a Mohr-Coulomb model with an associated flow rule is used, the strength reduction technique requires no more input data than limit-equilibrium methods because, for an associated flow rule, the elastic constants, the initial stress and the stress path have no effect on the collapse load.

It has been demonstrated, through a number of examples, that strength reduction can be applied to geotechnical stability problems involving slip surfaces as well as those involving deforming soil wedges. This allows the use of the same tool for both slope stability analysis and stability analysis of cantilever walls, braced excavations and retaining walls. These topics do not have to be artificially separated as they are in current practice. The method of slices is a specialized technique designed for a particular type of problem. In contrast, strength reduction is a general-purpose tool that can be applied to almost any geotechnical stability problem.

It should be noted that strength reduction gives a factor of safety with respect to soil shear strength. This is the traditional definition of the factor of safety for slope stability. However, for cantilever walls and retaining walls other definitions are commonly used. The strength reduction factor of safety is not equivalent to these and gives different numerical values. It is possible to compute these conventional factors of safety with a finite element or finite difference code by imposing an assumed failure mode and measuring the resulting soil pressures. However, this negates one of the main advantages of the strength reduction technique, the ability to automatically find the critical failure mechanism.

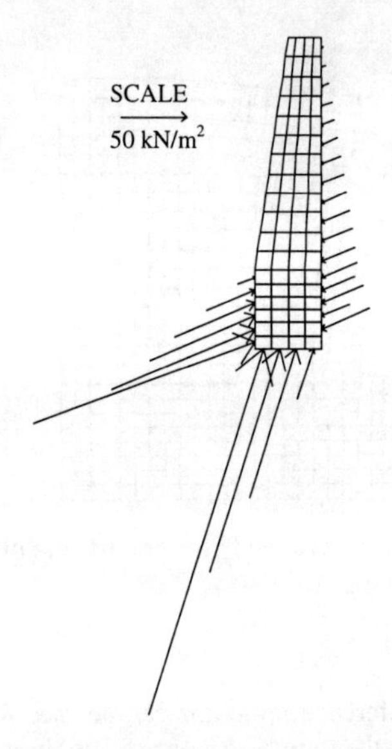

Figure 13. Soil pressures acting on retaining wall

References

Bishop, A.W. (1955). The use of the slip circle in the stability analysis of slopes. *Géotechnique* **5**, 7-17.

Chen, W.F. (1975). *Limit Analysis and Soil Plasticity*. Amsterdam, Elsevier.

Das, B.M. (1990). *Principles of Foundation Engineering,* Second Edition. Boston, PWS-Kent.

Dawson, E.M. and Roth, W.H. (1999), Slope stability analysis with FLAC, *Proceedings, FLAC and Numerical Modeling in Geomechanics*, Detournay, C. and Hart, R. (eds.) Minneapolis, September 1999, Balkema, pp. 3-9.

Dawson, E.M., Roth, W.H. & Drescher, A. (1999). Slope stability Analysis by strength reduction. *Géotechnique* **49**, No. 6, 835-840.

Derski, W., Izbicki, R., Kisiel, I. & Mróz, Z. (1989) *Rock and Soil Mechanics, Developments in Geotechnical Engineering, Vol. 48.* Elsevier and PWN – Polish Scientific Publishers, Warsaw.

Donald, I.B. & Giam, S.K. (1988). Application of the nodal displacement method to slope stability analysis. *Proc. Fifth Australia-New Zealand Conf. on Geomech.*, Sydney, Australia, 456-460.

Duncan, J.M. (1996). State of the art: limit equilibrium and finite-element analysis of slopes. *J. Geotech. Engng. Div. Am. Soc. Civ. Engrs.* **122**, No. 7, 577-596.

Itasca Consulting Group (1995). *FLAC, Fast Lagrangian Analysis of Continua, Version 3.3.* Itasca Consulting Group, Minneapolis, Minnesota, USA.

Matsui, T. & San, K.C. (1992). Finite element slope stability analysis by shear strength reduction technique. *Soils and Found.* **32**, No. 1, 59-70.

Naylor, D.J. (1982). Finite elements and slope stability. *Numer. Meth. in Geomech., Proc. NATO Advanced Study Institute.* Lisbon, Portugal, 1981, 229-244.

Otter, J.R.H., Cassell, A.C. & Hobbs, R.E. (1966). Dynamic relaxation (Paper No. 6986). *Proc. Inst. Civil Eng.* 35, 633-656.

Plaxis BV (1998) Plaxis Version 7.1 User's Manual. Plaxis BV, Delft, The Netherlands.

Ugai, K. (1989). A method of calculation of total factor of safety of slopes by elasto-plastic FEM. *Soils and Foundations* **29,** No. 2, 190-195 (in Japanese).

Ugai, K. & Leshchinsky, D. (1995). Three-dimensional limit equilibrium and finite element analyses: a comparison of results, *Soils and Foundations* **35,** No. 4, 1-7.

Zienkiewicz, O.C., Humpheson, C. & Lewis, R.W. (1975). Associated and non-associated visco-plasticity and plasticity in soil mechanics. *Géotechnique* **25**, No. 4, 671-689.

SHEAR STRENGTH OF SENSITIVE CLAY SLOPES IN MAINE

by Steven C. Devin[1] and Thomas C. Sandford[2], Members ASCE

Abstract

This paper presents the case history of a natural slope, about 10 meters in height, in a sensitive, low plasticity, glaciomarine clay near Portland, Maine. The area in the vicinity of the slope has a history of landslides. A field and laboratory testing program was conducted which included field and laboratory vane shear (VST), $\overline{CK_oUC}$ triaxial, and consolidation tests. Based upon regional geotechnical practice, the low undrained shear strengths measured behind the slope using the VST result in a stability factor of safety of 0.64 indicating failure for undrained conditions. However, the observed stable slope performance indicates that the actual factor of safety must be greater than unity. Regional geotechnical practice generally does not use corrections for measured s_u(FV) values in sensitive glaciomarine soils. Application of Bjerrum's (1973) anisotropic correction to the s_u(FV) measured at the site and considerations of stress history result in a computed factor of safety which is more consistent with the observed field performance of the slope. The application of VST anisotropic correction factors is recommended for future undrained stability analyses in this material.

Introduction

The current state of geotechnical engineering practice in northern New England does not adequately assess the stability of natural slopes in sensitive glaciomarine clays subject to undrained conditions. Oftentimes, the peak undrained shear strengths utilized in the undrained analysis of an intact slope yields factors of safety indicating that the slope is not stable. In an earlier paper (Devin and Sandford, 1995) the authors proposed prediction criteria to assess the likelihood of an initial single failure to evolve into a retrogressive or massive flowslide. Such a flowslide occurred along the Presumpscot River near Portland, Maine in 1868. Slide evolution predictions are valuable for land use planning and risk assessment of the consequences of failure, however, accurate assessment of initial instability can be elusive, particularly in sensitive and quick clays. As will be seen, standard analytical methods as currently applied in the region can provide misleading results in assessing the stability of natural slopes susceptible to flowslide behavior.

This paper presents a case history for one such slope located along the banks of the Presumpscot River in Westbrook, Maine. The slope is in the vicinity of a number of historical landslides, including the 1868 flowslide which encompassed an area estimated at 14 hectares (Morse, 1869; Devin and Sandford, 1995). Common practice in the area utilizes the peak undrained shear strength obtained from field vane shear tests (VST) behind the slope to calculate the factor of safety. This approach yielded a factor of safety of 0.64 on a slope which has been standing, and apparently stable, for years. A change in loading to a natural slope in a sensitive or quick clay, such as a steepening or an added embankment, can prove critical to the stability of the slope. Therefore it is important to be able to

[1] Consulting Engineer, P.O. Box 1782, Quincy, California 95971

[2] Associate Professor of Civil Engineering, University of Maine, Orono, Maine 04473

assess the factor of safety for this limiting condition. Such an analysis must consider: 1) the geology and stress history of the deposit; 2) the state of stress and consequent consolidation prior to loading; 3) shear strength anisotropy; 4) the drainage conditions during shear; 5) strain rate to failure; and 6) sensitivity. The methodology developed for assessing the shear strength and stability factor of safety at this site may prove valuable for sites with similar geology and geotechnical properties in the absence of comprehensive laboratory test data.

Geology and Stratigraphy

The site is underlain by glaciomarine sediments known as the Presumpscot Formation. These sediments consist predominately of late Pleistocene clays, silts, and fine sands deposited in a marine environment as the Laurentide ice sheet receded. The deposit at Westbrook has been interpreted as belonging to geologic unit GM-P (Belknap et al., 1988) which consists of a massive, relatively homogenous deposit of silts, clays, and fine sands showing little evidence of stratification (Devin and Sandford, 1995). The Presumpscot River has downcut through this relatively level ponded GM-P deposit and created banks which are susceptible to landslides. The location and extent of the 1868 flowslide (Morse, 1869) are shown in Figure 1.

The terrain at the site can be classified into three zones for further discussion of strengths. Zone I represents the intact ponded deposit and is located behind the crest. Zone II represents the portion of the deposit acted upon by the shear stresses imposed by the slope and encompasses the soil beneath the slope as well as parts of the crest and the toe affected by the shear stress of the slope. Zone III represents the downcut portion on the floodplain and beneath the river which is unaffected by the shear stresses of the slope.

Uplift and leaching of these sediments after deposition has potentially had a significant impact on their geotechnical properties such as has been documented for Eastern Canadian and Scandinavian clays (Bjerrum, 1954, LaRochelle, et al, 1970, Penner and Burn, 1978). Uplift above sea level was caused by crustal rebound which outpaced rising sea levels following glacier retreat. Pore water salinity, as measured by Mayer (1988) from samples obtained at the site, varied from 0.28 *g/l* to 1.87 *g/l* which indicates the probable leaching of saline pore water. Such leaching has been shown to be accompanied by a reduction in liquid limit and consequent increase in sensitivity since the time of

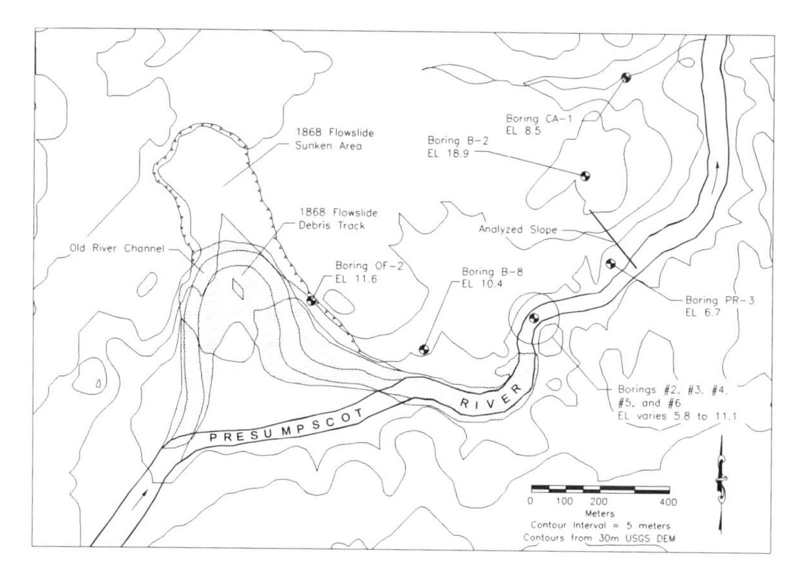

Figure 1, The study area showing the location of all borings and the extent of the 1868 flowslide.

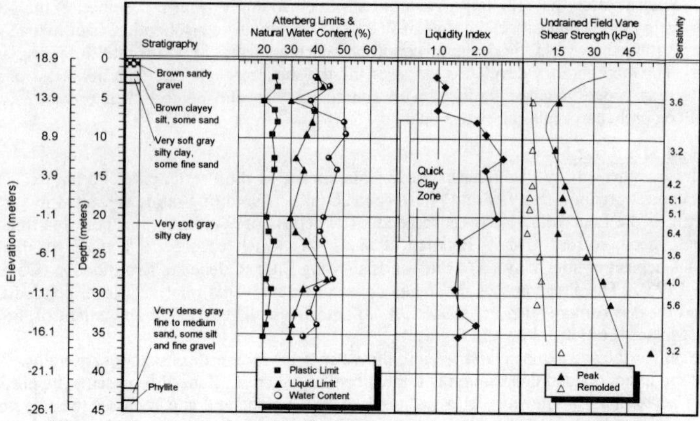

Figure 2, Summary of test data for Boring B-2 located within Zone I behind the crest of the slope.

deposition as suggested by Bjerrum (1954) for normally consolidated marine clays. Thus, changes in pore water chemistry have likely contributed to the evolution of a meta-stable soil structure exhibited by undisturbed samples obtained at the site.

The Zone I stratigraphy, as determined in Boring B-2 made in conjunction with this project, consists of about 0.75 to 1 meter of brown sandy gravel overlying a 1 to 1.2 meter thick weathered layer of brown clayey silt. Below this depth the Presumpscot sediments generally consist of very soft gray silty clay extending to a depth of 37.5 meters (EL –18.6). A quick clay layer extends from a depth of about 8 meters to 23 meters. Below the silty clay, a 1.4-meter thick layer of glacial till overlies bedrock. Unpublished data from borings made in connection with earlier pipeline and sewer projects in the vicinity and supervised by others (TAMS, 1965; Jordon-Gorrill, 1973; and S.D. Warren, 1972) show a similar stratigraphy with the possible addition of a surficial layer of Holocene alluvium. The location and surface elevation of all borings is shown in Figure 1 while a subsurface log and testing summary for Boring B-2 is presented in Figure 2.

As can be seen in Figure 1, the earlier borings are all generally located closer to the river (Zone III) or are located along the slope (Zone II) which rises from the river's edge. Surface elevations of the borings range from a high of 18.9 ± meters at Boring B-2 to an elevation of 5.8 ± meters at the river level for Boring #2, #3, and #4 which were drilled within the river channel and supervised by TAMS.

Geotechnical Properties

Stress History

Based upon limited consolidation test data, strength data, and natural water contents, the Zone I deposit behind the crest of the slope, appears to be normally consolidated with an overconsolidated upper layer. An overconsolidation ratio (OCR) of 1.03 was measured at a depth of 21.9 meters (EL –3.0) in Boring B-2 while an OCR of 1.44 was measured at a depth of 4.6 meters (EL 14.3) in the same boring. The consolidation test data show the significant compressibility of the normally consolidated sample with the secant compression index C_c' increasing from 0.32 for the overconsolidated sample at a depth of 4.6 meters to 1.11 for the normally consolidated sample at a depth of 21.9 meters. The lower value of C_c' at a depth of 4.6 meters may reflect an alteration of the more highly structured fabric evident in samples obtained at greater depths. The oxidation and lower water contents of the upper 3 to 4.6 meters in Boring B-2 suggest that the overconsolidation in the upper layer may be the result of chemical weathering, desiccation, and fluctuations in the groundwater level. The slight overconsolidation present at depth may be attributed to secondary compression. The field value of OCR is likely slightly higher than that measured in laboratory testing due to sample disturbance, and therefore the OCR below the weathered layer can be expected to be constant with depth (Bjerrum, 1967). The normal consolidation and the lack of distinct silt or fine sand laminae indicate that this deposit is geologic unit GM-P. The normally consolidated stress history of this deposit is also reflected in low values of the ratio of the peak undrained shear strength measured by the field vane to the effective overburden pressure $s_u(FV)/\sigma'_v$ of 0.15 (see Figure 2) in Boring B-2 behind the crest of the slope.

As will be seen, higher values of the $s_u(FV)/\sigma'_v$ ratio from the borings made along the slope (Zone II) and within the river (Zone III) suggest that the overconsolidation increases as one proceeds from the crest of the slope to the river. This may be attributable to overburden removal from erosion/fluvial processes, fluctuations in the groundwater level, and/or chemical weathering and desiccation. Shear stresses induced by the presence of the slope may also increase the consolidation beneath the slope and near the toe. The consequent increase in K_o (Terzaghi et al., 1996) along the slope and at the toe will decrease undrained shear strength anisotropy and will increase the measured field vane shear strength $s_u(FV)$ which effectively estimates the shear strength on a vertical plane. Principal stress rotation along the slope will affect the measured $s_u(FV)$ relative to that measured for a deposit without a slope. Consideration of these factors in the evaluation of the available undrained shear strength can yield factors of safety which are more consistent with slope performance.

Peak Undrained Shear Strength

The peak undrained shear strength measured by the field vane $s_u(FV)$ (Figure 2) in Boring B-2 behind the crest (Zone I) falls into three distinct layers. The upper layer, which is basically from the surface down to about 7 meters, shows lower water contents, lower liquidity indices, and evidence of weathering, oxidation, and desiccation which results in overconsolidation and an increase in $s_u(FV)/\sigma'_v$ in this upper layer relative to the lower layers.

Below the weathered layer, the very soft gray clay has two distinct layers, a quick clay layer and a less sensitive clay layer. The quick clay layer extends from a depth of 8 meters (EL 11.0) to a depth of 23 meters (EL –4.3) in Boring B-2. The liquidity index is generally greater than 2.0 in the quick clay while the liquidity index decreases to around 1.5 in the lower sensitive clay layer. The designation of quick clay is made with reference to the failure pattern prediction criteria presented by the authors (Devin and Sandford, 1995) in which flowslides are predicted in geologic unit GM-P where the liquidity index exceeds 1.5 and the $s_u(FV)/\sigma'_v$ ratio is less than 0.15.

The near constant measured shear strength in the quick clay layer does not correspond to the decrease in water content with depth. This borehole was advanced using casing and water. There is anecdotal evidence in the region that explorations conducted with casing and drilling mud yield a higher $s_u(FV)$ in very sensitive and quick clays than those conducted with casing and water. The $s_u(FV)/\sigma'_v$ ratio about 0.15 shown on Figure 2 represents an estimated upper limit for undisturbed strength in the quick clay. Some disturbance may have occurred at depth in the quick clay as a result of local yielding and movement of the clay near the bottom of the borehole with a consequent reduction in $s_u(FV)$.

Peak undrained shear strengths were obtained from VST and $\overline{CIUC}$ testing both beneath the slope (Zone II) and beneath or beyond the toe (Zone III). These data show a weathered layer with higher strengths in the upper 5 meters or so of each boring, irrespective of the elevation of the ground surface as shown in Figure 3. Below the weathered layer, the $s_u(FV)$ strengths nearer the toe are generally higher for a given depth than the $s_u(FV)$ strengths measured in Boring B-2 from behind the crest. The $s_u(FV)/\sigma'_v$ ratio beneath the weathered layer in Zone II and III generally is in the range of 0.20 to 0.50 approaching with depth the value of 0.15 found in Boring B-2 of Zone I.

The past history and current topography can account for the differences in the measured $s_u(FV)$ near the toe compared to those at the same depth beneath the uneroded plain behind the crest. Since the samples below the toe and below the river were presumably subject to the same overburden pressure as samples at the same elevation behind the crest, the water contents at the toe likely reflect the previous overburden pressure. Consequently, this material is overconsolidated with respect to its present overburden. Additionally, the weight of the soil in the slope imposes shear stresses which lead to further consolidation and a consequent increase in shear strength.

Since the river valley in this area reflects the down cutting of a massive deposit, the variation in strengths across the site can be plotted versus elevation as is done in Figure 3 to indicate the possible effects of past consolidation. Aside from the higher strengths due to weathered layers, the strengths show a pattern which trends toward the strengths found in Boring B-2, especially with increasing

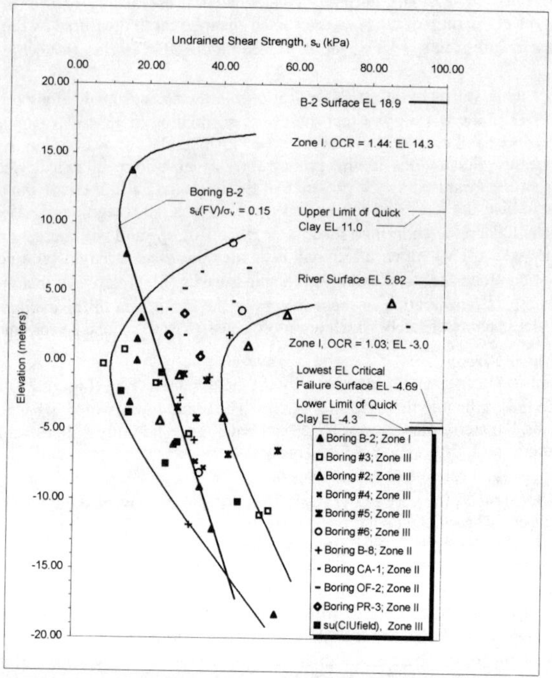

Figure 3, Undrained shear strength, $s_u(FV)$ and $s_u(CIU)_{field}$ vs. elevation across the site.

depth. This is to be expected if sediments within the river valley of Zone II and III have been consolidated to the same overburden as those in Zone I behind the crest.

Differences in measured strengths from Zone II and III relative to that of Boring B-2 in Zone I can be attributed to weathering near the surface irrespective of the elevation, swelling with associated strength loss from overburden removal, and strength gain from shear stress induced by the slope. Additionally, there were multiple explorations at different times with different exploration crews and equipment which also contribute to the differences.

Thus, the data presented in Figure 3 reinforce the concept that changes in the state of stress account for differences in strength across the site.

Remolded Shear Strength and Sensitivity

The remolded strength, while not critical to the initial stability of a natural slope, has a profound effect on post-failure behavior and serves as an indicator of stress-strain behavior and soil structure meta-stability. It is also an indicator that the calculated factor of safety for undrained shear cannot be close to unity, since local yielding will lower the available strength on the stability surface.

The sensitivity as measured by the field vane in Boring B-2 ranged from 3.2 to 6.4. These sensitivity values correspond to a material described as medium sensitive to very sensitive by Rosenqvist's (1953) criteria. A quick condition with a likelihood of viscous flow was confirmed for a sample taken at a depth of 9.8 meters (EL 9.1) with a liquidity index of 2.15 which, when remolded in a beaker, was poured onto a glass plate. This behavior suggests much more sensitivity in the deposit than indicated by the field vane. The high values of liquidity index, which Bjerrum (1954) has correlated to sensitivity in Norwegian clays, also suggest that the Presumpscot has much higher sensitivity than indicated by the field vane.

Sensitivity measured in the other borings varied from as low as 1.7 at a depth of 6.4 meters in Boring CA-1 (EL 2.1) to as high as 14 at a depth of 11.1 meters (EL –0.7) in Boring B-8. The high values of both the peak and remolded s_u(FV), along with the low value of sensitivity, may reflect the presence of shells in Boring CA-1.

A laboratory VST performed on an undisturbed sample taken at a depth of 28.0 meters (EL –9.1) in Boring B-2 yielded a sensitivity of 21. This compares with a sensitivity of 4 measured by the field vane at nearly the same depth. A similar result was observed for an undisturbed sample taken at the nearby 1983 Gorham landslide (Amos and Sandford, 1987). Thus, there are several indications that the VST, as typically performed in Maine, may underestimate the sensitivity of the Presumpscot clay (Devin, 1990).

Undrained Shear Strength Anisotropy

In regional practice, adjustments to measured s_u(FV) values for the possible effects of anisotropy are generally not made. Undrained shear strength anisotropy can be *inherent* due to depositional directional differences (Mitchell, 1993) and also can be the result of an anisotropic *initial state of stress* (Brinch-Hansen and Gibson, 1949; Jamiolkowski et al, 1985). In this case history the soil appears to primarily have an *initial state of stress* anisotropy since the deposit shows little evidence of stratification.

Bjerrum (1973) showed that undrained shear strength anisotropy is more pronounced in clays of low plasticity. Using Hvorslev's (1937) parameters as a basis, Bjerrum (1973) proposed a theoretical relation quantifying the variation in s_u with failure plane inclination. The model was calibrated against Triaxial Compression (TC), Triaxial Extension (TE), and Direct Simple Shear (DSS) tests run on clays of various plasticity and stress history. By assuming that the field vane measures s_u on a vertical plane ($\alpha = 90^{\circ}$), Bjerrum developed correction curves for use with the field vane for low, medium, and high plasticity clays. The majority of data presented in this paper were developed using a 50 mm x 180 mm field vane. Bjerrum's curve for an aged normally consolidated low plastic clay ($I_p = 10$) with an OCR = 1.2, and $K_{o(NC)} = 0.5$ is presented in Figure 4. Also plotted in Figure 4 are curves which the authors computed using Bjerrum's theoretical relationship for OCR = 1.2 and $K_{o(NC)} = 0.50$, $K_{o(OC)} = 0.58$, and $K_{o(OC)} = 0.75$ for a low plasticity clay.

The most important aspect of Figure 4 is the significant change in strength related to the orientation of the failure plane with respect to the principal directions of the stress. Thus for a vertical failure surface ($\alpha = 90^{\circ}$) the correction is 1.0 since the VST measures the strength on this plane. A stability surface may have an inclination close to vertical ($\alpha = 90^{\circ}$) at the start of the active area near the crest, then will transition to a horizontal surface ($\alpha = 0^{\circ}$). Most stability surfaces will have a passive area where the failure plane inclination is negative. These curves also demonstrate that as K_o increases, the undrained shear strength anisotropy decreases with an isotropic strength state reached at

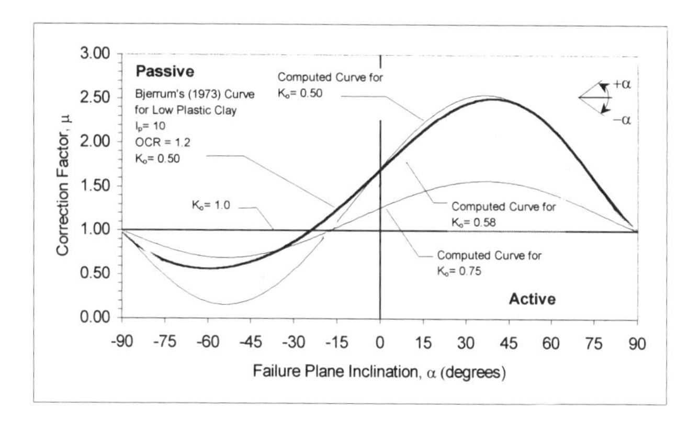

Figure 4, Field vane shear test (VST) correction for anisotropy.

$K_o = 1.0$. Thus, when overburden is removed as the result of down cutting by the river, the K_o will change, and consequently the correction for failure plane inclination will also change. The type of soil that Bjerrum used to establish the field vane correction in Figure 4 is similar to the normally consolidated Presumpscot Formation of Figure 2.

Effective Shear Strength and Stress-Strain Behavior

Anisotropically consolidated undrained triaxial compression, or $\overline{CK_oUC}$, tests were performed on an undisturbed sample taken at a depth of 27.4 to 28.0 meters (EL –8.5 to –9.1) in Boring B-2 using the Recompression Technique (Bjerrum, 1973). Samples were not consolidated beyond their *in situ* stress state so as to avoid destruction of the soil fabric. All samples were sheared at a rate of strain of 2 percent per hour.

Axial stress-strain curves normalized with the principal effective consolidation stress, σ_{1c}, are presented in Figure 5. The brittle behavior exhibited by the normally consolidated sample which reaches its peak strength at 0.2 percent axial strain has great significance to the stability of natural slopes. This brittle behavior indicates that the peak shear strength will be mobilized at a very small strain, the critical strain, beyond which significant structural breakdown and remolding of the soil fabric will occur. As can be seen in Figure 5, the measured TC axial strain at failure increases with increasing overconsolidation.

While only TC tests were run on samples from Boring B-2, it can be inferred from data presented by Lefebvre et al. (1983) and others (Koutsoftas and Ladd, 1985; Ladd, 1991) that the critical strain required to mobilize the peak strength will vary with failure plane inclination, mode of failure, and the initial state of stress. These data suggest that the critical strain will be greater for DSS and possibly for TE modes of failure. Thus, the peak strength will not be mobilized simultaneously along a potential failure surface and the factor of safety computed in a limit equilibrium analysis will be unconservative for low plastic clays, since the available shear strength on portions of the failure surface will be a post peak strength. While Koutsoftas and Ladd (1985) suggest a strain compatibility method to account for progressive failure along a potential surface of rupture, these methods are only applicable to strengths obtained from TC, TE, and DSS testing and not to an *in situ* test, such as the VST, which does not provide stress strain data. In addition, the time to failure is not accounted for in either the strain compatibility method or with the use of VST, although a field correction factor for time to failure for both laboratory (Terzaghi et al., 1996) and field vane (Bjerrum, 1972; Leroueil et al., 1983; and Terzaghi et al., 1996) tests is now widely used. These corrections attempt to account for rate of strain effects which show lower mobilized undrained shear strengths with increasing time to failure, particularly with increasing plasticity. The correction for low plasticity clays, such as the Presumpscot Formation, is close to unity and therefore regional practice does not routinely make corrections to $s_u(FV)$ for rate of strain effects.

The effective stress paths (Lambe and Whitman, 1969) for each sample are shown in Figure 6. The effective angle of internal friction at maximum obliquity, ϕ', determined in these tests was 36.9°

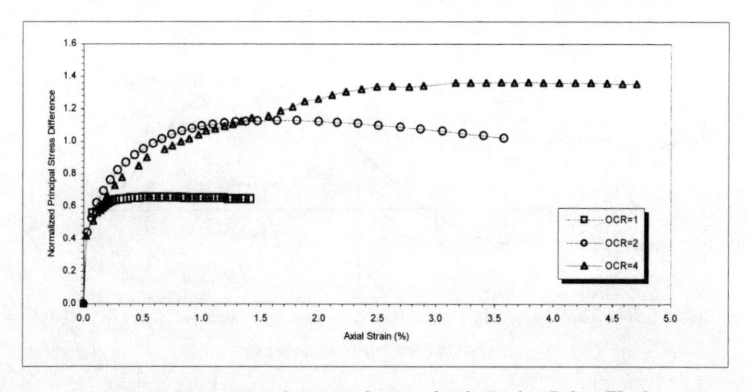

Figure 5, Axial stress-strain curves from testing in Boring B-2 at EL -9.

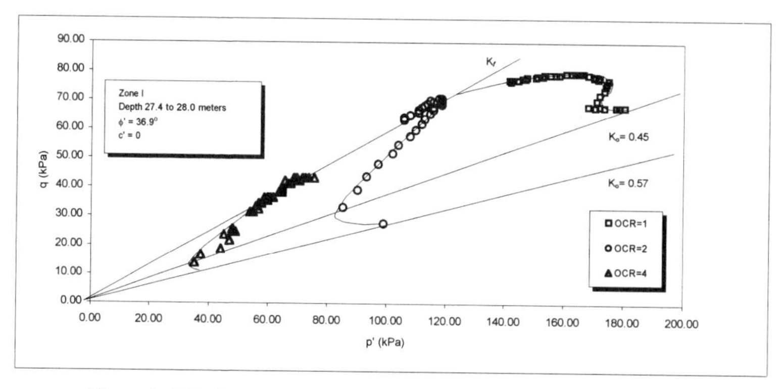

Figure 6, Effective stress paths from testing in Boring B-2 at EL -9.

which closely agrees with the friction angle determined by Amos and Sandford (1987) for the Presumpscot clay at the nearby Gorham landslide. The peak angle of internal friction, ϕ'_{peak}, measured at the peak shear stress, q_{max} , for the normally consolidated sample was determined to be 29.1°. No cohesion intercept was measured. The high value of ϕ' will yield a low K_0 and thus the effects of stress anisotropy will be important for this material. Skempton's (1954) pore pressure parameter at failure was measured for the normally consolidated sample as $A_{f(qmax)}$= 1.08 at q_{max} and $A_{f(mxoblq)}$= 2.67 at maximum obliquity. The overconsolidated samples showed much lower pore pressure generation with $A_{f(qmax)}$= 0.28 and $A_{f(mxoblq)}$= 0.40 for OCR=2 and $A_{f(qmax)}$= 0.12 and $A_{f(mxoblq)}$= 0.12 for OCR=4.

The values of the pore pressure parameter, A, suggest extreme sensitivity for the normally consolidated deposit behind the crest of the slope. For the normally consolidated sample at the peak undrained shear strength q_{max}, the value of $A_{f(qmax)}$ = 1.08 indicates that the pore pressure developed is approximately equal to the applied deviator stress. Beyond the peak strength and critical strain, during remolding, the pore pressures generated significantly exceed the magnitude of the applied deviator stress as the effective stress path approaches the K_f line. Pore pressure generation of this magnitude only occurs in soils with meta-stable skeletons which exhibit extreme sensitivity. As the breakdown of the soil fabric occurs and more and more effective stress is transferred from the soil skeleton to the pore water, induced pore pressures significantly exceed the applied load.

The low values of the pore water pressure parameter, A, measured for the overconsolidated samples reflect the stress history of these samples. Pore pressure generation in the overconsolidated samples is only one-tenth to about one-third of the applied axial load suggesting that the overall shearing resistance increases as frictional resistance is mobilized, *i.e.*, the inter-particle effective stress increases with strain.

Stability Analyses

In an area where there is historical evidence of retrogressive and flowslides, some slopes composed of highly sensitive clays have been apparently stable for many years. High liquidity indices and low $s_u(FV)/\sigma'_v$ ratios for this soil indicate potential retrogressive or flowslide behavior under undrained shear (Devin and Sandford, 1995). Although the slopes are composed of saturated clay, effective stress analyses for steady state seepage provide the appropriate long term factor of safety. Still, an undrained retrogressive or flowslide may be initiated with a single slide caused by rapid incremental stress increases from changes in slope geometry, slope loading, or cyclic shear strain caused by seismic ground motion. Thus the key to the determining the stability of an area subject to retrogressive or flowslides is the proper assessment of the undrained stability of the initial single slide. To obtain a realistic factor of safety for undrained conditions using the VST, stress history, shear strength anisotropy, and principal stress rotation must be considered in evaluating the available ground strength for undrained shear.

State of Stress Model

A history of the state of stress at the site using an engineering geology approach provides the framework for analyzing the shear strength to be used in a stability analysis. The Presumpscot River has cut a channel through a rather homogenous, massive silty clay deposit which still has intact plains which lie at nearly the same elevation and overlook the channel from both the north and south banks.

From this geologic and geomorphic basis and an understanding of the mechanics within the slope, we can establish a conceptual model of the state of stress across the site prior to any new loading. This model corresponds to the three previously designated zones. Zone I, located behind the crest of the slope, is characterized by vertical and horizontal major and minor principal stresses, respectively. Zone II, located beneath the slope and a portion of the crest and the toe, is characterized by preconsolidation resulting from removal of material which created the slope, spatial variation in the orientation of principal stresses, and the application of shear stresses to the soil. Zone III, located beneath the river channel and in the flood plain, is characterized by vertical and horizontal major and minor principal stresses, respectively. However, this zone is preconsolidated from overburden removal resulting from down cutting by the Presumpscot River.

For Zone I the principal stress orientation ($\delta = 0^\circ$) remains constant with depth. This zone consists of an aged normally consolidated clay with a weathered, overconsolidated surface layer with a thickness of about 5 meters. Overconsolidation within the weathered layer decreases with depth reaching a constant $OCR_{(NC)}$ in the normally consolidated layer located below the weathered layer. The constant $OCR_{(NC)}$ below this depth, in the probable field range of 1.10 to 1.20, is attributable to secondary compression. The ratio of the minor to the major principal stress, K_o, decreases with decreasing overconsolidation in the weathered layer and approaches a constant $K_{o(NC)}$ with depth in the normally consolidated layer. The estimated value of $K_{o(NC)}$ is 0.49 (Terzaghi, et al, 1996) based upon $OCR_{(NC)}$ equal to 1.15 and ϕ' equal to 33°.

Zone II is characterized by spatial variation in the orientation of the principal stresses ($\delta = 0^\circ$ to 90°) with $\delta \rightarrow 0^\circ$ with increasing depth. The spatial variation in δ is difficult to assess without sophisticated modeling. However, at failure, the orientation of the principal stresses along a surface of rupture is known. This can be used to estimate the magnitude and orientation of principal consolidation stresses along a trial failure surface. Preconsolidation from overburden removal in Zone II will also show spatial variation with an increase in $OCR_{(OC)}$ proceeding from Zone I towards Zone III and a decrease in $OCR_{(OC)}$ with depth ($z \rightarrow \infty$, $OCR_{(OC)} \rightarrow OCR_{(NC)}$). Consolidation, which will occur due to the imposition of shear stresses by the slope, has a spatial variation beneath the slope but is highest near the toe. Consequently, for a constant elevation, K_o will increase proceeding from Zone I towards Zone III while $K_{o(OC)}$ can be expected to approach $K_{o(NC)}$ with depth.

In Zone III the principal stress orientation ($\delta = 0^\circ$) remains constant with depth. This zone consists of a preconsolidated aged clay with a weathered surface layer of about 5 meters thickness which has been overconsolidated by weathering. Beneath the weathered layer, overconsolidation is attributable to overburden removal and secondary compression. The overconsolidation resulting from overburden removal decreases with depth and ultimately approaches $OCR_{(NC)}$ as $z \rightarrow \infty$. The magnitude of overconsolidation resulting from weathering will decrease with depth and transition to the overconsolidation attributable to overburden removal. Consequently, K_o in Zone III will rapidly decrease with depth through the weathered layer and then more gradually in the deposit below the weathered layer ultimately approaching $K_{o(NC)}$ as $z \rightarrow \infty$.

Undrained Shear Strength Model

Bjerrum (1972) summarized the research of many showing that shear strength varies depending upon the stress path to failure. It was recommended that the system of load application in testing should model the system of loads applied in the field to obtain corresponding shear strengths. For a stability evaluation, the field applied loads are modeled in the laboratory by TC, TE, and DSS. By experience, Bjerrum showed that VST with corrections correlated to embankment failures provide a shear strength intermediate between the extremes provided from the other tests. Bjerrum (1972) recommended that measured VST be corrected for the strain rate effect (μ_R), and for orientation of the shear surface with respect to anisotropy in the soil (μ_A) as follows:

$$(s_u)_{field} = s_u(FV) \cdot \mu_R \cdot \mu_A.$$

The strain rate correction is widely used, however, the low plasticity glaciomarine clays found in Maine have a strain rate correction close to 1.0 and thus this correction is generally not used locally with VST. In local practice, the correction for orientation of the failure surface is not considered with s_u(FV) in undrained analysis. As will be seen below, this can lead to misleading stability analysis results.

Both the initial state of stress and the undrained shear strength anisotropy (Bjerrum, 1973) have significant implications for the analysis of natural slopes using VST results. Consideration of the initial state of stress and anisotropy and reference to Figures 4 and 7 leads to the following field undrained shear strength model.

The VST effectively estimates s_u on a vertical failure plane. With the exception of Boring B-8 which used a field vane with $H/D = 2$, all s_u(FV) values were measured by a field vane with $H/D = 3.5$ where H is the height of the vane and D is the diameter of the vane. Thus, the assumption that s_u(FV) is measured primarily on the vertical $\alpha = 90^{\circ}$ plane is a reasonable one. Due to the lower values of K_o within Zone I, strength differences from the vertical plane to a failure surface orientation can be expected to be more significant in this zone than in Zones II and III. Consequently, corrections to measured s_u(FV) values from this zone become more significant than corrections in the other zones where the measured s_u(FV) on the vertical plane more closely approximates the available ground strength on the failure surface. In Zones II and III anisotropy increases with depth as $K_{o(OC)} \rightarrow K_{o(NC)}$, thus measured s_u(FV) values may deviate further from the available failure surface ground strength with increasing depth. The initial state of stress in Zone II with the rotation of principal stresses (*i.e.* $\delta > 0^{\circ}$) means that measured s_u(FV) values on the vertical plane are inclined to the principal stress direction, and thus corrections to the failure surface will be usually less than in Zone I. However, as with the preconsolidation effect, this effect will diminish with depth as the principal stress direction approaches $\delta = 0^{\circ}$. Undrained shear strength anisotropy therefore becomes more significant as the depth of the failure surface increases and consequently less significant for toe circles or shallow failures.

The measured s_u(FV) for Boring B-2 behind the crest may be adjusted for the rate effect, for shear strength anisotropy, and stress history to estimate the undrained strength available on a trial failure surface for the slope. A key element in using strengths from Zone I is the recognition that soil in Zone II and III has been consolidated in the past to the same overburden as exists in Boring B-2. Therefore, below the weathered layer, strengths on the failure surface of Zone II essentially correspond to strengths at the same elevation in Boring B-2 of Zone I along with some additional strength gain from consolidation caused by the presence of the slope. Figure 4 can be used for the correction for the failure plane inclination.

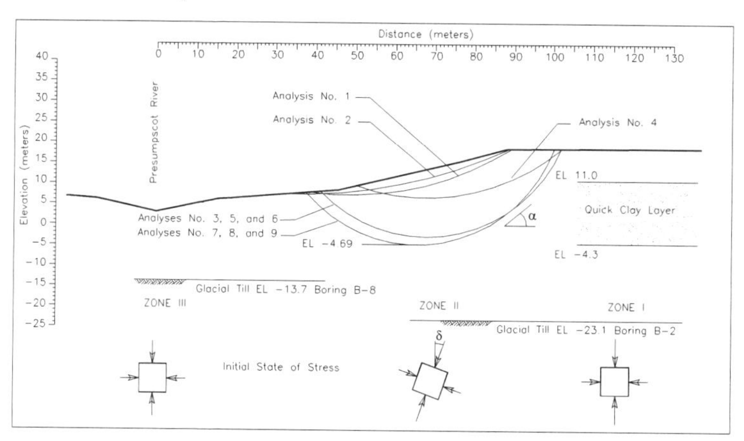

Figure 7, Initial state of stress model and critical failure surface locations.

The magnitude of the rotated principal stresses acting prior to a new loading on the slope can be estimated by conducting an effective stress stability analysis (Lowe and Karafiath, 1960) on the critical failure surface obtained from an undrained analysis using uncorrected s_u(FV). The results of the effective stress analyses give the mobilized shear strength and normal stress on the failure surface, and thus the magnitude of the principal stresses can be found at $(45^\circ - \phi'/2)$ to the failure surface for each slice. The computed major principal stress is treated as a consolidation stress and if it exceeds the preconsolidation stress at the same elevation behind the crest it is used with the $s_u(FV)/\sigma'_v$ ratio from the soil behind the crest to estimate the undrained shear strength. This is then corrected for rate (μ_R) and orientation (μ_A) of the failure surface relative to the orientation of the rotated principal stresses. This corrected shear strength estimates the available shear strength on the failure surface.

Analyses Results and Discussion

Both drained and undrained stability analyses were run on the slope shown in Figure 7 utilizing the measured shear strength data and the state of stress and undrained shear strength models presented above. All analyses were run using the Modified Bishop Method of Slices. The results of all analyses are summarized in Table 1 while the critical failure surfaces are plotted in Figure 7.

Drained analyses were run assuming a steady state seepage condition and both $\phi' = 36.9^\circ$ at maximum obliquity (Analysis No. 1) and $\phi'_{peak} = 29.1^\circ$ at q_{max} (No. 2). The computed factors of safety were 2.08 and 1.79, respectively. These show that the existing slope is stable as indicated by the field performance. The critical surfaces for these analyses can be seen in Figure 7 which also demonstrates how shallow these surfaces are relative to those computed in an undrained analysis.

Undrained analyses were initially run utilizing the values of s_u(FV) measured in Boring B-2 uncorrected for strain rate (Bjerrum, 1972) and with the common assumption in practice of isotropic behavior, *i.e.*, the same strength on all orientations. It was also assumed that the stratigraphy as indicated by Boring B-2 extended laterally into and beyond the slope, so that shear strengths of Boring B-2 were utilized at the same elevation on the failure surface. This commonly used approach implicitly assumes that consolidation beneath the slope is the same as behind the crest. The computed factor of safety for these conditions was 0.64 suggesting that the slope should have failed. However, the field evidence obviously does not support these results since the slope *is* standing and apparently stable in an area which has experienced numerous historical landslides (Morse, 1869; Devin and Sandford, 1995).

The previous undrained analysis was rerun with strengths reflecting a constant $s_u(FV)/\sigma'_v$ ratio of 0.15 for the silty clay (No. 4). This constant strength ratio with depth corresponds to an estimate of strength in the quick clay layer as shown in Figure 2 which corrects for possible disturbance. This increases the factor of safety to 0.79, which still indicates failure.

Even for a homogeneous deposit like this site, field measurements along the failure surface are more representative than the s_u(FV) values measured in Boring B-2, located some 180 to 215 meters behind the crest of the slope. These locations will better reflect variations in s_u due to spatial variation in principal stress magnitude and direction as well as consolidation from shear stresses induced by the slope. Since the soil at these locations is overconsolidated, there will be less tendency to experience disturbance during drilling. With reference to Figure 3 and the location of the critical failure surface determined in Analysis No. 3, representative average s_u values were obtained directly from field measurements in Zone II and III. Factors of safety of 0.94 and 1.13 were computed using an s_u of 25 kPa and 30 kPa, respectively (Nos. 5 and 6). These factors of safety imply failure or near failure in a material that may have local yielding at low factors of safety. Considering the proposed initial stress state and undrained shear strength models, the strengths utilized in this analysis may better represent the average available ground strength along the critical failure surface than the measured s_u(FV) from Zone I. These values had not been corrected for strain rate or failure surface inclination. Although there were extensive explorations from a number of different sources at this site, it was still not possible to obtain measured strengths along the entire trial failure surface. However, if such data were available, it must be recognized that these strengths would still be measured on the vertical plane in each VST with an unknown K_o and σ_h' caused by overburden removal from river down cutting and stresses imposed by the slope. Thus the measured strengths would still be different than the ground strength available along a failure surface of different orientation.

The s_u(FV) values from Boring B-2 which had been used directly in Analysis No. 3 without adjustments for possible disturbance were corrected for rate of strain and orientation of the failure plane with Bjerrum's correction curve for K_o = 0.50 (Figure 4) and rerun in Analysis No. 7 to obtain a factor of safety of 1.24. However, this analysis does not include strength gain from consolidation caused by shear stresses induced by the slope. The factor of safety of 1.24 is more consistent with the observed stable field performance of the slope than the factor of safety of 0.64 when measured strengths from Zone I are used directly. However, it is expected that the undrained stability factor of safety should be closer to the drained factor of safety since mobilized strengths in the drained case represent initial conditions for undrained failure.

The measured s_u(FV) values from Boring B-2 were also corrected for rate of strain, the existing consolidation stress, which includes the effects of shear stresses induced by the slope, and the orientation of the shear surface. The consolidation stress from the slope, principal stress direction, and K_o were estimated by conducting a drained, steady-state seepage, effective stress analysis on the critical failure surface found for Analysis No. 3. If the computed major principal stress was larger than the stress at the corresponding elevation in Boring B-2, the computed major principal stress was then used to find s_u(FV) at the same overburden stress in Boring B-2. The strength was then corrected

Table 1, Summary of Stability Analyses

No.	Type of Analysis	Description of Analysis	Stress State Zone s_u(FV)	Computed *FS*
1	Effective Stress	Drained conditions; Steady State Seepage $\phi' = 36.9^\circ$, $c' = 0$	N/A	2.08
2	Effective Stress	Drained conditions; Steady State Seepage $\phi' = 29.1^\circ$, $c' = 0$	N/A	1.79
3	Total Stress	Undrained conditions; Measured isotropic s_u varying with depth; uncorrected for rate effect; varying $s_u(FV)/\sigma'_v$	I	0.64
4	Total Stress	Undrained conditions; Isotropic s_u varying with depth; uncorrected for rate effect; constant $s_u(FV)/\sigma'_v = 0.15$ below weathered layer	I	0.79
5	Total Stress	Undrained conditions; Isotropic s_u = 25 kPa constant with depth	II or III	0.94
6	Total Stress	Undrained conditions; Isotropic s_u = 30 kPa constant with depth	II or III	1.13
7	Total Stress	Undrained conditions; Anisotropic s_u varying with depth and failure plane inclination, α, and corrected for rate effect; varying $s_u(FV)/\sigma'_v$	I	1.24
8	Total Stress	Undrained conditions; σ_{1c}' and K_o computed in Zone II; anisotropic s_u varying with depth and failure plane inclination, α, and corrected for rate effect; varying $s_u(FV)/\sigma'_v$	I corrected to II	1.31
9	Total Stress	Undrained conditions; σ_{1c}' and K_o computed in Zone II; anisotropic s_u varying with depth and failure plane inclination, α, corrected for rate effect; constant $s_u(FV)/\sigma'_v = 0.15$ below weathered layer	I corrected to II	1.69

All *FS* computed with Modified Bishop Method of Slices

for rate effects and anisotropy using Bjerrum's K_o = 0.50 curve from Figure 4 and an orientation of the failure surface of (45°- $\phi'/2$) relative to the direction of the major principal stress. This Analysis (No. 8) gives a factor of safety of 1.31 as compared to a factor of safety of 1.24 when the effects of the slope are not considered. Therefore the consolidation effects of the slope do impact the available ground strength and the computed factor of safety at this site.

The strengths from Boring B-2 as used in Analysis No. 4, *i.e.,* constant $s_u(FV)/\sigma'_v = 0.15$, were also used in Analysis No. 9 together with the shear stress induced consolidation approach outlined for Analysis No.8. This gives a factor of safety of 1.69 which seems most consistent with the observed performance of the slope.

Obviously for a stability analysis to be reliable, the strengths utilized in the analysis must be representative of those available along the potential surface of sliding in the field. While the shortcomings of a limit equilibrium analysis have been discussed, particularly in the absence of representative stress-strain data for the anticipated modes of failure, the variation in strength with failure plane inclination, *i.e.* undrained shear strength anisotropy, has been neglected in regional practice and is quite important when using VST strengths. In the absence of extensive and expensive laboratory test data, Bjerrum's (1973) VST corrections for an anisotropic analysis as given in Figure 4 and the strain rate effect appear to address the shortcomings in results obtained by regional practice. However, the magnitude of the correction is sensitive to K_o which is not normally measured directly in the field. The use of drilling mud with casing instead of water with casing to advance the borehole appears justified to reduce disturbance in very sensitive soils. Investigations for slope stability analyses should preferably be conducted at the toe, within the slope, and at the crest of a slope taking into account stress history and principal stress orientation at each VST location and along the failure surface.

Conclusions

An apparently stable slope, about 10 meters in height, in a massive, relatively homogenous deposit of sensitive glaciomarine clay in Westbrook, Maine, affords the opportunity to examine some regional geotechnical practices related to the evaluation of its stability. When the slope is analyzed under long-term, drained effective stress conditions, the high effective angle of friction, which ranges from 29.1° to 36.9° depending upon failure criteria, gives an adequate factor of safety of 1.79 to 2.08. The undrained shear strengths measured behind the slope using vane shear tests (VST) result in a stability factor of safety of 0.64 indicating failure for undrained conditions. The stable slope performance indicates that the actual factor of safety must be greater than unity.

As a result of the value of ϕ' and the normally consolidated stress history of the deposit, the coefficient of earth pressure at rest, K_o , will be low ($K_o \approx 0.5$) which results in significant undrained shear strength anisotropy induced by the applied stress system. This is significant since the VST measures undrained shear strength, $s_u(FV)$, on the vertical plane. For a normally consolidated deposit with a horizontal surface, the vertical plane has the lowest confining pressure, while a failure surface mobilizes undrained shear strength on other orientations with higher confining pressures.

Regional geotechnical practice generally does not use corrections for measured $s_u(FV)$ values in sensitive glaciomarine soils. However, Bjerrum (1972; 1973) has summarized research using the VST and recommends corrections for rate of strain and for failure plane inclination in anisotropic soil. For this soil the correction for the rate of strain is close to unity, but the correction for the failure plane inclination is important. Based upon Bjerrum's results, a correction to the VST for failure plane inclination was developed for the low plasticity material at this site. When the failure plane inclination corrections are applied to the $s_u(FV)$ measured behind the crest at the site, a factor of safety of 1.24 is computed which better reflects the observed field performance of the slope. However, this factor of safety is still low when one considers that the mobilized shear for drained conditions, with a much higher factor of safety, are the initial conditions for undrained shear. Possible disturbance from borehole advancement in the quick clay layer behind the crest was suggested by lack of correspondence between water contents and strengths and by the use of water instead of drilling mud in advancing the borehole. An estimated adjustment for disturbance, as well as failure plane inclination and strength gain from shear stress induced consolidation within the slope, increases the computed factor of safety to 1.69. This level of factor of safety is more consistent with the performance of the slope.

The initial state of stress was important at this site for assessing the available shear strength on a potential failure surface. Preconsolidation of the soil beneath the slope and at the toe was determined from geologic and geomorphic evidence. Allowance was made for this preconsolidation in estimating the appropriate consolidation stress along the failure surface for evaluating the available undrained shear strength. Additionally, to estimate the effect of induced shear stresses on the magnitude and direction of the principal stresses, an effective stress stability analysis was conducted to represent stress conditions on the failure surface just prior to undrained loading. If these principal stress values exceeded preconsolidation stresses, then they were used to determine undrained strength along the failure surface. The strengths corrected for the presence of the slope and failure surface orientation gave a factor of safety of 1.31 compared to the 1.24 determined with corrections for failure surface orientation alone.

Soil strengths measured along the failure surface are more representative of the available ground strength, but for this investigation of failure plane orientation effects, they were difficult to correct for anisotropy due to difficulties in estimating principal stress rotation and K_o related to overconsolidation. For this homogeneous deposit, values of s_u(FV) measured along the slope or beyond the toe provided a higher strength than s_u(FV) measured at the same elevations behind the crest. The overconsolidated soils on the slope are less susceptible to disturbance during drilling than the very sensitive soil behind the crest, and there is also some consolidation from the slope. Without correcting for orientation of the failure surface, estimated factors of safety of 0.94 to 1.13 were obtained for estimated average values of VST strength along the failure surface. Due to the higher K_o for the overconsolidated soil within the slope, the correction factor for orientation is less than that for the soil behind the crest.

Recommendations

Bjerrum's (1973) anisotropic correction of VST for orientation of the failure surface should be applied to natural slopes in regional practice to better predict shear stability for undrained conditions. However, the magnitude of the correction factor is sensitive to the value of K_o which is not readily measured. Therefore, engineering judgment must be exercised when estimating the appropriate value of K_o to be used in an analysis.

Appendix I. References

Amos, J. L., and Sandford, T.C., 1987, Landslides in the Presumpscot Formation: an engineering study, Open file report 87-4, Maine Geological Survey, Augusta, 68 pp.

Belknap, D.F., Shipp, R.C., Kelley, J.T., and Schnitker, D., 1988, Quaternary seismic stratigraphy of the Maine inner shelf, C.T. Jackson volume -- Maine Geological Survey Bulletin, Augusta.

Bjerrum, L., 1954, Geotechnical properties of Norwegian marine clays, Geotechnique, vol. IV, no. 2, pp. 49-69.

Bjerrum, L., 1967, Engineering geology of Norwegian normally-consolidated marine clays as related to settlements of buildings, Seventh Rankine Lecture, Geotechnique, Vol. 17, pp. 81-118.

Bjerrum, L., 1972, Embankments on soft ground, Proc. of the ASCE specialty conf. on earth and earth-supported structures, vol. II, pp. 1-54.

Bjerrum, L., 1973, Problems of soil mechanics and construction on soft clays and structurally unstable soils, General report, session 4, Proc. of the eighth international conf. on soil mechanics and foundation engineering, Moscow, vol. 3, pp. 111-159.

Brinch-Hansen, J., and Gibson, R.E., 1949, Undrained shear strengths of anisotropically consolidated clays, Geotechnique, Vol. I, p. 189.

Brinch-Hansen, J., 1962, Relationships between stability analyses with total and effective stresses, Danish Geotechnical Institute, Bull. No. 15, Copenhagen, 12 pp.

Devin, S.C., 1990, Flowslide potential of natural slopes in the Presumpscot Formation, M.S. Thesis, University of Maine, 127 pp.

Devin, S.C., and Sandford, T.C., 1995, Landslides in glaciomarine clays of coastal Maine, *in* Landslides Under Static and Dynamic Conditions – Analysis, Monitoring, and Mitigation, Geotechnical Special Publication No. 52, ASCE, New York.

Hvorslev, M.J., 1937 óber die festigkeitseigenschafften gest›rter bindiger b›den, kbn. (Gad), 159 pp.

Jamiolkowski, M., Ladd, C.C., Germaine, J.T., and Lancellotta, R., 1985, New developments in field and laboratory testing of soils, Proc. of the eleventh international conf. on soil mechanics and foundation eng., San Francisco, Vol. I, pp. 57-153.

Jordon-Gorrill Associates, 1973, Boring Logs for CA-1, OF-2, and PR-3, Presumpscot Interceptor, Westbrook, Maine Test Borings, Inc., *Unpublished.*

Koutsoftas, D.C., and Ladd, C.C., 1985, Design strengths for an offshore clay, J. Geotech. Eng., ASCE 111, No. 3, pp. 337-355.
Ladd, C.C., 1991, Stability evaluation during staged construction, J. Geotech. Eng., ASCE, 117, No. 4, pp.540-615.
Lambe, T.W., and Whitman, R.V., 1969, Soil Mechanics, John Wiley and Sons, New York, 553 pp.
LaRochelle, P., Chagnon, J.Y., and Lefebvre, G., 1970, Regional geology and landslides in the marine clay deposits of eastern Canada, Can. Geotech. J., 7, pp 145-156.
Lefebvre, G., Ladd, C.C., Mesri, G., and Tavenas, F., 1983, Report of the Subcommittee on Sampling and Laboratory Testing of the Committee of Specialists on Sensitive Clays of the NBR Complex, SEBJ, Montreal, Annexe III.
Leroueil, S., Collins, G., and Tavenas, F., 1983, Total and effective stress analyses of slopes in Champlain Sea clays, SGI Report No. 17, pp. 293-321.
Lowe III, J., and Karafiath, L., 1960, Stability of earth dams upon drawdown, Proc. of the first Pan-American conf. on soil mech. and foundation eng., Mexico City, Vol II, pp 537-552.
Mayer, L.M., 1988, Final report on mineralogy and pore water chemistry of Presumpscot clays, Maine Geological Survey, Augusta.
Mitchell, J.K., 1993, Fundamentals of Soil Behavior, 2nd Ed., John Wiley and Sons, New York, 437 pp.
Morse, E.S., 1869, On the landslides in the vicinity of Portland, Maine, Boston society of natural history proceedings, vol. 12.
Penner, E., and Burn, K.N., 1978, Review of engineering behaviour of marine clays in eastern Canada, Can. Geotech. J., 15, pp 269-282.
Rosenqvist, I. Th., 1953, Considerations on the sensitivity of Norwegian quick-clays, Geotechnique, vol. IV, no. 2, pp. 49-69.
S.D. Warren Co., 1972, Boring Log for Boring B-8, Westbrook, Maine, Maine Test Borings, Inc., *Unpublished.*
Skempton, A.W., 1954, The pore pressure coefficients A and B, Geotechnique, Vol. 4, p. 143.
Terzaghi, K., Peck, R., and Mesri, G., 1996, Soil Mechanics in Engineering Practice, 3rd Ed., John Wiley and Sons, New York, 549 pp.
Tibbets, Abbett, McCarthy, & Stratton, (TAMS), 1965, Stability of Pipe Trench, Presumpscot River, Maine, Portland Pipeline Corporation, *Unpublished.*

Appendix II. Notation

A, A_f	Pore pressure parameter and pore pressure parameter at failure
c'	Effective cohesion
$\overline{CIUC}$	Consolidated isotropically undrained triaxial compression test
$\overline{CK_oUC}$	Consolidated anisotropically undrained triaxial compression test
GM-P	Glaciomarine ponded; massive geologic unit (*facies*) of the Presumpscot Formation
I_p	Plasticity Index
$K_{o(NC)}$	Normally consolidated coefficient of lateral earth pressure at rest
$K_{o(OC)}$	Overconsolidated coefficient of lateral earth pressure at rest
$OCR_{(NC)}$	Overconsolidation ratio of aged normally consolidated clay
$OCR_{(OC)}$	Overconsolidation ratio of aged overconsolidated clay
$s_u(CIU)_{field}$	Undrained shear strength calculated from $\overline{CIUC}$ tests where $(s_u/\sigma_v')_{field} = \frac{2}{3}(s_u/\sigma_c')_{CIU}$
$s_u(FV)$	Undrained shear strength measured by field vane shear test
α	Inclination of failure plane measured from horizontal
δ	Rotation of major principal stress from vertical
ϕ'	Effective angle of internal friction at maximum obliquity
ϕ'_{peak}	Effective angle of internal friction at q_{max}
μ_A	Field vane correction factor for shear strength anisotropy
μ_R	Field vane correction factor for rate of strain
σ_h'	Horizontal effective stress
σ_v', σ_{vc}'	Vertical effective stress and consolidation stress
σ_{1c}'	Major principal effective consolidation stress

Effects of Sinkholes on Earth Dams

Mandar M. Dewoolkar[1], Associate Member, ASCE
Kitidech Santichaianant[2],
Hon-Yim Ko[3], Member, ASCE
Ton Goddery[4]

Abstract

Two centrifuge experiments investigating the effects of sinkhole formations on the stability of earth dams are discussed in this paper. A special container and a trap door assembly were designed to simulate formation of sinkholes. Calibration tests on level ground sand models were also conducted. The tests on embankment models provided insights into the mechanisms involved in the cavity formation inside embankments, as a result of sinkhole development. The two embankment models of compacted Bonnie silt were identical except the presence of water reservoir in the second model. However, the water reservoir in this test was contained in an impermeable latex membrane bag, so no seepage was developed. The sinkhole was about 3.6 m deep and 11 m in diameter at the base of a 35.9 m tall embankment in terms of prototype dimensions. In the test with no water reservoir, the surface of the embankment did not suffer major deformations; although a large cavity was formed above the sinkhole. In the test with water reservoir, the shear zone reached the dam surface which created a 3.6 m deep and 11 m diameter depression in the dam surface in terms of prototype dimensions.

Introduction

The possibility of sinkhole development within a dam, its foundation and the reservoir base poses a serious threat to the safety and integrity of the dam. In the dam-reservoir environment, sinkholes may develop as a result of (i) presence of natural or man-made surficial and subsurface voids, (ii) presence of incompatible gradations at boundaries of different materials, (iii) presence of cracks, (iv) presence of dispersive or collapsible materials, or (v) increase of hydr-

[1]Research Associate, University of Colorado, Boulder, CO 80309-0428
[2]Graduate Student, University of Colorado, Boulder, CO 80309-0428
[3]Professor, University of Colorado, Boulder, CO 80309-0428
[4]Professional Research Assistant, University of Colorado, Boulder, CO 80309-0428

aulic gradients across critical zones associated with the impounding reservoir (Lo, 1984). When a sinkhole is developed in the foundation of a dam, questions such as how and when localized loss of foundation support can cause a sinkhole to reach the dam surface are of primary concern.

The consequences of loss of ground support in sinkhole formations have been associated with the phenomenon of active or positive arching in soils which leads to the classical trap door problem. Effects of sinkholes on dams could probably best be studied through field testing which is very expensive, at times dangerous, and inherently difficult. Uncertainties in the field investigation involve accurate prediction of the location of a sinkhole in an existing dam to arrange for appropriate instrumentation, inhomogeneity in soil conditions and, of course, safety issues. The alternatives to field testing have been laboratory experiments conducted under earth's gravity or under elevated acceleration levels in a centrifuge.

Small scale model tests under earth's gravity suffer from a lack of similitude of stress levels between models and any realistic large scale structures. This deficiency can be overcome with the use of centrifuge modeling technique in which scaled models are subjected to predetermined, high acceleration levels to produce similitude of stresses and strains in the model and the real or hypothetical prototype that it represents.

As far as is known, loss of ground due to sinkhole formation in embankments has not been studied in a centrifuge. Some centrifuge studies investigated different forms of loss of ground support; for example, Stone and Wood (1988), Iglesia (1991) and White, et al. (1994) studied active trap door problem with a controlled movement of trap doors at specified g-levels; Craig (1990) and Abdulla and Goodings (1994) studied effects of existing voids on a clay layer and a cemented sand layer, respectively, with increasing g-levels to failure; Allersma (1998) studied loss of ground under layered soils; Iglesia, et al. (1991) studied loss of ground under simulated jointed rock; Stone and Brown (1993) studied loss of ground relevant to mining operations; and Kutter, et al. (1994) studied collapse of sand into spherical cavities.

Two centrifuge experiments on models of earth embankments investigating the effects of sinkhole formations on the stability of dams are discussed in this paper. A special container and a trap door assembly were designed to simulate formation of sinkholes in a displacement controlled fashion. Significant efforts were invested in the initial design and subsequent modifications of the trap door assemblies in order to measure correctly soil load in its undisturbed state, i.e. before the trap door is lowered. This was confirmed by conducting calibration tests on dry sand models with level ground surface.

Model Configuration

The conceptual design of the container and the trap door assembly is shown in Figure 1. The inner container is housed in a larger outside container and is supported by six 27.94 cm long, 0.9525 cm thick tubes of 13.97 cm outer diameter. This clearance is used to accommodate an actuator and an electrical motor for the operation of the trap door. The entire trap door assembly including the motor and actuator was supported by the inner container bottom plate preventing any relative movement between the trap door and the inner container bottom. It was envisioned that if the container floor is to deflect with increasing g-level, the trap door will deflect with it.

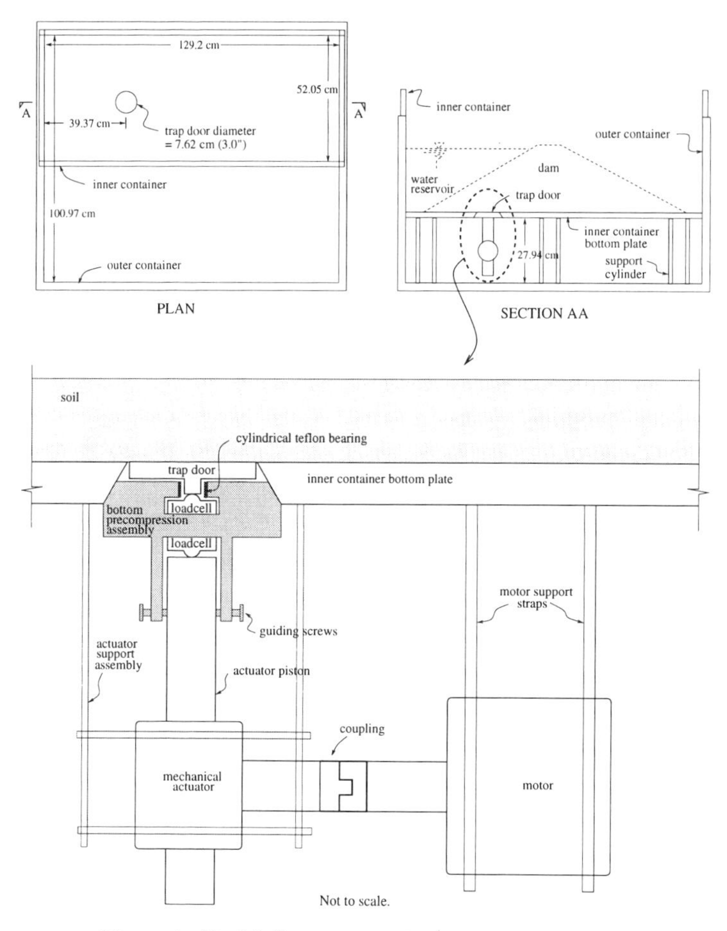

Figure 1: Model Container and Trapdoor Assembly

The details of the trap door assembly was finalized after incorporating two sets of modifications so as to prevent the separation of the trap door from the inner container bottom prematurely. The trap door assembly consisted of the suspended trap door which was completely separated from the rest of the components. The diameter of the trap door (D) was 7.62cm (3"). It was resting only on the top loadcell. The top loadcell was attached to the bottom precompression assembly which was used to apply a precompression between the assembly and the container bottom plate to prevent premature separation. The bottom loadcell was attached to the bottom precompression assembly and rested on the actuator piston. The guiding screws helped in preventing the assembly from tilting. By raising the actuator piston, a desired amount of precompression could be applied to prevent premature separation of the assembly from the container bottom plate. The measurement of the top loadcell was unaffected by the precompression. It measured only the load from the soil and the weight of the trap door. Possible tilting of the trap door was prevented by the tight fit between the door and the bottom precompression assembly. In order to minimize the friction between the two, a cylindrical teflon bearing was introduced. This system was tested by conducting two tests. In the first test, a known weight was placed on the trap door at the center. The centrifuge was spun to 150g. This load was correctly measured by the top loadcell throughout the spin-up of the centrifuge. In the second test, the same weight was placed at an eccentricity of 6 mm from the center. Again, the centrifuge was spun to 150g. The top loadcell measured the same load which indicated that a small eccentricity in the load application did not affect the measurement. However, a tilted trap door would result in redistribution of stresses in the soil, thus affecting the loadcell measurement. These abovementioned tests also indicated that the friction between the trap door and the teflon bearing was insignificant.

All the tests presented here were conducted at 150g measured at the bottom of the outer container using the 400 g-ton centrifuge at the University of Colorado at Boulder. At the mid-height of the soil models, the g-level was about 141g. In all the tests, the trap door was lowered at a rate of 0.01 mm/sec. Before the model preparation, the position of the trap door was adjusted by operating the motor-actuator assembly such that a contact force of about 450 N between the bottom precompression assembly and the inner container bottom was generated. The soil model was then constructed. A minimum amount of contact force of about 450 N was maintained throughout the centrifuge spinup to prevent the premature separation of the trap door from the inner container bottom. By using the two loadcell measurements and simple calculations, the load from the soil supported by the trap door before and during the lowering of the trap door and the instance at which the trap door separated from the inner container bottom were determined.

Level Ground Calibration Tests

As a calibration of the trap door system, tests on level ground, uniform, dry sand models were conducted. Based on the density of the sand, the stress distribution on the trap door before it was lowered could be calculated. The resulting soil force could then be calculated and compared with the measurements. The model configurations of tests LevelSand1 and LevelSand2 were identical and is shown in Figure 2a. These two tests were conducted to establish repeatability of the measurements and testing techniques. Model was prepared inside an open-ended cylindrical container which was placed around the trap door as shown in Figure 2a. A circular piece of latex membrane of about 12 cm diameter was placed on top of the trap door to prevent the sand from getting into the trap door assembly once the trap door was lowered and a gap was created. The influence of the latex piece on the model behavior and measurements was believed to be insignificant, especially at high g-levels. Nevada No. 100 sand (uniform, subrounded, medium fine to fine sand, $D_{50} = 0.1$ mm, specific gravity of 2.67, and maximum and minimum dry unit weights of 17.33 and 13.87 kN/m^3, respectively) was placed at a dry unit weight of 15.86 kN/m^3 by pluviation.

The soil forces supported by the trap door from the two tests plotted against the downward trap door displacement (Δ) are shown in the left side plot of Figure 2b. In the right side plot, the soil force and the trap door displacement are normalized with the undisturbed soil force and the diameter of the trap door, respectively. The undisturbed soil force was calculated to be 1207 N which compared very well with the measured soil force before the trap door was lowered

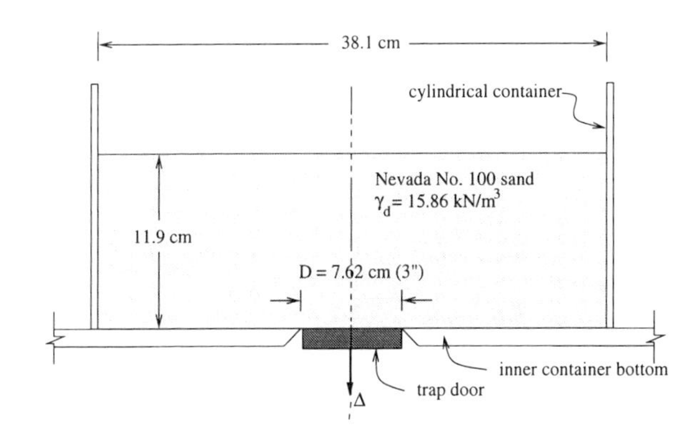

(a) Model configuration of tests LevelSand1 and LevelSand2

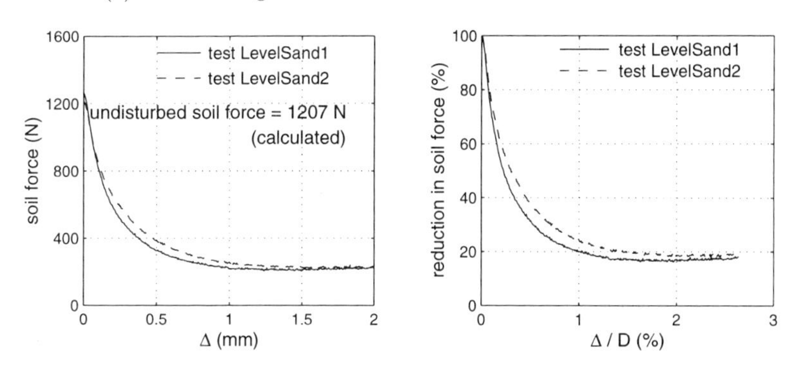

(b) Measured soil force versus downward displacement of the trap door

Figure 2: Model Configuration and Soil Force Measurements of the Trap Door Experiments on Level Sand Models at Model Scale

in both the tests. Once the trap door was lowered, the soil force reduced by about 80% in the downward displacement of the trap door of about 1% of its diameter. The measurements from the two tests compared well with each other. The results from the calibration tests conducted on level sand models indicated that the trap door assembly functioned appropriately and prevented premature separation of the trap door from the container. Also, the measurements were reliable and repeatable.

In comparison to previous centrifuge studies on level ground models with trap doors, the following remarks could be noted. Craig (1990) and Abdulla and Goodings (1994) studied effects of existing voids with increasing g-levels in layers of clay and in weakly cemented sands, respectively. Craig (1990) also studied the loss of ground simulated by draining dry sand from underneath clay layers at a constant g-level. No direct measurements of load or stresses were made in these studies. Stone and Wood (1988) studied controlled movement of a trap door underneath a layer of sand as well as clay. White, et al. (1994) also studied active trap door movement underneath a layer of sand. The main focus of these studies was to investigate mechanisms of deformation within the soil layers. The load or stresses on the trap door were not measured. The results from the trap door experiments on level sand models in axi-symmetric conditions presented here can be compared only with Iglesia's (1991) results. He simulated controlled downward movement of a trapdoor underneath a layer of sand in plane strain condition. He presented load-displacement curves similar to the ones shown in Figure 2b based on direct measurements of load and displacement measured on the trap door. Iglesia's measurements are qualitatively similar to the measurements presented here. However, his results were affected by premature relative movements between the trap door and the adjacent base (the earlier being stiffer than the latter) with increasing g-level resulting in higher (by about 100%) than expected geostatic "undisturbed" load before the trap door operation was initiated. This problem of premature separation of the trap door assembly and the adjacent base was resolved in the study presented in this paper.

Tests on Embankment Models

Two tests on models of earth dams were conducted. Models were prepared using Bonnie silt whose gradation curve, standard Proctor compaction data and other properties are shown in Figure 3.

As shown in Figure 4, models of tests damNW1 and damWW1 were very similar, except that a water reservoir existed in the latter. Water was contained in an impermeable latex membrane bag in test damWW1. Therefore, flow of water from the reservoir into the embankment did not take place. Four LVDTs were used to monitor vertical displacements of the dam surface. The 25.4 cm tall model embankments were prepared in ten lifts of 2.54 cm thick layers of compacted Bonnie silt at about 16% water content. A special wooden mold was used for this purpose. For each lift of the model, the final volume of the compacted soil was calculated. An appropriate amount of moist Bonnie silt was compacted in this volume so as to achieve a dry density of 90% of the maximum standard Proctor density. After each compacted silt layer, a very thin layer of white powder was introduced to indicate the failure surfaces during the post-

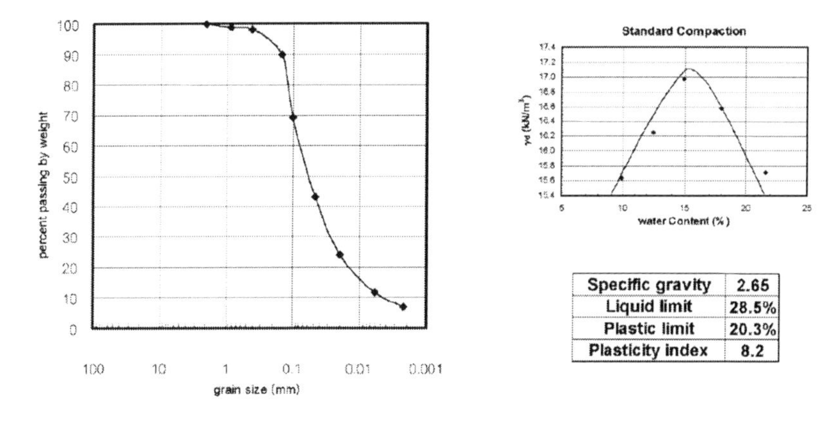

Figure 3: Properties of Bonnie Silt

test investigation. The container was weighed before and after the model was prepared. The actual dry density was calculated to be 89% of the maximum standard Proctor density.

The tests were conducted at 141g and the trap door was lowered up to 25.4 mm, i.e. $\Delta/D = 33\%$. The measured soil forces are plotted in plots (a) and (b) of Figure 5 for up to 2.5 mm and 25 mm of trap door movement, respectively.

Unlike the case of level ground, the stress distribution on top of the trap door is unknown because of the inclined geometry of the model. As an approximation, the weight of the soil column on top of the trap door was estimated to be 1440 N. The measured undisturbed soil force was 1590 N in test damNW1. As seen in Figures 5a and b, the soil force dropped to 100 N which is only 6.3% of the undisturbed value. The displacement of the door at this point was only about 1 mm. As the door continued to displace downward, more soil separated from the embankment creating a bigger cavity. As a result, the soil force increased. The surface displacements of the embankment as indicated by the measurement of lv-A were very small as shown in Figure 5c.

The post-test investigation revealed a failure pattern as shown in the photographs in Figure 6. Photograph-a shows the view along section A1-B (Figure 4). Photograph-b shows the view along section A1-A2. As indicated by white layers that are visible in the photographs, the dam above the cavity was intact even after the door displaced by 25 mm. It is speculated that in the beginning, when the door displaced by about 1 mm and when the soil force was at its minimum, the cavity was small. As the door continued to move downward, the cavity continued to grow. The diameter of the cavity was about the same as the trap door diameter.

The weight of the soil resting on the trap door was estimated to be about 845 N at 141g. However, the measured soil force at the trap door displacement of 25.4 mm was 500 N (Figure 5b). The difference of 345 N was speculated to be balanced by the friction between the failed soil and the rest of the dam along the bottom half portion of the cavity.

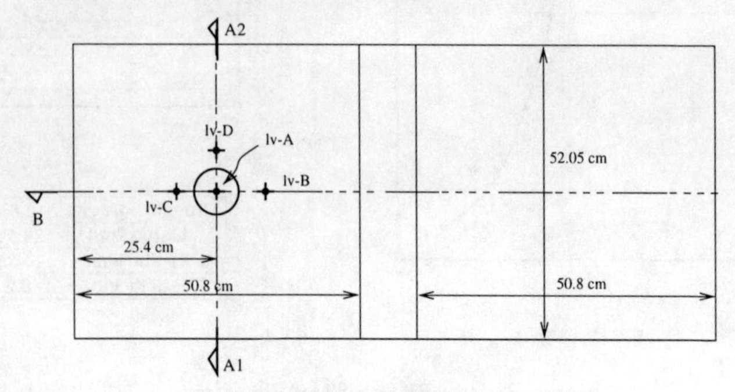

PLAN view of tests damNW1 and damWW1

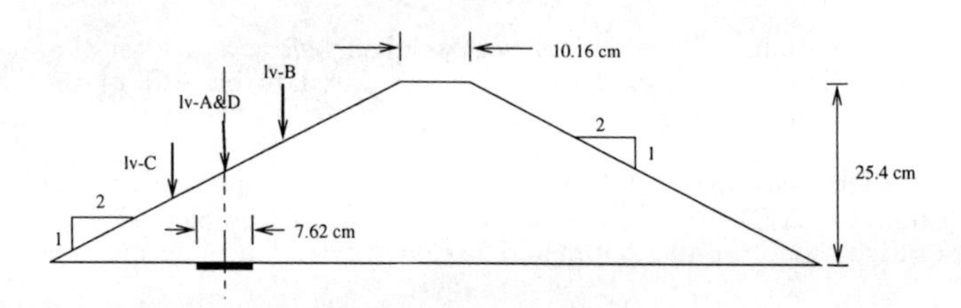

ELEVATION view of test damNW1

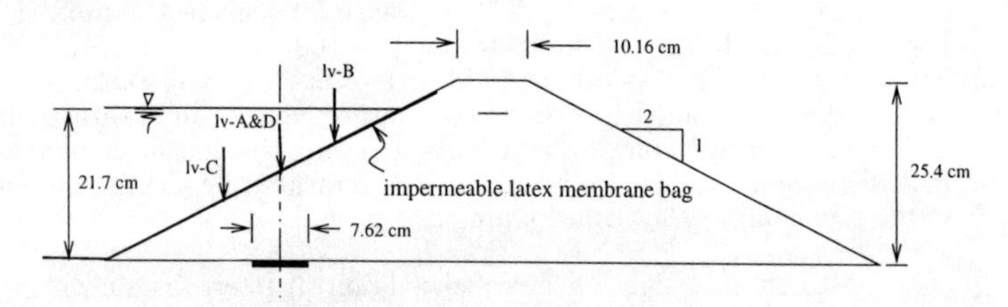

ELEVATION view of test damWW1

Figure 4: Embankment Tests Model Configurations at model scale

In test damWW1, because of the presence of water reservoir, the undisturbed soil force was 2040 N, about 450 N higher than that in test damNW1. It should be noted that the water reservoir was contained in an impermeable latex membrane bag, so seepage forces were not present. As seen in Figure 5, the soil force dropped to 100 N. The displacement of the door at this point was only about 1.5 mm. The displacement of the embankment surface was very small.

This reduction in soil force, the required door displacement to cause the reduction, and the surface settlements compared very well with those in test damNW1. This indicated that although a water reservoir existed on top of the embankment, the same volume of cavity was formed on top of the trap door when the displacement was about 1.5 mm. Thus, the additional load due to water reservoir was completely supported by the soil arch formed above the trap door.

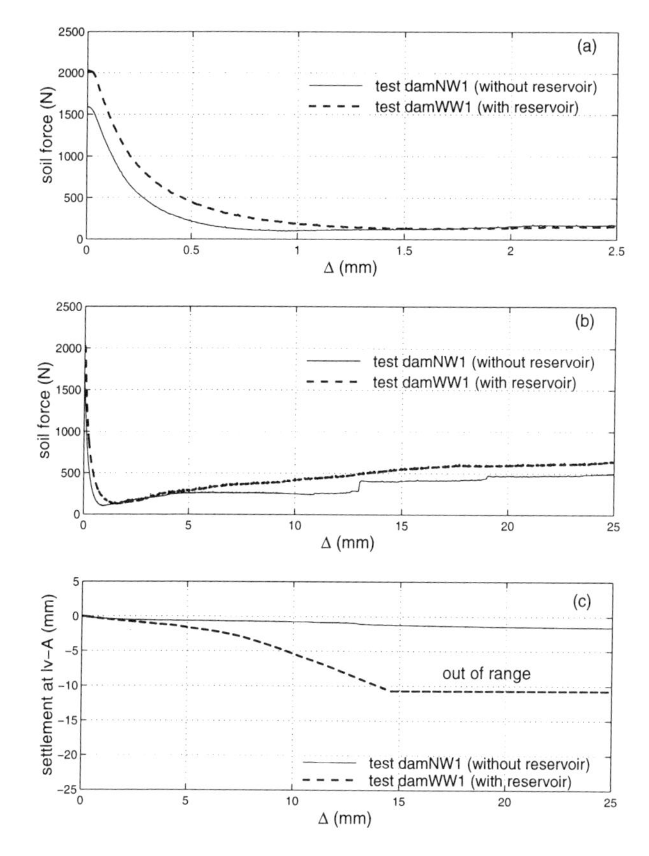

Figure 5: Embankment Test Results at model scale

However, as the trap door continued to move downward, the behavior of the embankment changed drastically from that in test damNW1. The top view of the dam after the test is shown in Figure 7a. With the additional door movement, the cavity continued to grow and reached the surface of the dam. Photograph-b shows the view along section A1-A2 (Figure 4). As seen, the failure surface is cylindrical with the diameter approximately equal to the trap door diameter. The displacement measured by lv-A (Figure 5c) reached 11 mm after which the transducer went out of range. The total displacement measured during the post test investigation was about 25 mm. The rest of the LVDTs were outside the failure zone and hence measured relatively small displacements.

At the end, when the total trap door displacement was 25.4 mm, the soil force was 660 N. The weight of the soil and water on top of the door was estimated to be about 2100 N at 141g. The difference of about 1440 N was speculated to be balanced by the friction along the failure surface.

Because of the presence of the water reservoir, the undisturbed soil force in test damWW1 was higher than that in test damNW1. As the trap door was lowered, the soil force reduced to 100 N in both the tests in a door movement of about 1.5 mm. As the trap door continued to move downward, the soil force increased. The increase in the soil force was higher in test damWW1 than in test damNW1 because of the water reservoir. In test damNW1, without reservoir, the surface of the embankment did not suffer major deformations, although a large cavity was formed above the trap door. On the other hand, in test damWW1 with reservoir, with greater trap door displacement, the shear zone eventually reached the dam surface creating a large depression in the dam surface for the same size sinkhole. In both cases, cavities or failure planes were formed just above the trap door. The rest of the dam suffered minor distortions.

The models in tests damNW1 and damWW1 represented a 35.9 m tall prototype dam with 2H:1V side slopes. The sinkhole formation would be about 3.6 m deep and 11 m in diameter. The results from test damNW1 indicated that such a sinkhole would create a large cavity inside the dam; however, the surface of the dam would not suffer significant deformations. On the other hand, if the water reservoir was present, as in test damWW1, the formation of the same size sinkhole would eventually force the shear zone to reach the surface of the dam. The depression on the surface would be about 3.6 m deep and 11 m in diameter. If seepage forces were present, the failure could be even worse.

Discussion and Conclusion

Effects of sinkoles on the stability of embankment dams were investigated. A special container and a trap door assembly were designed to simulate formation of sinkholes in a centrifuge.

The results from the calibration tests conducted on level sand models indicated that the trap door assembly functioned appropriately and prevented premature separation of the trap door from the container base at increased g-levels in a centrifuge. The soil force measurement made on the trap door was reliable and repeatable.

The tests on embankment models provided insights into the mechanisms invloved in the cavity formation inside embankments, as a result of sinkhole development. The two embankment models of compacted Bonnie silt were identical except the presence of water reservoir in the second model. However, the water reservoir in the second test was contained in an impermeable latex membrane bag, so no seepage was developed. The sinkhole was about 3.6 m deep and 11 m in diameter in terms of prototype dimensions. For the model configuration studied, in the test with no water reservoir, the surface of the embankment did not suffer major deformations. About 1050 m^3 volume of soil was disturbed above the sinkhole and a large cavity was formed. In the test with water reservoir, the failure surfaces reached the dam surface which created a 3.6 m deep and 11 m diameter depression in the dam surface in terms of prototype scale. In both cases, the portion of the dam just above the trap door was affected. The rest of the dam suffered insignificant disturbance.

It would be interesting to examine a more realistic situation in which the water reservoir is not contained in a latex bag. In such a case, seepage forces could play a key role producing more devastating effects.

Acknowledgements

The US Bureau of Reclamation, Denver provided a research grant for this work which is gratefully acknowledged.

References

Abdulla, L. and Goodings, D. J. (1994), "Study of sinkholes on weakly cemented sand", Centrifuge 94, Lee & Tan (eds), Balkema, Rotterdam, pp. 797-802.

Allersma, H. G. B. (1998), "Development of cheap equipment for small centrifuge", Centrifuge 98, Kimura, Kusakabe & Takemura (eds), Balkema, Rotterdam, pp. 85-90.

Craig, W. H. (1990),"Collapse of cohesive overburden following removal of support", Canadian Geotechnical Journal, Vol. 27, pp. 355-364.

Iglesia, G. (1991), "Trapdoor Experiments on the centrifuge: A study of arching in geomaterials and similitude in geotechnical models", Ph.D. Thesis, Department of Civil Engineering, Massachusetts Institute of Technology.

Iglesia, G., Einstein, H. H., Whitman, R. V. Jessberger, H. L. and Güttler, U. (1991), "Trapdoor experiments with simulated jointed rock", Centrifuge 91, Ko & McLean (eds), Balkema, Rotterdam, pp. 561-567.

Kutter, B. L., Chang, J. and Davis, B. C. (1994), "Collapse of cavities in sand and particle size effects", Centrifuge 94, Lee & Tan (eds), Balkema, Rotterdam, pp. 817-822.

Lo, R. C. (1984), "Sinkhole problem related to dam engineering", Sinkholes: Their Geology, Engineering and Environmental Impact, Proc. of the First Multi-disciplinary Conf. on Sinkholes, Beck B. F. (ed), Balkema, Rotterdam, Orlando, Florida, October, pp. 267-272.

Stone, K. J. L. and Brown, T. A. (1993), "Simulation of ground loss in centrifuge model tests", Geotechnical Testing Journal, GTJODJ, Vol. 16, No. 2, pp. 253-258.

Stone, K. J. L. and Wood, D. M. (1988), "Model studies of soil deformations over a moving basement". Eng. Geology of Underground Movements, Geological Society Eng. Geology Special Publication, No. 5, pp. 159-165.

White, R. J., Stone, K. J. L. and Jewell, R. (1994), "Effect of particle size on localisation development in model test on sand", Centrifuge 94, Lee & Tan (eds), Balkema, Rotterdam, pp. 817-822.

(a)

(b)

Figure 6: Excavated Model of Test damNW1

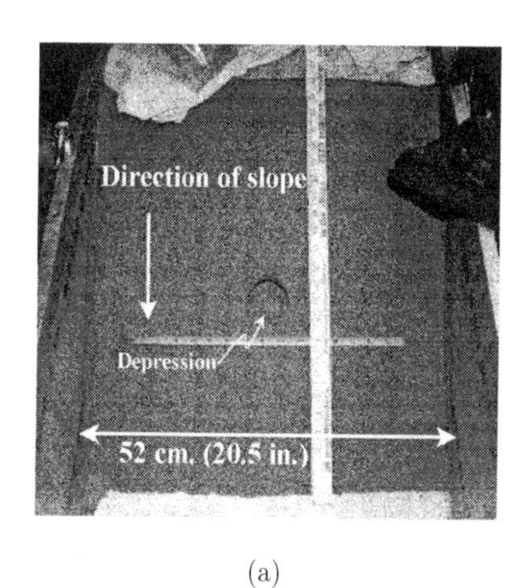

(a)

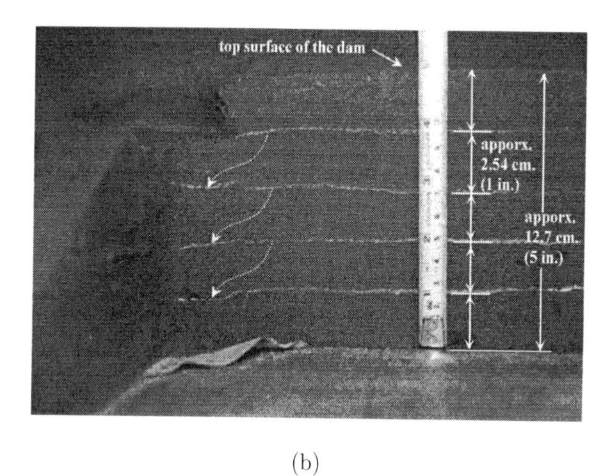

(b)

Figure 7: Excavated Model of Test damWW1

STABILIZATION OF A SLIDE USING A TIEBACK PILE WALL

Vishnu Diyaljee[1], Fellow, ASCE, Murthy Pariti[2] and Roy Callioux[3]

Abstract

The use of drilled shaft concrete piles is generally not a widespread method of stabilizing landslides along the Alberta Highway Infrastructure as a result of perceived high costs, the uncertainty of performance of these structures in the long term and the ready availability of land for realignment, offloading and toe berm construction. This paper presents a case history of a roadway embankment slide which was stabilized using a tieback drilled shaft concrete pile retaining wall. Although the piles were not installed to the desirable depth, the wall has performed satisfactorily since its construction some 10 years ago with none or minimal maintenance to the surface of the roadway.

INTRODUCTION

Highway 33, a Provincial Primary Highway in the North Central Section of the Province of Alberta, Canada, was constructed during the1950s. The highway crosses a small creek located in between two ridges about 8 km north from the Town of Swan Hills and 250 km north-west of the City of Edmonton. The location of the project site is shown on Figure 1.

Since construction, slope instability conditions have been experienced in a short section of roadway embankment fill crossing a creek channel. Initial investigations indicated the existence of subsurface springs in the creek banks which were blocked by the embankment fill. Drainage improvements were made through the construction of subsurface and surface drains along the highway ditch in the backslope and the

---

[1]Managing Director, GAEA Engineering Ltd., Edmonton, Alberta

[2]Geotechnical Engineer, GAEA Engineering Ltd., Edmonton, Alberta

[3]Ex-Construction Manager, Peace Region & Director of Consistency in Construction, Alberta Infrastructure

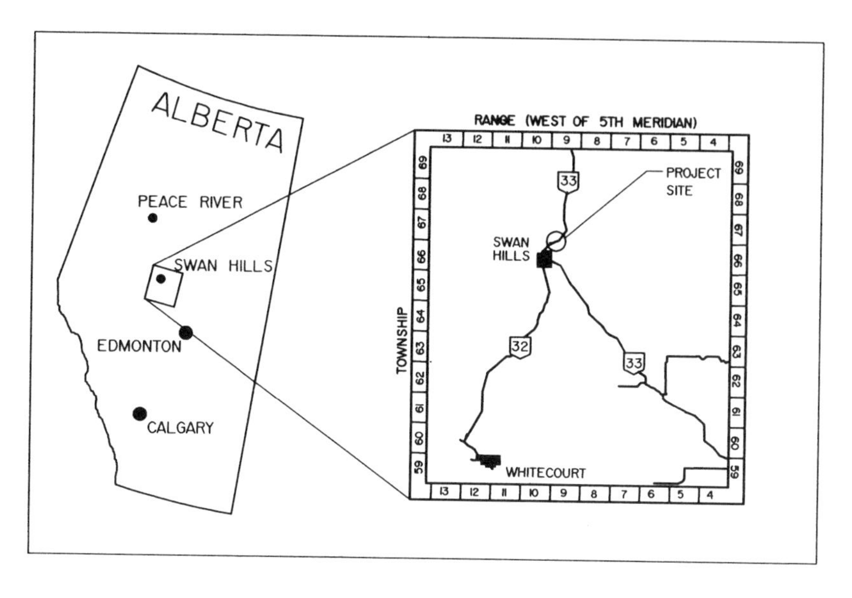

Figure 1. Project Location

installation of horizontal drains drilled through the slide area from the toe of the side slope. These measures, however, proved only partially successful in preventing further sliding of the embankment.

During the latter half of 1980, improvements were planned for this section of highway in a two stage process - grade widening in 1989, and a raise of grade in 1990. These improvements necessitated more permanent slide stabilization measures as any additional height of grade would require some form of lateral restraint.

This was accomplished by the construction in 1989 of a drilled straight shaft concrete pile wall consisting of 82 piles, each 760 mm in diameter, drilled at 1 m center-to-center spacing. Tieback anchors were subsequently installed through the pile wall in 1991 to enhance the stability of the slide area. The slope on the downhill side of the pile wall was reconstructed with wood chips, and capped with a silty clay soil. A gravel layer was installed at the bottom of the wood chips to improve the subsurface drainage downside of the pile wall. On completion of the wall and tieback installation, slope indicators were installed to monitor the immediate and long term performance of the pile wall.

The objective of this case history paper is to review the remedial measures implemented and to discuss their performance to date.

SITE CONDITIONS

As indicated earlier, the highway crossed a small creek at the slide location. The terrain adjacent to the highway formed a basin which gently sloped towards the creek. The slopes of the basin were littered with fallen and tilting trees which indicated a generally progressive slope movement pattern towards the creek. The formation of ice blocks in the winter time over much of the slopes indicated the existence of possible seepage paths. A centreline culvert under the highway at about 120 m north of the slide area handled the creek flow across the highway from the east to west.

A major backslope was located at about 30 m from the south end of the pile wall. Ice build up was noticeable during the winter months at several locations along this backslope indicating the existence of active springs within the slide area.

SUBSURFACE DRAINAGE MEASURES INSTALLED PRIOR TO PILE WALL INSTALLATION

Prior to the installation of the pile wall in 1989, horizontal drains were installed at the toe of the sideslope in the slide area on the west side of the road to improve the stability of the embankment fill. These drains flowed free for a year or two and eventually became silted up, broken, or non-functional. Five horizontal drains and three shallow trench drains with perforated pipes were installed in 1988 as an immediate response to sliding instability occurring during that period. These horizontal drains, drilled through the embankment fill under the road to reach the backslope, yielded an average groundwater flow of about 1.5 to 2 litres/min.

PILE WALL INSTALLATION - 1989

Installation of drilled straight shaft concrete piles was undertaken in July 1989 ahead of the proposed grade re-construction schedule. In view of the slide conditions, it was agreed to undertake the grade revision work in the slide area over the 1990 and 1991 construction seasons.

A pad was prepared on the side slope about 2.5 m below the elevation of the road as an access for the drilling equipment. In total, 82 piles, each 760 mm in diameter, were installed from Sta. 9+620 to 9+700 at an average center-to-center spacing of 1 m. Although these piles would be conventionally anchored into bedrock or hard shale in typical piling projects, the depths of piles in this project were constrained by the location of horizontal drains installed within the slide area in the previous years. The location of the pile wall in relation to the roadway embankment is shown in Figure 2.

The piles were extended 2.3 m above the drill pad by using cardboard forms (sonotubes), which were subsequently removed after the concrete was cured. The piles were extended above the pad elevation to facilitate reconstruction of the roadway after the completion

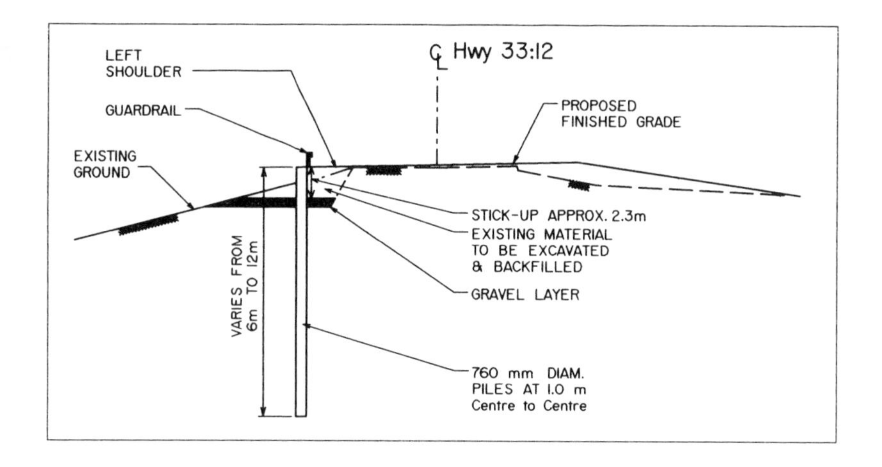

Figure 2. Location of Pile Wall along Roadway

of pile wall installation. The pile holes were nominally reinforced and backfilled with 25 MPa strength, Type 10, Portland Cement Concrete. The pile wall installation was carried out by North American Construction Ltd of Edmonton, Alberta.

Following the completion of the pile installation, the areas on the downslope side of the pile wall and between the piles and the west shoulder of the road were backfilled with native soil that had been excavated earlier. The road grade was brought back to match the elevation of the adjacent roadway surface. Two slope indicators (SI #A and B) installed adjacent to the pile wall in the reconstructed shoulder area of the roadway indicated movements at depths of 14.5 and 17.5 m, respectively, along the contact with stiff clay till and shale. These slope indicators were subsequently destroyed during the winter of 1989/90 by maintenance equipment. A general layout of the slope indicators installed is shown on Figure 3.

The highway grade was raised by about 1.0 m during the summers of 1990 and 1991. Prior to the construction of the grade raise, a site inspection was done in May 1990 to check the performance of the pile wall. This inspection revealed that little or no lateral movement was evident in the roadway embankment fill in the area of the concrete piles, but the road had settled considerably south of the pile wall close to Sta. 9+560.

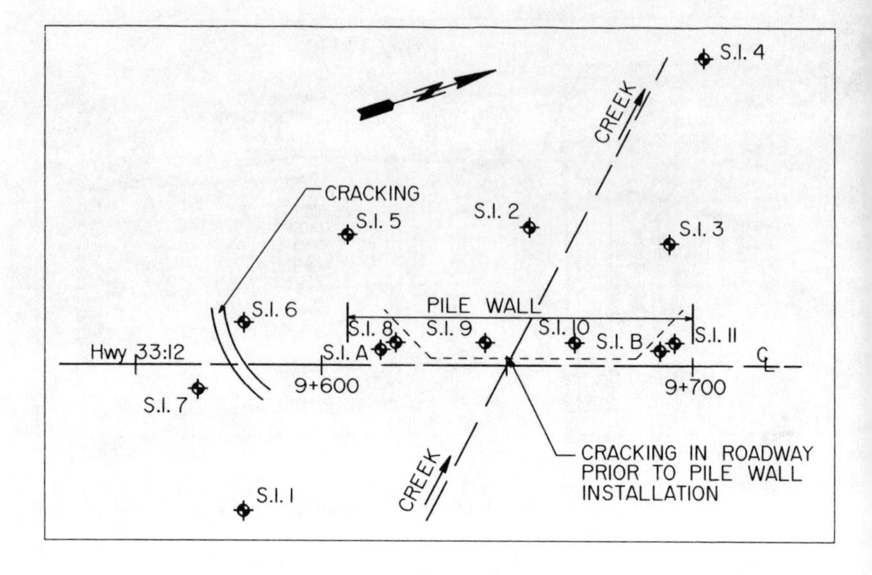

Figure 3. Layout of Slope Indicators and Crack Patterns

During the time interval between the phases of the pile wall installation and the grade raise construction, four slope indicators (SI #1 to 4) were installed in March 1990 along the creek banks on either side of the roadway embankment to monitor subsurface movements occurring therein on either side of the highway. Readings from these slope indicators indicated that progressive movements were occurring along the creek banks on either side of the highway. No slope indicators were installed to replace SI #A and B in the pile wall area in view of the grade raise construction scheduled for the summer of 1990.

TIEBACK INSTALLATION - 1991

Based on the indication that slow movements were still occurring and the fact that the piles were not drilled into bedrock, it was considered more appropriate from a long term stability consideration to strengthen the free standing piles by installing a tieback system. The tieback system was installed in September 1991 by Beck Drilling and Environmental Services Inc. of Calgary, Alberta.

The design of the tiebacks was undertaken using laboratory obtained shear strength parameters of $c'=12$ kPa and $\phi'=25°$ for the native soil behind the pile wall. The design called for two rows of tiebacks consisting of 28 tiebacks in each row installed at 1 m and 3 m depth, respectively, below the top of the piles. Each waler beam was designed to be a composite section consisting of two of C380 x 60 kg channels laid back to back

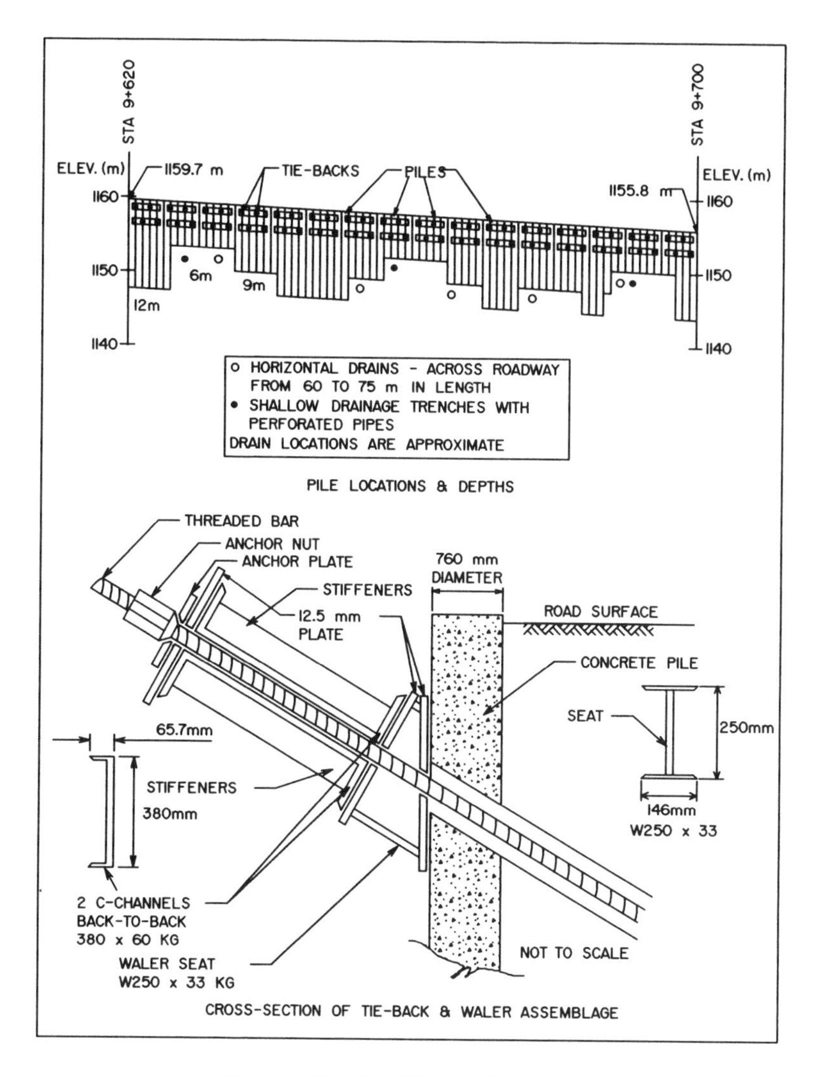

Figure 4. Details of Tieback Installation

and covered both at the top and bottom by steel plates of 12.5 mm nominal thickness. Typical details of the tieback installation are shown in Figure 4.

Two anchors were installed in each beam at an angle of 30° to horizontal. Each anchor consisted of a 36 mm diameter Dywidag Grade 150 steel rod inserted in a 125 mm predrilled hole and grouted for a length of 7.5 m into the intact shale zone. The length of each anchor in the native silty clay zone was about 36 m. Each anchor was designed to resist a tensile force of 600 kN. Fifty (50) MPa strength cement concrete was specified for the grout.

The top row of tiebacks was installed at 1 m below the top of piles. For the purpose of installing this row of tiebacks, a working bench was cut on the downslope of the pile wall to a depth of 2 m below the top of concrete piles to allow for access of equipment and machinery. After completion of the top row of tiebacks, excavation was done to a further depth of 2 m below the first bench and the lower tiebacks were installed at a depth of 3 m below the top of the piles.

In the process of the tieback installation, about six (6) anchors failed to achieve the desired strength, probably due to variable quality of rock. The failed anchors were replaced by new ones at an additional cost. Grout pipes were also placed in those failed holes and multi-stage grouting was done as necessary.

After the completion of the tieback installation, a gravel layer was placed on top of the lower drill pad to allow quick drainage of any seepage coming through the roadway embankment fill. The downslope of the pile wall was reconstructed in the summer of 1992 with wood chips and a silty clay cap.

PAVEMENT CRACKING SOUTH OF PILE WALL LOCATION

Cracking of the pavement at about 60 m south of the pile wall location (Sta. 9+560) was observed even before the installation of the wall in 1989 and became more pronounced in the summer of 1990. This cracking was almost diagonal to the centreline of the road and can still be observed today (Figure 3).

INSTRUMENTATION

General

Slope indicators were installed at this site periodically after the installation of the pile wall in 1989 to monitor the long term performance of the pile wall. The instrumentation is currently being monitored twice a year. Visual inspection of the site is also undertaken periodically to determine any unusual features which may affect the integrity of the roadway.

As described earlier, the first set of slope indicators, SI #A and B, was installed in the pile wall area in August 1989. The second set, SI #1 to 4, was installed in March 1990 to monitor slide movements along the creek on either side of the road. The third set, SI #5 to 11, was installed in September 1991 after the grade was raised to the design

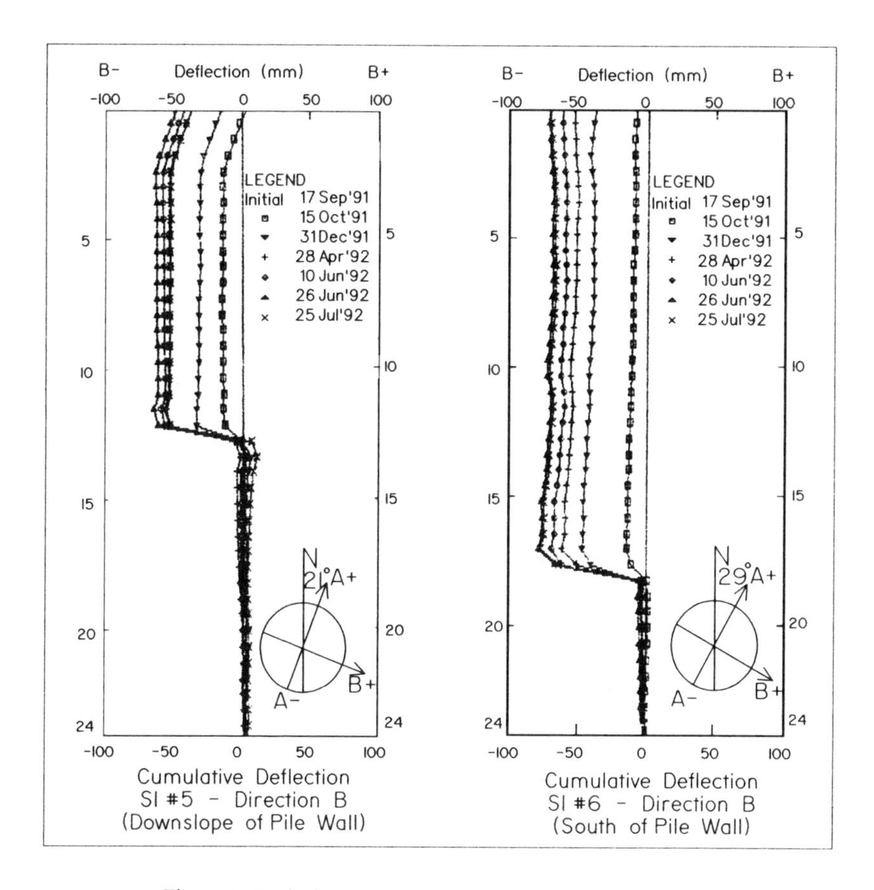

Figure 5. Typical Slope Indicator Plots Outside of Pile Wall

elevations and the new pavement completed. Of these, SI #5, 6 and 7 were installed on either side of the road at the crack location while SI #8 to 11 were installed between the outer edge of the south bound lane and the pile wall.

Unfortunately, many of the slope indicators have sheared or have been destroyed. SI #8 and 10 were destroyed by road maintenance equipment or by animals. Only two (SI #9 and 11) are currently functional.

Inference From Instrumentation Monitoring

Well-defined failure planes are indicated in the plots of the slope indicators located by the pile wall and south of the pile wall, as shown in the typical plots of Figures 5 and 6.

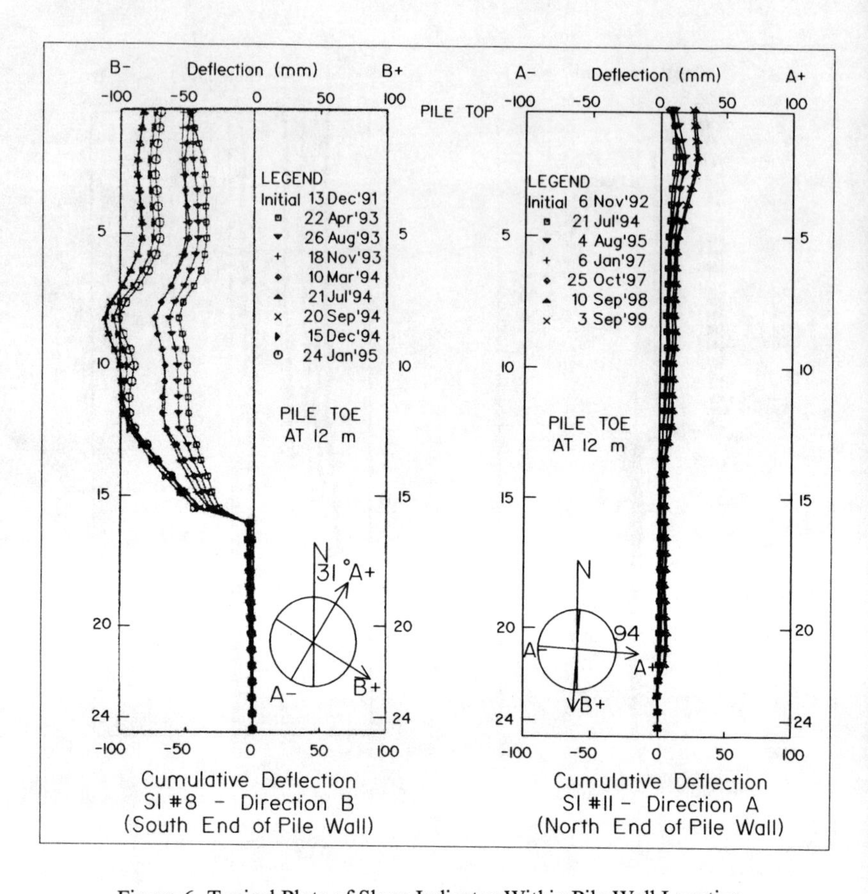

Figure 6. Typical Plots of Slope Indicator Within Pile Wall Location

Pile Wall Area

It is interesting to note from the plots of Slope Indicators (SI #8 and 11) within the pile wall area that slide movements are occurring in the roadway embankment at more than one depth. The depth of deepest failure plane varies from about 16 m at the south end (as shown by SI #8) to about 22 m (as shown by SI #11) at the north end. From this observation, it is clear that slide movements are occurring well below the tips of the deepest piles. A cross section of the roadway embankment showing typical soil stratigraphy and the ground movements is presented in Figure 7. As far as the rate of ground movement is concerned, the time-movement plots typically indicate continuing slow creep movements and no evidence of dramatic changes was noticed either in the plots or visually on the ground to date.

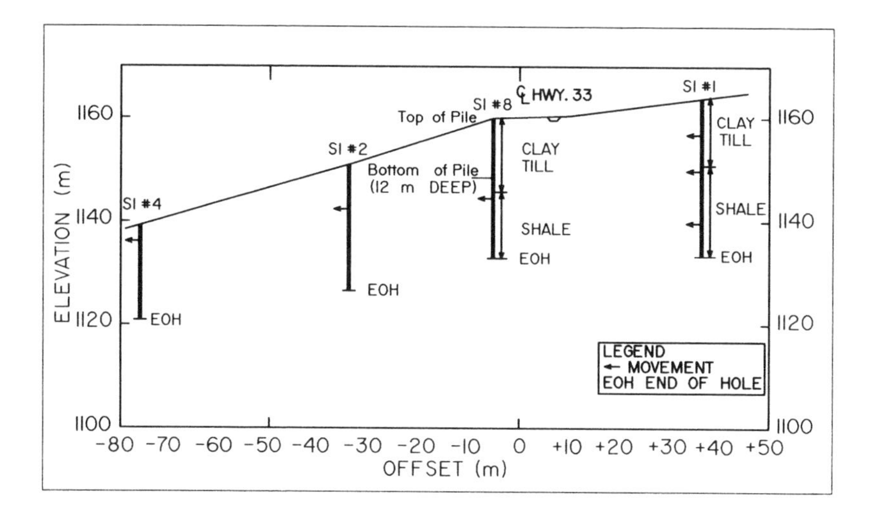

Figure 7. Cross-Section showing Stratigraphy and Movement Depths (Sta. 9+650)

A set of typical time- movement plots of SI #11 is presented in Figure 8. It is further noticed that the sub- surface movements are more restrained below the 12 m depth mark in comparison with the movements between the road elevation and the bottom of the piles.

Pavement Crack Location South of the Pile Wall

Slope indicators #5 to 7 installed in the crack area have sheared at different depths varying from 8 m in SI #7 to 18 m in SI #6. Generally, the average depth of movement is about 12 m below the road.

CURRENT STATUS OF THE SLIDE AREA

It is nearly 10 years since the pile wall was installed. Although the slope indicators show slow creep movements in the order of 30 to 50 mm at the road level over this time, the pile wall seems to be holding well, and there is no distress observed in the two driving lanes. Only a small settlement is visible in the shoulder area adjacent to the piles. A photograph taken recently of the road and the piles is shown in Figure 9.

Mixed success has been reported in the literature regarding the use of pile walls in slide stabilization. However, in the opinion of the authors, this form of stabilization has proved to be a successful solution for this site, considering the complex ground conditions and site constraints.

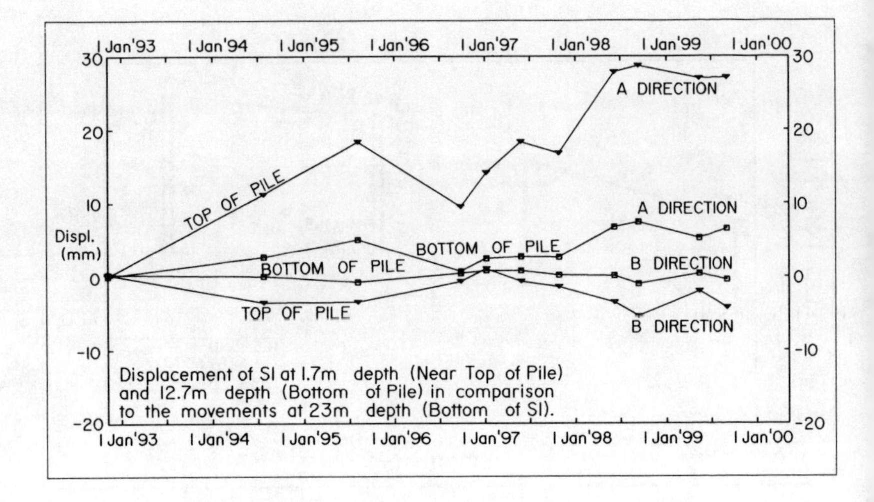

Figure 8. Time-Movement Plots For SI #11

The cracking south of the pile wall continues to be a matter of concern. Since the depth of movements are in the order of 12 to 18 m, a recommendation was made to extend the pile wall southward to encompass the pavement cracking location. However, no specific measures have been implemented to date. It is likely that the pile wall system may be extended sometime in future if continued maintenance of the roadway surface by patching is no longer considered to be a cost effective approach in keeping this section of roadway functional.

CONCLUDING REMARKS

The approach of using drilled straight shaft concrete piles to stabilize slides has not been the norm for stabilization of slope stability problems along the Alberta Highway Infrastructure. In many situations, relocation or the use of earthwork methods such as berms, slope flattening etc, has been utilized as a result of the availability of adjacent land. It has been noted, however, that many highways are located in what can be described as the "best alignment" routes, and very often the use of earth work methods of solution to slide problems is not feasible.

The stabilization approach described herein was the first of its kind to be used on the Alberta Highway Infrastructure. Since then, this approach has been adopted over the last 10 to 12 years in a few areas where site constraints and other factors precluded the use of other forms of stabilization.

Figure 9. Recent Photograph of Pile Wall and the Highway

In reviewing the infrastructure, we see that this type of remedial measure is probably very suitable to the resolution of many similar instability problems. The long term performance of this and other similar projects would no doubt be of importance to the utilization of this method of stabilization.

Over the last three (3) years, this and other sites have been placed on a twice-a-year monitoring schedule with an annual inspection undertaken in the spring of each year. This monitoring schedule will allow the performance of the stabilization measures to be evaluated and, where necessary, the implementation of additional measures to preserve the integrity of the highway.

ACKNOWLEDGMENTS

The authors would like to thank the authorities of Alberta Infrastructure, Peace Region, for allowing them to make use of the data in the Department files for preparing this paper. The opinions expressed in the paper are entirely those of the authors and may not necessarily constitute a position of the Department. Thanks are also due to P. Boos, P.Eng. and J. Donnelly of GAEA Engineering Ltd. who have helped in preparing the figures and typing the manuscript, respectively.

STABILIZATION OF MEIKLE RIVER SLIDE ON HIGHWAY 35

Vishnu Diyaljee[1], Fellow, ASCE, Murthy Pariti[2] and Roy Callioux[3]

Abstract

Stabilization of an unstable section of roadway along a major river valley experiencing deep seated movements was effected through the construction of a pile wall, simple drainage measures and a gabion wall. Since the construction of the pile wall in 1997, the effectiveness of these remedial measures has been evaluated through instrumentation monitoring and observation. So far the roadway is performing well.

INTRODUCTION

Northwestern Alberta has many geologically young river valleys which are very well noted for landslide activity along their valley walls. The Meikle River valley, one of such valleys, is located along Provincial Primary Highway 35 and situated as shown in Figure 1, about 900 kms north-west of the City of Edmonton, Alberta, Canada, in the geographical area of the Province known as the Peace Region. The area between the upper edge of the Meikle River valley wall and the flood-plain on the north side of the river has been affected by landslides at some time in the geologic past. As determined from aerial photographs, almost the entire stretch of highway between the Meikle River Bridge and the Canadian National Railway (CNR) tracks is within landslide terrain. The general opinion is that these slides have occurred as a result of the river cutting into the bank at the toe of the slope (Nasmith, 1964).

As shown on the aerial photograph, Figure 2, the highway follows a curved side hill alignment on the north side of the Meikle River, crossing the CNR tracks at the top of the valley which is located about 1.2 km north of the Meikle River Bridge. The highway has three lanes within this 1.2 km distance, one lane for the south bound travel and two

---

[1]Managing Director, GAEA Engineering Ltd., Edmonton, Alberta, Canada

[2]Geotechnical Engineer, GAEA Engineering Ltd., Edmonton, Alberta, Canada

[3]Ex-Construction Manager, Peace Region, Alberta Infrastructure, Alberta, Canada

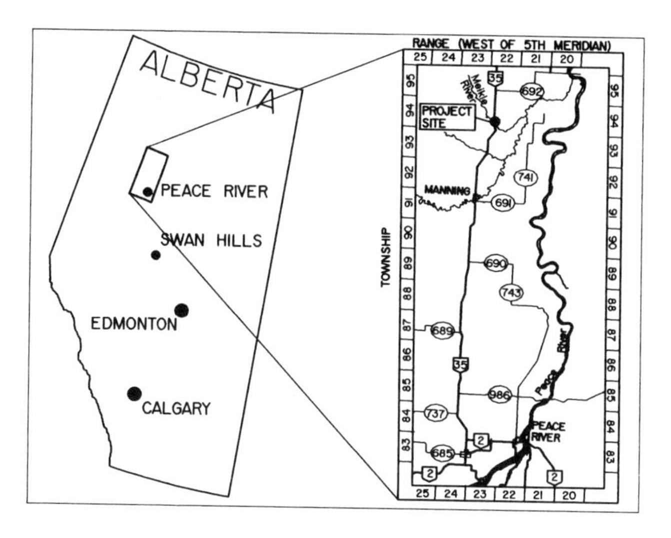

Figure 1. Location of Project Site

lanes for the north bound travel, with one of these being a climbing lane. The highway grade between the Meikle River Bridge and the CNR tracks is about 6 to 7% with an elevation difference of about 60 m from the top of the valley at the CNR tracks to the river at the bridge location.

Slide conditions and pavement distress were reported for the first time in 1991 at two sites (Site 1 and Site 2) identified in Figure 3 and located 0.5 and 0.9 kms, respectively, north of the Meikle River Bridge. At Site 1, sideslope failure occurred over a 30 metre stretch of the highway resulting in 4 to 5 slumped terraces towards the river. At Site 2, settlement of the roadway resulted in a sag at two locations within a distance of about 30 metres. The second site is located at a cut/fill transition of the highway at the top of the valley.

As an immediate short term response, drainage measures were carried out at the two sites by the Regional Maintenance Staff of Alberta Transportation & Utilities(AT&U), now Alberta Infrastructure. At Site 1, the surface water was diverted away from the slide location to minimize ingress of surface water into cracks in the slide area. At Site 2, a perforated pipe subsurface drain was installed in the backslope ditch with the outlet daylighted in a centreline culvert downhill of the slide area. This subsurface drainage measure was implemented based on the evidence of seepage on the sideslope which was inferred to be travelling through the highway from the backslope.

The primary objective of this case history paper is to review the remedial measures implemented at Site 2 and their performance to date.

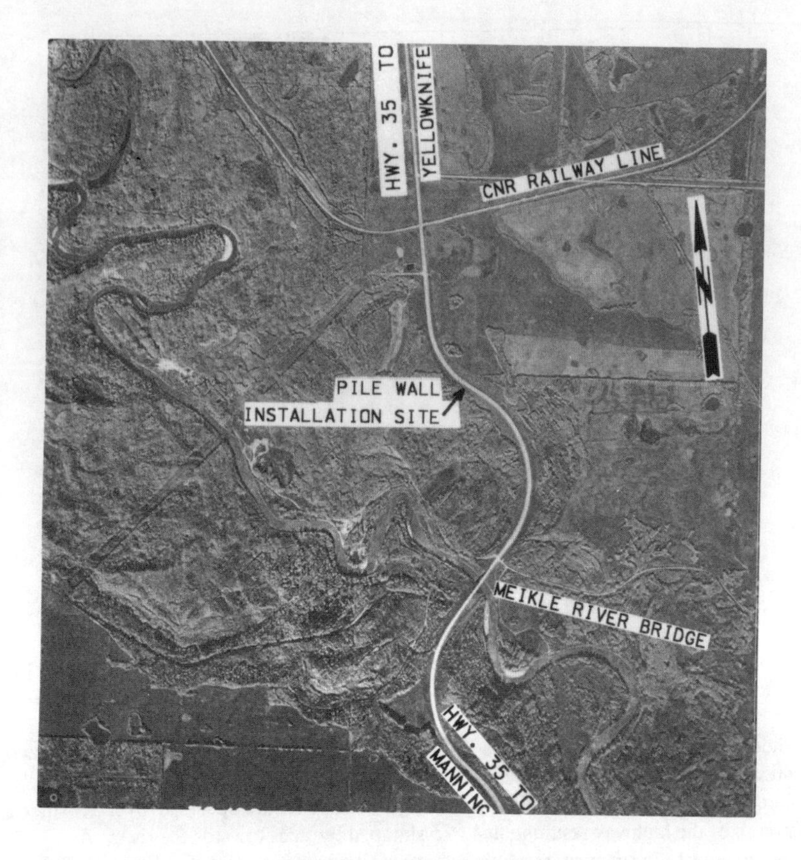

Figure 2. Aerial Photograph Showing Roadway Alignment

GEOTECHNICAL EVALUATIONS AND MEASURES IMPLEMENTED

Between 1991 and 1995

As a first step in identifying the depth of movements, five slope indicators (SI #1 to 5) were installed at both sites in 1992. The location of these slope indicators are shown in Figure 3. While the monitoring was still in progress, a sideslope slipout encroaching the south bound lane occurred at Site 1 in October 1993. Since the slide movements were noticed to be well below the toe of the side slope, a pile retaining wall was constructed as an emergency measure. The construction of this wall was undertaken by the Regional Maintenance Staff on recommendations from the Geotechnical Services Section of AT&U.

Eleven (11) slope indicators (SI #21 to 31) were installed during 1994 on both sides of the road to assess the global nature of the slide activity (Figure 3). Based on a visual observation of accelerating pavement distress and a review of the latest slope monitoring data, an internal AT&U recommendation was made in March 1996 by the Geotechnical Services Section to explore the feasibility of a realignment of the highway on the north side of the bridge without incurring relocation of the bridge.

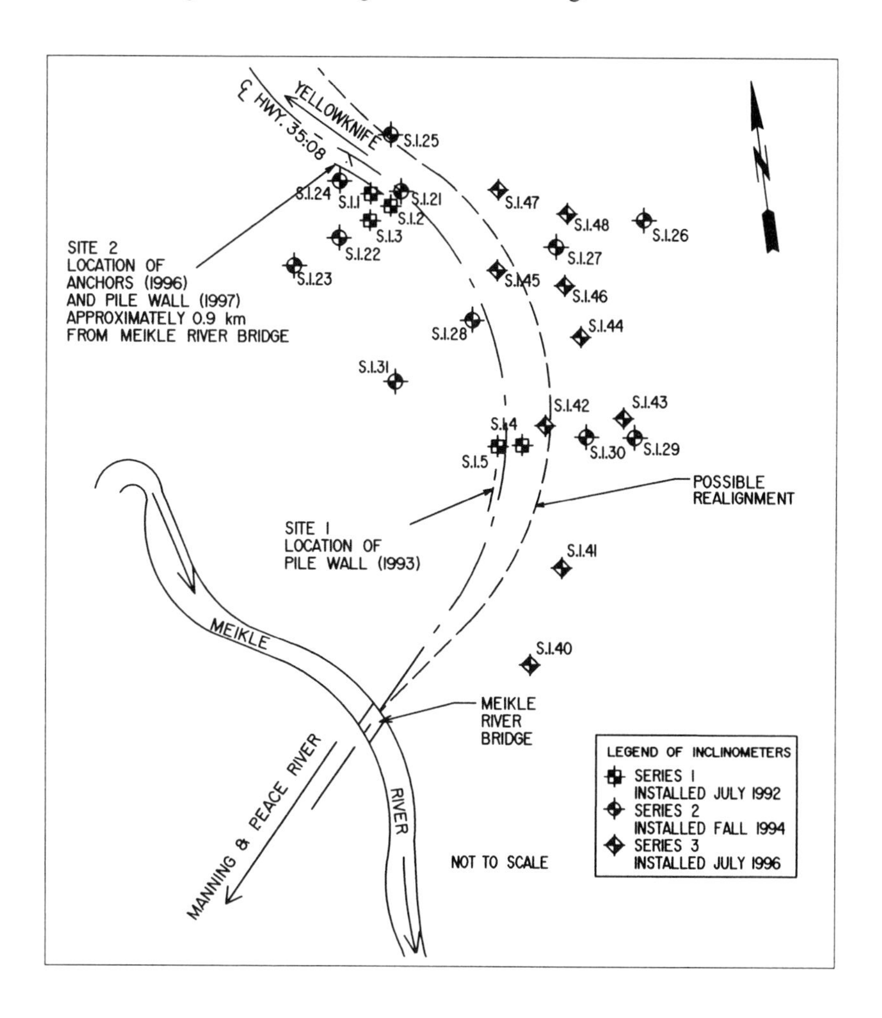

Figure 3. Plan of Instrumentation Locations

In May 1996 and Later

While the realignment feasibility was under study, differential settlement and pavement cracking started to re-appear in the south bound lane in May 1996 over a 120 m stretch of the highway at Site 2. This created an unsafe situation especially for the downhill traffic coming from the north. GAEA Engineering Ltd, a consulting civil engineering firm providing geotechnical services to AT&U on landslide problems in the Peace Region, was requested to inspect the site conditions and design appropriate remedial measures. Based on a visual inspection of the site conditions, screw anchors were installed in the south bound lane in July 1996 as a temporary remedial measure to improve the stability of the highway.

Installation of additional slope indicators, especially in the uphill side of the highway, was also recommended since any realignment option on the north side of the bridge would involve substantially deep cuts. Accordingly, nine (9) additional slope indicators (SI #40 to 48) were installed covering a wider area of the uphill portion and to depths varying between 30 and 60 m in few holes (Figure 3).

The generalized stratigraphy of the subsurface soils encountered consisted essentially of gravelly clay varying from soft to stiff in the upper 8 m and stiff to hard to a depth of 24 m. Very hard clay shale was encountered below a depth of 26 m.

INSTALLATION OF SCREW ANCHORS

As mentioned previously, screw anchors were installed in the south bound lane in July 1996 as a short term measure to improve the stability of the roadway in the south bound lane. The projected advantage of the screw anchors was two fold: (i) to allow traffic to utilize the existing roadway, and (ii) to avoid the possible widening of the highway towards the backslope.

Widening was considered as an option to shift the highway into the backslope to maintain the three lanes in the event of a closure of the southbound lane by the slide activity. The construction of this widening would have resulted in substantial costs, since a much longer length of highway than the length affected by the slide would have to be constructed to allow for proper horizontal geometrics, sight distances and stable cut slopes. In addition, this scheme would have necessitated the acquisition of private property which would have been time consuming and not in the best interests of the travelling public. As a result of these constraints it was decided not to pursue the widening option.

Fifty (50) helical screw anchors each consisting of a 7.6 m long by 114 mm diameter shaft and two 300 mm diameter helixes, were installed in the slide area by Alberta Anchors Inc. of Fort Saskatchewan. For the installation of the anchors, two notches were made in the road by excavating the asphalt and base course to a depth of about

1.5m below ground to facilitate the insertion of the anchors and to ensure that they were buried below ground. The anchors were installed in an inclined direction towards the backslope of the road at an angle of approximately 30° to the vertical (Figure 4). After the installation of the anchors was completed, the notches were backfilled with gravel and the surface was repaved.

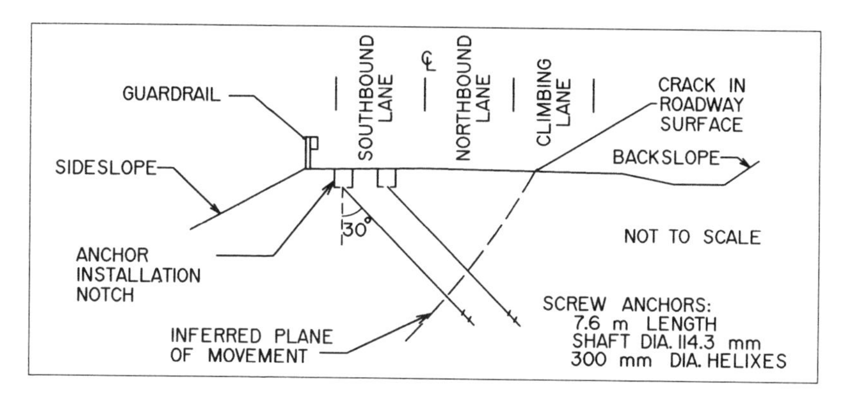

Figure 4. Installation of Screw Anchors

RENEWED SLIDE ACTIVITY IN THE SUMMER OF 1997

Following the installation of the screw anchors, the road surface did not exhibit any major distress through the winter of 1996/97. However, during the months of April to June 1997, the pavement surface started to show renewed differential settlements and sloughing of material on the sideslope on a more severe scale in the section where the screw anchors were installed earlier. These developments were considered to be the result of unusually heavy snow fall that occurred during the winter of 1996/97. As a result of this renewed activity, immediate remedial measures were considered warranted to avoid a total loss of the south bound lane. The realignment option was once again reviewed more closely in the light of the latest slope indicator data and rejected because of the following drawbacks:

1. A realigned section would still be exposed to slide terrain as may be inferred from the deep seated nature of movements shown by the slope indicators installed in the area considered feasible for realignment (Figure 5).

2. The realignment would have to be done through privately owned land and hence could be subject to long delays due to likelihood of expropriation measures to acquire the necessary right-of-way.

3. Shifting the alignment of the highway would also involve steeper grades, longer

realigned length of the highway and substantial cut volumes before reaching the top of the north valley. The disposal of excess material would also involve long haul distances leading to high costs.

4. The Meikle River Bridge could not be relocated any further to the east of its present location as a result of a massive slide, which was present along the north bank of the river only a short distance downstream of the bridge.

As a result of these observations and findings, it was considered more prudent and cost-effective to maintain the road along its existing alignment and implement appropriate remedial measures.

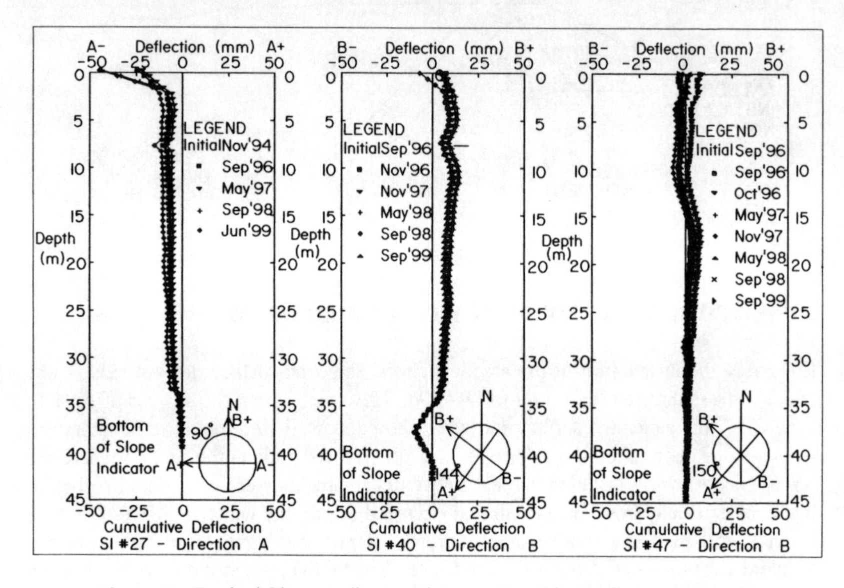

Figure 5. Typical Slope Indicator Plots on Possible Realignment Location

STABILIZATION CHOICE

A careful review of the site was undertaken to determine the most appropriate slide remedial measure. Two alternates were considered in principle, viz., (i) construction of a toe berm in the flood plain to provide the necessary lateral restraint to the highway and the valley walls, and (ii) installation of drilled straight shaft concrete piles.

Toe Berm

Construction of an earth toe berm in the flood plain would literally involve filling in the area that was scoured by the Meikle River in geologic past. While this was a possible solution to implement, this construction would entail the movement of a large volume of earth material. This concept was referred to Alberta Environment for formal approval as part of the procedure that is mandatory when construction activities influence land disturbance outside of the existing highway right-of-way.

From the referral, it was determined that the Meikle River valley is a prime habitat for a wide range of wild life and the river carries three important varieties of fish. Any highway construction in the valley will disrupt the wild life and increase the risk of erosion of valley slopes and siltation in the river which could be detrimental to the fish. The referral also pointed out that an Historical Resources Impact Assessment of the project would be needed since the physiographic features of the site were considered to have high potential for the discovery of archeological resources.

Since these investigations take a considerable amount of time, it was decided to utilize the pile wall concept which would allow construction activities to be undertaken close to the existing roadway and without any time delay, thereby minimizing the continuing and escalating public complaint of the danger posed to commuter traffic of an unstable highway with a steep gradient.

Drilled Straight Shaft Concrete Piles

The approach of using drilled straight concrete piles to stabilize slide areas has been successfully used on a selective basis within the last 10 to 12 years in a few problem areas along the Alberta Highway Infrastructure, especially in the Peace Region.

While the installation of drilled straight shaft piles was considered a feasible alternative in principle, it was also recognized that the slide activity along this 1 km stretch of roadway was deep seated, as inferred from Figure 5, wherein existence of multiple slide zones is generally noticeable at depths varying from 5 to 40 m. However, in the immediate vicinity of Site 2, the depth of slide zones vary from 8 to 28 metres. Figure 6 shows typical monitoring data of two slope indicators of Site 2.

The presence of multiple slide zones is typical at many sites in the Peace Region, where rivers are responsible for valley formation. The various slide zones would correspond generally to depths at which various stages of river down-cutting or toe cutting occurred as a result of the river progressing from a youthful to a mature stage. The overall picture is one of retrogressive sliding activity which, from an aerial photograph review, influenced the movement of land beyond the immediate top of the valley crest parallel to the roadway. This emphasizes the importance of a careful aerial photograph review of the valley slopes since, very often, realignments are taken just beyond the valley crest without recognizing that this area may also contain slide zones.

Based on the understanding of the sliding mechanism, it was determined that the installation of piles well below the lowest perceptible slide zone would be very costly. Hence, engineering judgement was exercised to limit the depth of piles to 24 metres, which was considered to be a reasonable compromise from a cost-effective point of view and the variation in depths of sliding observed.

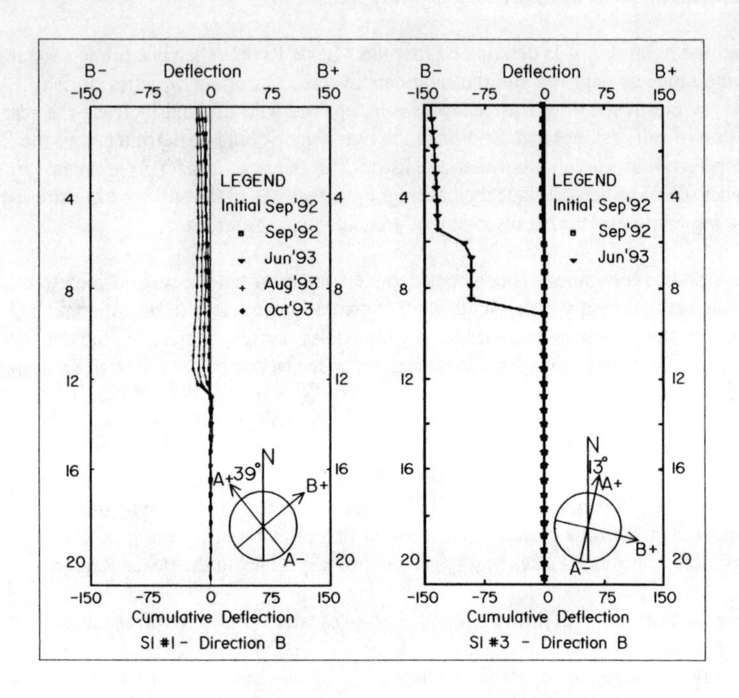

Figure 6. Slope Indicator Plots In the Vicinity of Site 2.

STABILIZATION SCHEME

The stabilization scheme proposed and implemented was the immediate construction of a row of drilled straight shaft concrete piles to allow the retention of the roadway and thus prevent the loss of the south bound lane. This was to be followed by the installation of a capping beam to provide rigidity to top of the piles, prevent rotation of the pile tops, and allow any lateral movement of the pile wall to occur with the piles acting as a unit rather than individually. Beyond the pile wall, the sloughing sideslope was to be retained by a gabion wall to prevent material from behind and in front of the piles from slipping out and exposing the piles thereby reducing the soil lateral restraint against the piles. Drainage measures were also proposed to remove the surface run off from the road and subsurface seepage within the roadway embankment to improve the over-all stability of the slide area.

Installation of tiebacks was also considered as an option to enhance the stability of the wall and the slide area. However, the tiebacks were not favoured since their installation costs would have been quite expensive and there was uncertainty about the performance of the site. Rather, it was decided to proceed with the less costlier option of the construction of a capping beam, gabion wall and surface drainage measures. Installation of tiebacks was, however, still a consideration should performance show that there was a need to preserve the integrity of the pile wall based on observations of site monitoring and that the site conditions would still support this as a feasible option.

To accommodate the availability of funds, installation of drilled straight shaft piles was first undertaken in the fall of 1997 to retain the road in position for the convenience of the travelling public. The construction of the capping beam, gabion wall and drainage measures was done later in the summer months of 1998.

PILE WALL INSTALLATION

The pile wall installation was undertaken by North American Construction Inc. of Edmonton, Alberta, during the fall of 1997. Seventy seven (77), 760 mm diameter by 24 m deep piles were drilled with a Texoma 900 drill at 1.5 m centre to centre spacing along the location of the guardrail. H-piles, 310 mm x 96 kg in size, were installed with their flanges parallel to the roadway centre line and the pile holes backfilled with 30 MPa strength concrete. The settled portion of the south bound lane was then brought back to grade with native soil to match the level of the asphalt pavement of the north bound lanes and left unpaved to check for any settlements during the spring/summer of 1998.

Selection of the size of the piles, their spacing and depth, was done based on a general interpretation of the slope indicators monitoring information in the neighborhood of the slide area, overall cost implications, past experience and engineering judgement.

Three slope indicator tubes were installed along the length of the pile wall, one at the centre and the two near the ends of the pile wall. The slope indicators were embedded in the piles by affixing these tubes to the H piles at the junction of the web and flange of the H piles through a rectangular slot running along the length of the pile.

CAPPING BEAM CONSTRUCTION

As explained earlier, a reinforced concrete capping beam was utilized on this project to provide fixed support conditions for the pile tops. The cross sectional dimensions of the capping beam were maintained as 1.5 m wide by 1 m deep for about 20 m length of the beam at the outer ends of the pile wall, and 1.75 m wide by 2 m deep for a distance of 70 m in the middle of the pile wall. The 2 m depth was designed to prevent movement of roadway embankment material in between the piles, due to seepage and internal erosion occurring in the middle part of the pile wall. Weep holes were also provided in this section to prevent the build up of pore pressure behind the wall.

The capping beam design was considered as a cost effective approach as the massive size of the beam used on this project was expected to provide an equally satisfactory rigidity to the top of piles. Another indirect advantage of the capping beam was that the guardrail could be positioned on top of the capping beam.

The capping beam construction contract was undertaken between August 20 and September 5, 1998 by Ruel Concrete Ltd of Peace River, Alberta. Concrete was supplied by Szmata Concrete and Aggregates Ltd, of Grimshaw, Alberta. The guardrail was mounted on top of the capping beam after its completion, Figure 7. The traffic lane adjacent to the pile wall was restored with an additional gravel base course and paved with new asphalt concrete pavement.

Figure 7. Completed Capping Beam Installation

GABION WALL INSTALLATION AND DRAINAGE MEASURES ON SIDESLOPE

The gabion wall installation and drainage measures were undertaken on the sideslope during the month of November 1998. The work was carried out by Kauri Contracting Ltd of Grimshaw. These measures were instituted to preserve the integrity of the sideslope on the downhill side of the pile wall.

Four shallow gravel filled finger drains with perforated pipe were installed in the slumped material of the sideslope between the pile wall and the gabion wall. These were installed perpendicular to the centre line of the roadway to tap into the seepage below the road elevation coming from the uphill terrain. These finger drains were then connected to a longitudinal collector drainage pipe system installed behind the gabion wall. A 150 mm diameter non-perforated corrugated plastic pipe was then connected to the collector pipe and daylighted away from the gabion wall further down the slope.

Following the completion of the gabion wall and the subsurface drainage system, the sideslope between the gabion wall and the pile wall was reshaped to a uniform slope using discarded tires and wood chips as part of the fill material to reduce the lateral pressure on the gabion wall. A non-woven geotextile was then laid on top of the tires/wood chips and then capped with clay material. A typical drawing related to the gabion wall construction and drainage measures is shown in Figure 8.

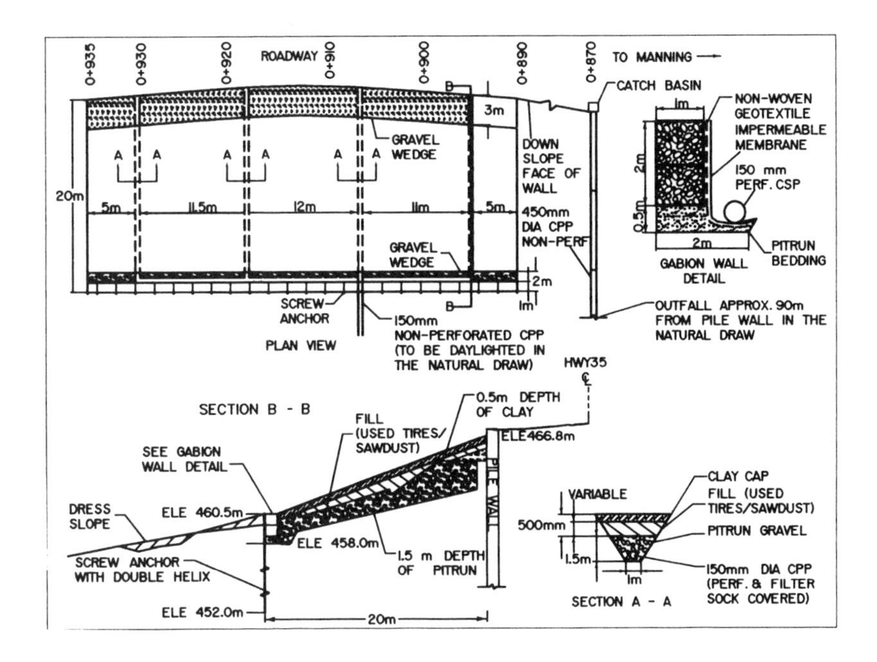

Figure 8. Gabion Wall Installation and Drainage Measures

As part of surface drainage improvement measures, a catch basin was constructed at the lower end of the pile wall to collect the surface runoff from the roadway. A 450 mm diameter non-perforated corrugated plastic downdrain pipe was connected to the catch basin and was laid along the sideslope to a distance of about 150 metres, where it was daylighted in a treed area. The purpose of the catch basin system and the downdrain was to minimize the seepage of surface runoff into the soil on the downhill side of the pile wall and to prevent the occurrence of shallow soil slipouts along the sideslope within the rehabilitated slide area.

CURRENT STATUS OF THE PROJECT SITE

To date, the roadway is performing well, except for minor settlement of the material downslope of the gabion wall. It should be noted that since the gabion wall construction and drainage measures were undertaken in winter months, compaction of the material might not have been perfect at that time. Hence, the settlement being observed currently may be the result of readjustment within the reshaped material.

The slope monitoring instrumentation is being read on a semi-annual basis. The three slope indicators installed in the pile wall are showing slow creep movements generally within the top 8-12 metre depth range (Figure 9). Below that depth, the movements are practically negligible. It is also interesting to note that there is a kink in the deflection plots at about 2 m depth below the road. This kink generally coincides with the depth of surficial sloughing, noticed prior to the pile wall installation, on the side slope of the highway. Typical time-movement plots corresponding to both A and B grooves of SI #50, which was installed in the middle of the pile wall, are shown in Figure 10. The rate of movement is generally in the order of 0.02 mm/day and there is no indication of drastic changes observed in the displacement plots.

CONCLUDING REMARKS

The total cost of the various measures described in this paper was about $1 Million Canadian dollars. Although this order of expenditure may seem high, the remedial measures implemented have proved quite helpful and cost-effective in maintaining the highway through this major river valley. Considerable engineering judgements had to be exercised to decide on the most logical and practically viable slide remedial measure, while at the same time to be cost effective.

The final choice of stabilization measure was governed not only by the technical appreciation of the problem, but as well, by social, economic, and environmental factors, and by constraints imposed by topography and geometrics of the alignment. Although it is often desirable to install piles well below the deepest slide plane with tiebacks to retain them, preference was given to the principle of observational approach and engineering judgement to make the project reasonably viable and cost effective in maintaining the integrity of the road for the convenience of the travelling public.

The installation of drilled straight shaft concrete pile walls in this project, is the third of its kind used along the Alberta Highway Infrastructure in the Peace Region to stabilize landslides in complex geologic conditions. The first installation was undertaken in 1988 to stabilize the Judah Hill landslide which has been previously reported in the literature

(Diyaljee,1992). The second installation was done near the Town of Swan Hills. A detailed description of the second project has been reported in another paper submitted to this Conference (Diyaljee et al, 2000).

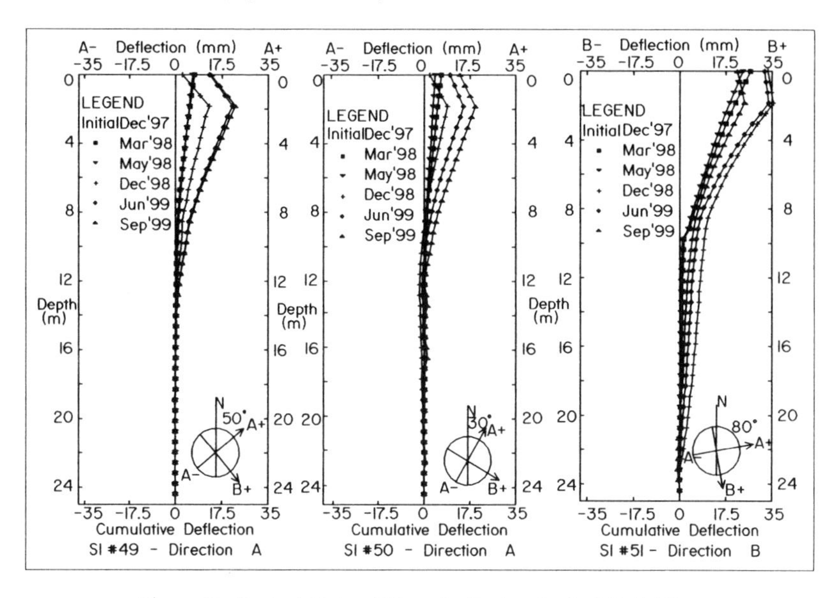

Figure 9. Typical Plots of Slope Indicators in the Pile Wall

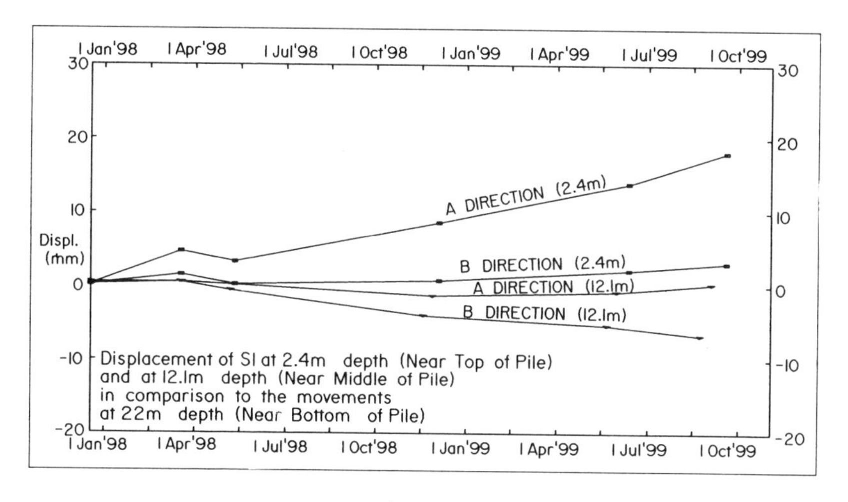

Figure 10. Typical Time-Movement Plots of SI #50

So far all of these installations are performing satisfactorily thereby allowing the highways to perform their intended function of providing safe, efficient and effective movement of goods and people within and through Alberta.

Over the last three (3) years, this and other sites have been placed on a twice-a-year monitoring schedule with an annual inspection undertaken in the spring of each year. This monitoring schedule would allow the performance of the stabilization measures to be evaluated and where necessary, the implementation of additional measures to preserve the integrity of the highway.

ACKNOWLEDGMENTS

The authors would like to thank the authorities of Alberta Infrastructure, Peace Region, for allowing them to make use of the data in the Department files for preparing this paper. The opinions expressed in the paper are entirely those of the authors and may not necessarily constitute a position of the Department. Thanks are also due to P. Boos, P.Eng. and J. Donnelly of GAEA Engineering Ltd. for their help in preparing the figures and typing the manuscript, respectively.

REFERENCES

Diyaljee, V.A. (1992). "Stabilization of Judah Hill Landslide." *Proceedings of the Sixth International Symposium on Landslides,* Christchurch, New Zealand.

Diyaljee, V.A., Pariti, M. and Callioux, R. (2000). "Long Term Performance of A Slide Retained By A Tie Back Pile Wall", *GeoDenver 2000 Symposium,* Denver, CO.

Nasmith, Hugh (1964). "Landslides And Pleistocene Deposits In The Meikle River Valley Of Northern Alberta." *Canadian Geotechnical Journal,* Vol 1. No. 3, 155 - 166.

NUMERICAL ANALYSIS AS A PRACTICAL DESIGN TOOL IN GEO ENGINEERING

Charles Fairhurst[1], Branko Damjanac[1] and Roger Hart[1]

Abstract

With the advances in computer technology and the increased accessibility and flexibility of computer software developed for geo-engineering applications, numerical analysis has gained wide recognition as a qualitative tool to increase understanding and guide prediction of anticipated behavior during underground construction or mining operations. Successful practical problem solving in geo-engineering depends on formulation of a concise conceptual model that correctly incorporates the key factors for the situation at hand. Historically, this conceptual model was built on empirical knowledge and existed only in the mind of the experienced and well-informed engineer. Numerical analysis provides a way to present the conceptual model explicitly, such that it can be reviewed critically, and the effect of possible assumptions and design changes examined and discussed. This paper reviews four case examples to illustrate some of the many ways in which numerical analysis can and has been used to provide valuable assistance in improving practical geo-engineering design.

Introduction

The *observational method* (Terzaghi and Peck, 1968) is a well established geo-engineering procedure to allow the knowledge of subsurface conditions as revealed during excavation to be used to modify the design during construction. The field observations made during excavation allow the engineer to refine her/his conceptual image or *model* of the critical characteristics and mechanisms that affect the design. Predictions of the anticipated behavior are then calculated or estimated based on this mental image/model, and the design is altered as needed as a consequence of these predictions.

----

[1]Itasca Consulting Group, Inc., 708 South 3rd Street, Suite 310, Minneapolis, MN 55415

This procedure has been used successfully in many construction and mine development projects in which design changes are made as required based on direct observations of response in the field using visual inspection, instrumentation, empirical knowledge and, in some cases, back-analysis. A conceptual model is first developed based on observations and forecasts of behavior then made. The model and the design are revised or refined as appears appropriate based on the observed response to the initial design. The procedure is frequently referred to as a "design-as-you-go" approach to tunnel excavation/support or to mine development. Designs that can accommodate significant changes in ground behavior from those assumed are desirable. In tunneling, for example, support systems that involve yielding supports and/or the so-called New Austrian Tunneling Method (NATM), when well applied, are examples of such flexible design systems.

While the insight and guidance of the experienced engineer/consultant are invaluable components of successful geotechnical design, the inherently implicit and unstated nature of at least some of the assumptions used in arriving at design recommendations can be a limitation – and sometimes a barrier to communication within the design team. Numerical analyses, when well applied, can serve as an additional investigative tool to assist in overcoming this limitation, to make the observational approach more effective. By conducting numerical analyses on the conceptual model, it is possible to show explicitly the mechanisms assumed to represent the actual behavior. The results of these analyses can lead to a better understanding of the key mechanisms and improved predictions of response in comparable situations and design challenges in the future. The effect of changes in design and consequences of uncertainty in assumed parameters or mechanisms can be shown graphically (and dynamically, if useful) for communication and discussion with members of the design team (site geologists, construction engineers, designers, consultants, etc.). Changes can be made to the model based on the discussions, and the predicted consequences assessed in further discussion. Numerical modeling can show to particular advantage in complex and/or unusual situations where the appropriate course of action is not readily apparent, or for which there is no "standard response." The case histories discussed illustrate this attribute of numerical modeling.

This paper describes three case studies in which numerical analyses were used to gain confidence in the conceptual model developed from field observations. Two of the studies are related to underground mining operations and one is related to surface excavation. In all three cases, numerical analyses are performed to assess the viability of the conceptual model. Finally, a fourth case is presented that describes a systematic approach in which numerical analysis can be incorporated into an operational procedure for tunnel construction.

Failure of a Large Hangingwall Wedge

Difficult access to a site can sometimes preclude direct observation and prevent assessment of the possible mechanism involved in a problem situation. In the first case example, large ground settlements observed during a mining operation caused shut-down of the mine. The mine management was concerned that additional mining in the area would reactivate dangerous further displacement. A conceptual model was formulated based on available observations, and numerical analysis provided an insight into the factors that must be considered in the model in order to make confident predictions. This case example is described in detail by Board, et. al., 1999.

The Kidd operation (Timmins, Ontario, Canada) by Falconbridge, Ltd. mines roughly 10,000 tpd of copper and zinc ore. In the late summer and fall of 1997, major movements in the proximity of the east wall of the Kidd open pit were observed. The Kidd pit had been mined from 1965 to 1977 to depths of 260 m followed by subsequent mining by underground methods to current depths in excess of 1830 m. Although initial concerns centered on the stability of the pit walls, the potential ramifications of the movements to the underground were recognized and measures were implemented to ensure the safety of personnel as well as to minimize the potential damage to existing operations. In late October, damage to underground operations was observed as deep as the 2000 Level (610 m below surface). The global mobilized mass was assessed to be a "wedge"-shaped hangingwall block extending from surface to about 610 m below surface. The wedge boundaries are defined by a fault (North B Fault) and shear zone (East-West Shear) which have continuous lengths in excess of 1070 m; its base by shearing through the host rock mass; and its eastern face by deep tension cracks. Figure 1 shows an aerial view of the pit and denotes the wedge boundaries. Mined and filled excavations at the western face of the wedge allowed mobilization of a mass in excess of 30 M tons. Movements in excess of 75 cm per day were recorded at the surface stations that preceded the shutdown of complete operations on October 29, 1997.

The initial proposed failure mechanism involved a complex, three-dimensional interaction of mined voids, slip along pre-existing planes of weakness, shear failure through the rock mass, and compaction of backfill. Representation of these phenomena using a numerical model is not a trivial exercise. The approach used was to make as simple a model as possible that represented the three-dimensional geometry of the mining and faults with reasonable accuracy. The model was then used to conduct sensitivity studies of the mining sequence while varying the rock mass strength until the model best matched the observations made underground and on the surface. The primary idea was to determine the factors that control initiation and development of this failure. Once some confidence was gained in the model as a tool, it could be used to explore the impact of various future mining scenarios on the response of the block system.

Numerical analysis was performed with *3DEC* (Itasca, 1998), which is a three-dimensional discontinuum code. The numerical model allows the rock mass to be represented as a series of deformable blocks separated by fault planes that can shear and/or open. It was necessary to construct a 3D model of the geology and stoping blocks for the model. A model of the geology and stoping for the mine was developed from level plans and CAD drawings. To this geologic and mining geometry, the major faults and shear zones were added by cutting the model based on mapping, thus forming additional blocks. Figure 2 shows a view of the resulting *3DEC* model for comparison to Figure 1.

Figure 1. Aerial view of the pit from the southwest following failure (from Board, et. al., 1999)

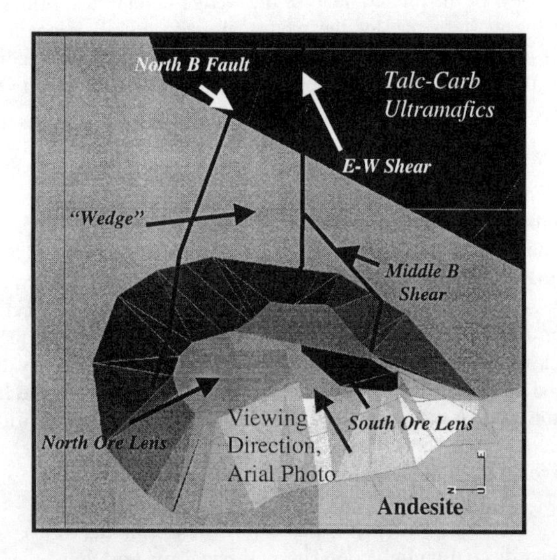

Figure 2. Plan view of the *3DEC* model looking into the open pit (from Board, et. al., 1999)

An example of the results of the simulations is shown in Figure 3, in which velocity vectors are presented on a longitudinal section through the center of the wedge. It was found that, for the best guess estimate of properties, mobilization of a complex hangingwall block occurred when stope panels (pillars) in the wedge region from 2000 Level and above were removed from buttressing the toe area as occurred in 1997. Once these stope panels were removed, the remaining pillars and fill in the waste lens were insufficient to resist the shearing forces at the base of the toe. The toe area fails, followed by slip on the faults. The velocity of the blocks that comprise this model show that a large body is mobilized that is bounded by: a) a deep tension crack behind the crest of the slope, b) a basal shearing surface that interconnects the tension cracks with the toe of the wedge at the 2000 Level, c) the mined and filled stopes to the west; and d) the bounding North B and E-W Shear on the north and south sides. The velocity vectors at the pit crest are oriented at approximately 35° down from horizontal – this is the same as monitored on the pit crest. It is also noted that the dip and direction of the velocities in the model at the 1600 Level are roughly the same as those observed underground on the North B Fault. The region of most intensive shearing and instability in the model is the basal region that extends from the 2000 Level to around the 1800 Level at an angle roughly parallel to the plunge of the intersection line of the North B Fault and E-W Shear.

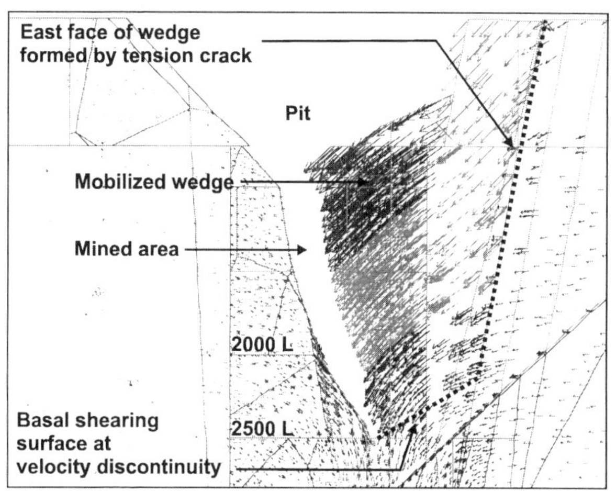

Figure 3. Longitudinal section through *3DEC* model showing velocity vectors associated with wedge movement (from Board, et. al., 1999)

The *3DEC* model reproduced in-situ observations very well and provided clear indication of the likely failure mechanism. Consequently, the model indicated that further mining would not cause additional movement of already destabilized block nor would there be destabilization of new regions of the rock mass. The mining operation was continued and to date (approximately 40% extraction) only minor movements of the block have been observed.

Mine Tremors Induced by Coal Mining Near Major Faults

In catastrophic situations, the testing of a possible scenario may be necessary to determine the cause of the problem in order to develop appropriate preventive measures. In this case study, numerical analyses were conducted to verify the hypothesis that major faults were the likely source of mine tremors induced by coal mining. The investigation focused on the consequences of a rockburst that occurred on September 11, 1995 at the Nowy Wirek Colliery of the Upper Silesia Coal Basin, Poland. A detailed description of this study is presented by Kwaśniewski and Wang, 1999.

Multi-seam coal extraction is conducted at the Nowy Wirek Colliery in proximity to several major faults. Mining in the vicinity of one fault, Fault III, was believed to have triggered the rock mass tremor and resulting rockburst. The energy from the seismic event was recorded as 80 MJ, and the associated rockburst resulted in the destruction of a setup entry to a working face at a depth of approximately 670 m and the death of five miners.

Observations of the seismic data recorded at the site indicated a long history of high seismic activity in the area of Fault III. The results of analyses of seismic source parameters, specifically high values of stress drop and seismic moment, also suggested that activity was related to motion along Fault III. For example, the direction, dip and type of focal plane corresponded to the orientation and location of Fault III.

A three-dimensional numerical analysis was conducted with *FLAC3D* (Itasca, 1997) in order to assess if the conceptual model of motion along major faults is a rational explanation of the rock tremors resulting in rockbursts. The boundaries of the *FLAC3D* model were selected to encompass the two major faults, Fault III and Fault IV, within the mining region, as well as the source location of the September 11, 1995 tremor. Figure 4 shows a plan view of the site including the two faults, the source location, and the lateral boundaries of the *FLAC3D* model. Figure 5 shows a schematic view of the model including the two major faults and nine coal seams located within the model region.

The geology of the region consists of hard carboniferous rocks covered by series of weak soil sediments. More than 50 strata of materials were represented in the model. The extraction of the coal seams was simulated by deleting elements in a sequence that corresponded to the mining history. Goaf material in caved zones and hydraulic backfill were simulated as materials with time-dependent properties to approximate the effect of compaction and deformability changes with time. The major faults were represented by interface elements along which slip and separation can occur in accordance with a Coulomb slip criterion and tensile strength limit.

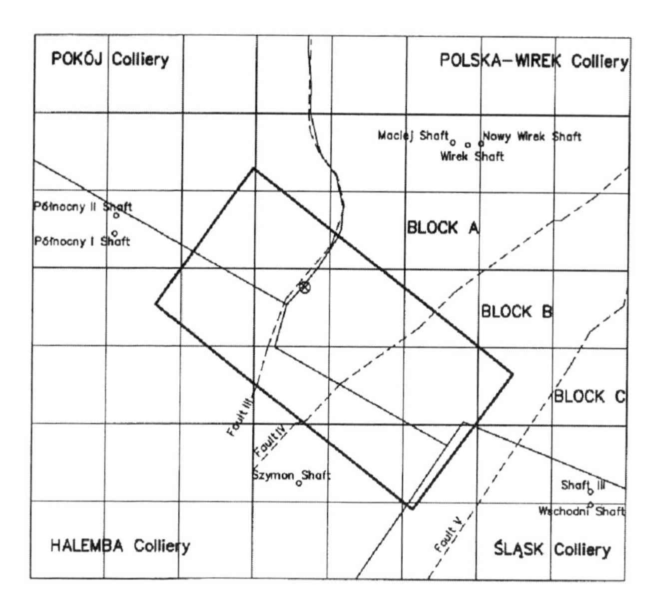

Figure 4. Plan view of studied mining area showing lateral boundaries of *FLAC3D* model (from Kwaśniewski and Wang, 1999)

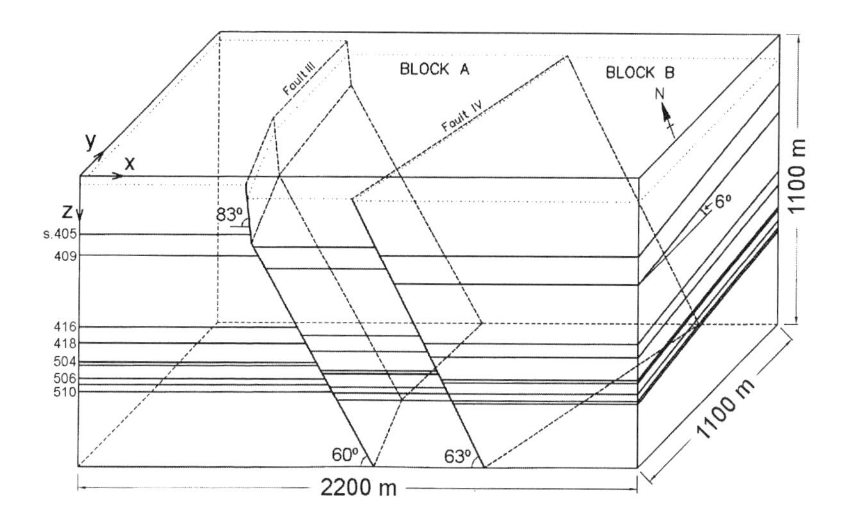

Figure 5. Schematic view of *FLAC3D* model showing two faults and nine coal seams (from Kwaśniewski and Wang, 1999)

The explicit finite-difference modeling approach used in *FLAC3D* permits both static and dynamic analysis. The model is first brought to a static force equilibrium state for the initial stress-state conditions. Then, the mining activity is simulated, and the dynamic response is monitored at selected locations in the model. Figure 6 displays one such response, a vertical acceleration located at a position corresponding to the location of the 1995 tremor and rockburst. A very strong vibration was calculated at this location, with a peak acceleration greater than 80 g. The region of strong vibration in the model was found to coincide with the location of the observed tremors. In addition, the analysis showed that high stress concentrations developed on Fault III and within remnant parts of coal seams as a result of the mining pattern, reaching particularly high values at the depths coincident with the rockburst location. The zones of high stress concentrations developed during the course of mining, rising dramatically prior to the time of the catastrophic rockburst.

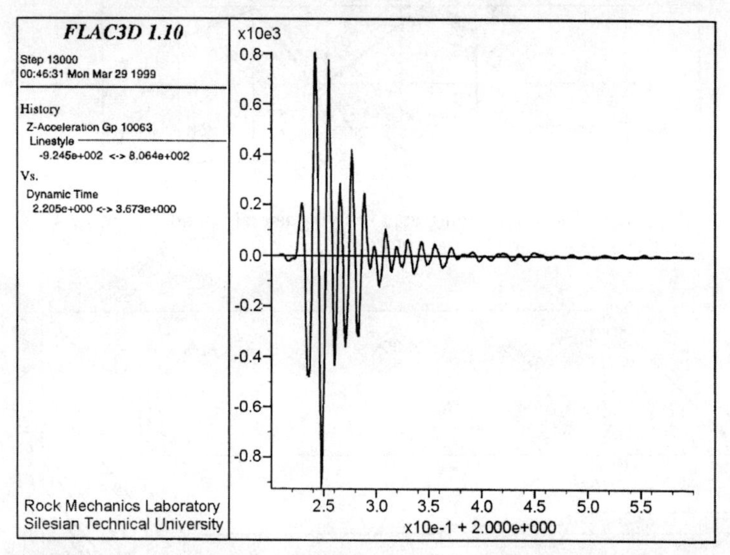

Figure 6. Vertical acceleration predicted by *FLAC3D* model at a position along Fault III corresponding to the location of 1997 tremor (from Kwaśniewski and Wang, 1999)

The numerical analysis thus showed that high stress concentrations on the fault, occurring as a result of disorganized mining and the presence of coal pillars left along the fault, could have triggered a rock mass tremor following extraction, and caused the catastrophic rockburst observed on the site. As a result, the analysis identified regions of potential rockburst hazard in the course of current planned mining activity. Additional analyses have been conducted to assist with the design of future mining patterns. This work is ongoing.

Failure of a High Cut in Clay Shales

Although a conceptual model may be useful in prediction making, it is often of prime importance to assess the limit of application of the model for accurate projection. This case study analyzes the deformation and failure mechanism of a high cut in clay shale. In this instance, numerical analysis incorporating both continuum and discontinuum material behavior indicated that the observed displacements could be explained only when excavation-induced shear bands in the clay slope were taken into account. This study was performed for the staged mining cut at the Santa Barbara open-pit mine located in Arezzo, Italy. The publication by D'Elia et al., 1999 discusses the case study in detail.

The mining of the brown coal seam at the Santa Barbara open pit required cutting of slopes with heights progressively increasing up to 200 m. The excavation is made in highly non-homogeneous structurally complex formations of clay shale mixtures with rock inclusions. Figure 7 shows a cross-section of a representative geological setting including the coal seams. The figure also denotes various heights of excavation cuts.

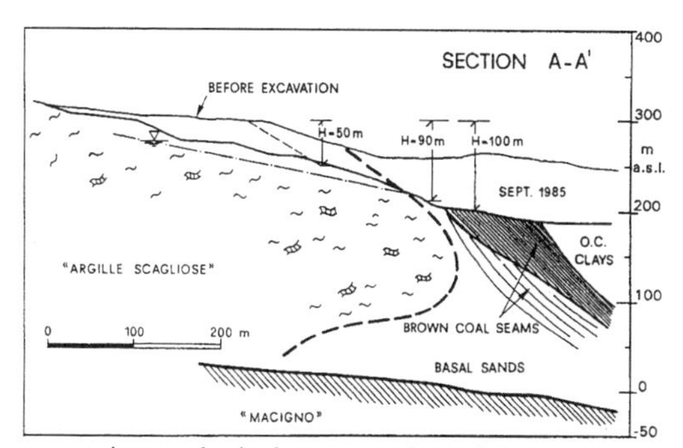

Figure 7. Representative geological cross-section of the study region (from D'Elia et al., 1999)

The cut design followed an observational method with field monitoring and in-situ and laboratory testing conducted throughout the excavation period. A three-stage conceptual model was developed to explain the deformational behavior of the cut. In the first stage, for cut heights up to 90 m, an extensional state developed in the upper layer of ground as evidenced by surface displacements and formation of tension cracks. These observations indicated that the state of stress in this phase was close to an active failure state. In the second stage, for heights up to 150 m, profiles of displacement displayed sharp counter-rotations suggesting that shear bands were forming along the cracks created in the previous phase. Sharp counter-scarps were also observed at the ground surface in this phase. In the third stage, as the cut height reached 200 m, the development of a downslope-facing landslide scarp indicated that a general slope failure surface was forming. A sketch of the state of deformation at the third phase is shown in Figure 8.

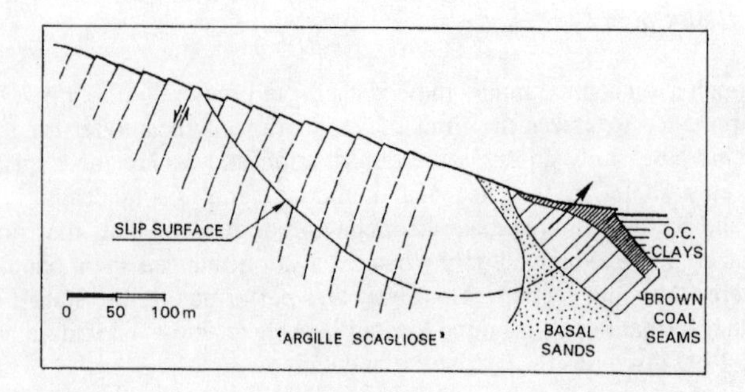

Figure 8. Sketch of the deformation state at the formation of the slip surface (from D'Elia et al., 1999)

Numerical analyses were conducted to evaluate and confirm the conceptual model hypothesized from the observations and measurements. The analyses were designed to replicate the evolution of the stability conditions so that results could be compared to the observed phenomenon at each phase. A fundamental component of the numerical analyses was the simulation of the effect of the sub-vertical tension cracks that developed at the end of the first phase.

The numerical simulations were performed with *FLAC* (Itasca, 1998) to compare a continuous-material to a discontinuous-material representation of the behavior of the clay shale. An elastic-plastic Mohr-Coulomb constitutive model was used to represent the continuous-material behavior, while discontinuous-material behavior was simulated with a ubiquitous-joint model. The ubiquitous-joint model is an anisotropic plasticity model that includes weak planes of specific orientation embedded in a Mohr-Coulomb solid. The *FLAC* model configuration and properties are illustrated in Figure 9; the same configuration was used to compare the response for the clay shale represented as a Mohr-Coulomb material versus a ubiquitous-joint material.

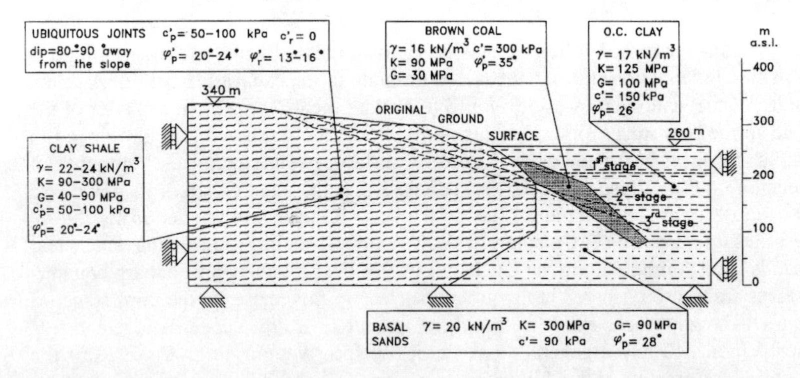

Figure 9. *FLAC* model configuration and properties (from D'Elia et al., 1999)

Figures 10 and 11 compare *FLAC* simulations at the third stage of cutting (160 m excavation height). For the continuous-material model, the analysis produces a continuous shear band that extends from the upper limit of the cut to the bottom of the excavation, as shown in Figure 10a. The deformational pattern corresponds to a rotational slide and does not correlate to the observed behavior.

For the discontinuous-material model, the progressive excavation first causes slip to occur along the sub-vertical ubiquitous joints and finally a fairly continuous slip surface develops within the clay shale, as shown in Figure 11a. This simulation produces characteristics of flexural toppling that correspond to the field observations, and suggests a general failure mechanism that is in accordance with that drawn from the conceptual model.

The conclusions of this study are that progressive updating of the material model representation should include discontinuous behavior as well as continuous material behavior. A continuum material model is not sufficient to reflect the observed behavior. Through an observational approach and appropriate modification to the material model implemented in the numerical analysis, an increased understanding of the deformation and failure process can be obtained. Consequently, excavation design for the high cuts can be improved based on the increased confidence in the conceptual model.

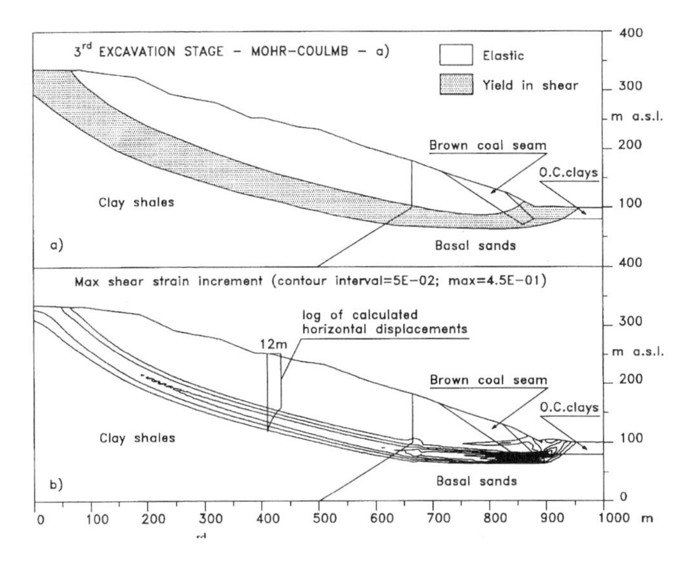

Figure 10. Results of continuum-material analysis at third excavation stage (from D'Elia et al., 1999)

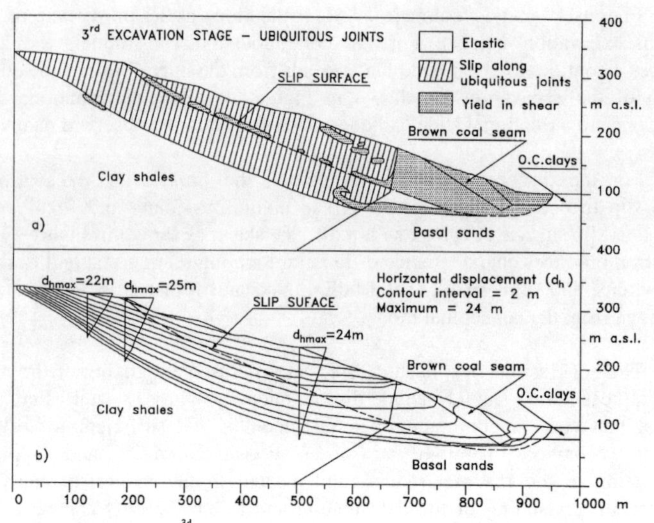

Figure 11. Results of discontinuum-material analysis at third excavation stage (from D'Elia et al., 1999)

Continuous Evaluation of Compensation Grouting Required for an Advancing Tunnel

Numerical analysis is usually applied to evaluate *a priori* the construction technique used to limit soil movement during tunnel construction. However, as tunnel excavation progresses, unforeseen features may be uncovered that call for a revision of the conceptual model and the predictions. This case study describes a general design technique that combines the flexibility of the observational method with the planning capabilities provided by numerical analysis. In this approach, numerical modeling is used in the initial design stage to provide a first estimate of the behavior resulting from the construction approach. As construction proceeds, new data collected on site are used to readjust or validate the numerical predictions. The daily updating computations are then used to define the construction program for the following day. The publication by Buchet and Van Cotthem, 1999 describes this approach when using the construction technique, *compensation grouting*, during advancement of a tunnel.

The construction of underground infrastructures offers an environmentally viable solution to the problem of transportation in dense urban areas. In these circumstances the tunneling process is often performed in soft ground and at shallow depth. Inevitably, ground movements of various magnitude can result, which can threaten the integrity of overlying structures. Compensation grouting is a construction technique that is more and more widely used, in conjunction with traditional support systems, to control undesirable ground movements. The conceptual model underlying this technique is simple: local volumetric expansion of the soil is generated by grout injection. The resulting magnitude of ground heave is a function of the grout volume. In theory, ground heave can be adjusted by controlling the grout injection to compensate for the settlement caused by the tunneling process. The soil volumetric expansion can be calibrated on laboratory samples for grouting material of specific properties and injection rate.

Although the mechanism involved is simple and can easily be predicted for simple geometry and structure (i.e., laboratory samples), the global effect of grout injection may be difficult to assess in complex geometrical and structural environments such as those arising at the tunneling site. In general, the location and amount of grout injection are not known *a priori*. Instead, compensation-grouting measures are applied in response to soil and structure displacement data recorded during tunnel construction. In the proposed approach, numerical analysis is first used in the design stage to predict ground movements induced by tunneling. Computer simulations are performed for grouting schemes involving various grouting injection points and volumes until one scheme is found to produce acceptable surface and structure settlements. A typical mesh used in a *FLAC3D* (Itasca, 1997) numerical analysis is presented in Figure 12. An example calculation of ground heave induced by grouting is illustrated in Figure 13.

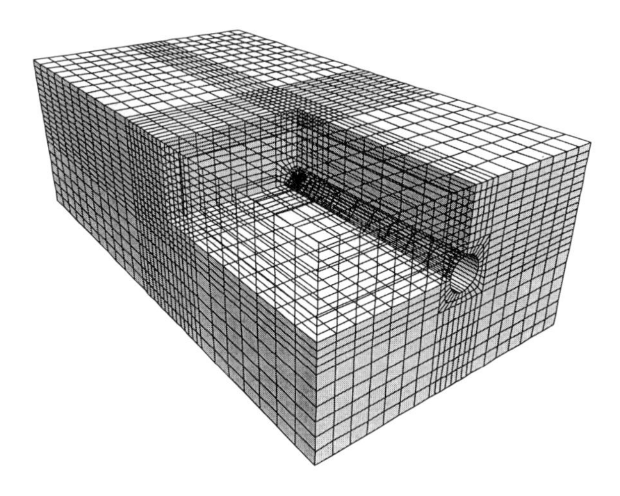

Figure 12. Typical *FLAC3D* grid used in the compensation-grouting analysis (from Buchet and Van Cotthem, 1999)

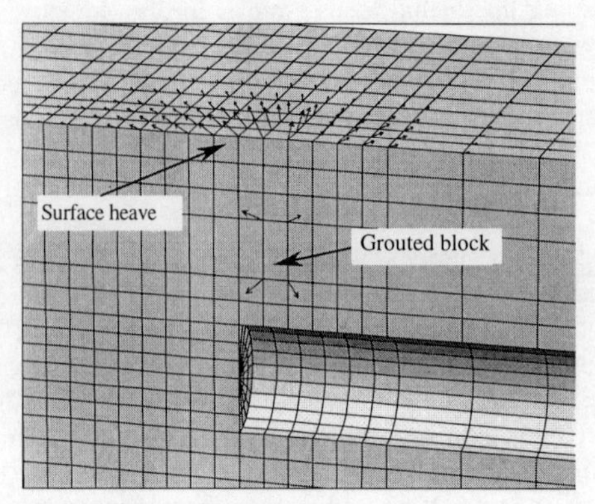

Figure 13. Surface displacement results from compensation-grouting simulation (from Buchet and Van Cotthem, 1999)

The results of the numerical simulations for the initial design are used on site by the grout plant to plan the layout of grouting boreholes and grout volumes. Numerical modeling is then applied to assist in the adjustment of the compensation-grouting operations during the tunnel construction phase. The monitoring of ground surface and structure displacements during tunnel advancement and grouting operations is an important factor to the success of the grouting operation. Daily corrective numerical simulations are conducted to adjust to unanticipated site conditions and provide modifications to the grouting program for the following days.

This study illustrates the procedure in which numerical simulations can assist in applying the mechanism defined by a simple geo-mechanical conceptual model in a complex geometrical and structural environment. This "extended" conceptual model is first used in the project planning stage (initial design) and then reviewed and updated in the construction stage (observational method).

Concluding Remarks

As interaction between geotechnical design engineers and the developers of practical numerical modeling techniques increases, it is inevitable that more "user-friendly" procedures will evolve. Computers will become progressively as much an integral and indispensable part of the geotechnical design process as they are already in many other sectors of engineering and general daily activities.

While encouraging fully such developments, it is emphasized that the uncertainty and vicissitudes of the geotechnical environment remain, and the judgement of the experienced and well-informed geotechnical engineer are still as essential as ever. Numerical modeling simply provides a new, potentially valuable tool, which when used intelligently, can lead to even sounder practical judgement and decisions. There will still be situations in which good decisions can be reached without need of this tool. In recommending numerical modeling as a practical tool in geotechnical design, the authors recall the engineering maxim: *Theory without practice is sterile, Practice without theory is blind!*

References

Board M., Brummer R., Seldon S. and Pakalnis R. (1999), "Analysis of the Failure of a Large Hangingwall Wedge Kidd Mining Division, Falconbridge, Ltd.," submitted to *CIM Bulletin*.

Buchet B. and Van Cotthem A. (1999), "3D 'Steady State' numerical modeling of tunneling and compensation grouting," *Proceedings of the International FLAC Symposium on Numerical Modeling in Geomechanics,* Minneapolis, MN USA, 1-3 September 1999. Detournay & Hart (eds.), Balkema, Rotterdam, pp. 255-261.

D'Elia B., Esu F., Tommasi P. and Utzeri L. (1999), "*FLAC* modeling of the deformation and failure mechanism of a high cut in clay shales," *Proceedings of the International FLAC Symposium on Numerical Modeling in Geomechanics,* Minneapolis, MN USA, 1-3 September 1999. Detournay & Hart (eds.), Balkema, Rotterdam, pp. 47-54.

Itasca (1998), *FLAC,* Fast Lagrangian Analysis of Continua, Version 3.4, Itasca Consulting Group, Inc., Minneapolis, MN.

Itasca (1997), *FLAC3D,* Fast Lagrangian Analysis of Continua in 3 Dimensions, Version 2.0, Itasca Consulting Group, Inc., Minneapolis, MN.

Itasca (1998), *3DEC,* 3 Dimensional Distinct Element Code, Version 2.0, Itasca Consulting Group, Inc., Minneapolis, MN.

Kwaśniewski M.A. and Wang J.-A. (1999), "3D numerical modeling and study of mine tremors induced by coal mining in the vicinity of major faults – A case study," *Proceedings of the International FLAC Symposium on Numerical Modeling in Geomechanics,* Minneapolis, MN USA, 1-3 September 1999. Detournay & Hart (eds.), Balkema, Rotterdam, pp. 379-388.

Terzaghi K. and Peck R. B. *Soil Mechanics in Engineering Practice,* 2nd Edition, John Wiley & Sons Inc., New York, March 1968.

Influence of soil strength spatial variability on the stability of an undrained clay slope by finite elements

D. V. Griffiths [1] and Gordon A. Fenton [2]

Abstract

An investigation has been performed into the stability of an undrained clay slope having spatially randomly varying shear strength. The results of the study lead to a direct comparison between the probability of slope failure and the traditional Factor of Safety for a range of statistically defined input shear strength properties. The results highlight the influence of the spatial correlation length, a variable which is routinely omitted from conventional probabilistic studies in geotechnics.

Introduction

The paper presents results obtained using a program developed by the authors which merges nonlinear elasto-plastic finite element analysis (e.g. Smith and Griffiths 1998) with random field theory (e.g. Fenton 1990, Vanmarcke 1984). Some initial work using this approach has been reported by Paice and Griffiths (1997), however the problem to be considered in this paper is an undrained clay slope ($\phi_u = 0$) of height H with a gradient of 2:1 resting on a foundation layer, also of depth H. A typical finite element mesh is shown in Figure 1.

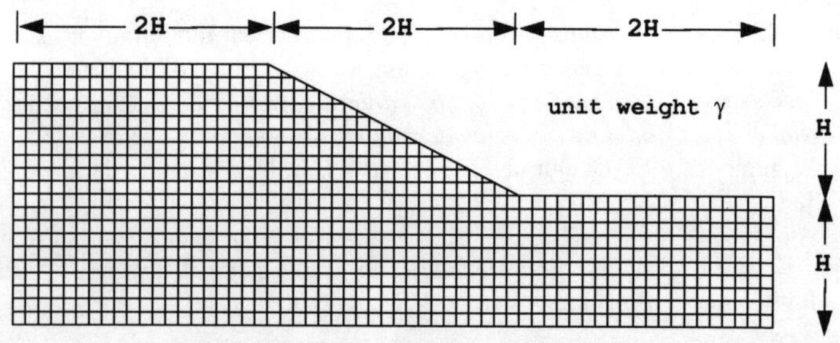

Figure 1: Mesh used for slope stability analyses.

[1] Professor, Geomechanics Research Center, Colorado School of Mines, Golden, CO 80401
[2] Professor, Department of Engineering Mathematics, Dalhousie University, Canada

In this study, the variability of the undrained shear strength (c_u) is assumed to be characterized by a lognormal distribution with the following three parameters:

		Units
Mean	μ_{c_u}	Stress
Standard Deviation	σ_{c_u}	Stress
Spatial Correlation Length	$\theta_{\ln c_u}$	Length

The mean and standard deviation can conveniently be expressed in terms of the dimensionless coefficient of variation defined as

$$C.O.V._{c_u} = \frac{\sigma_{c_u}}{\mu_{c_u}} \tag{1}$$

Since the actual undrained shear field is assumed lognormally distributed, taking its logarithm yields an "underlying" normally distributed (or Gaussian) field. The spatial correlation length is measured with respect to this underlying field, that is, with respect to $\ln c_u$. In particular, the spatial correlation length ($\theta_{\ln c_u}$) describes the distance over which the spatially random values will tend to be significantly correlated in the underlying Gaussian field. Thus, a large value of $\theta_{\ln c_u}$ will imply a smoothly varying field, while a small value will imply a ragged field. The spatial correlation length can be estimated from a set of shear strength data taken over some spatial region simply by performing the statistical analyses on the log-data. In practice, however, $\theta_{\ln c_u}$ is not much different in magnitude from the correlation length in real space and, for most purposes, θ_{c_u} and $\theta_{\ln c_u}$ are interchangeable given their inherent uncertainty in the first place. In the current study, the spatial correlation length has been non-dimensionalized by dividing it by the height of the embankment H.

It should be emphasised that the spatial correlation length is rarely taken into account in routine probabilistic studies relating to geotechnical engineering. In the majority of these cases, a Single Random Variable approach (e.g. Harr 1987, Duncan 2000) is used, which is equivalent to setting $\theta_{\ln c_u} = \infty$.

It has been suggested (see e.g. Lee *et al* 1983, Kulhawy *et al* 1991 and Duncan 2000) that typical $C.O.V._{c_u}$ values for the undrained shear strength lie in the range 0.1-0.5, however the spatial correlation length is less well documented, especially in the horizontal direction, and may well exhibit anisotropy. While the analysis tools used in this study have the capability of modeling an anisotropic spatial correlation field, all the results presented in this paper assume that $\theta_{\ln c_u}$ is isotropic. This is not a severe restriction, since the geometry can often be scaled to achieve the desired spatial correlation structure

Brief description of the finite element method used

The slope stability analyses use an elastic-perfectly plastic stress-strain law with a Tresca failure criterion. Plastic stress redistribution is accomplished using a viscoplastic algorithm which uses 8-node quadrilateral elements and reduced integration in both the stiffness and stress redistribution parts of the algorithm. The theoretical basis of the method is described more fully in Chapter 6 of the text by Smith and Griffiths (1998), and for a discussion of the method applied to slope stability analysis, the reader is referred to Griffiths and Lane (1999).

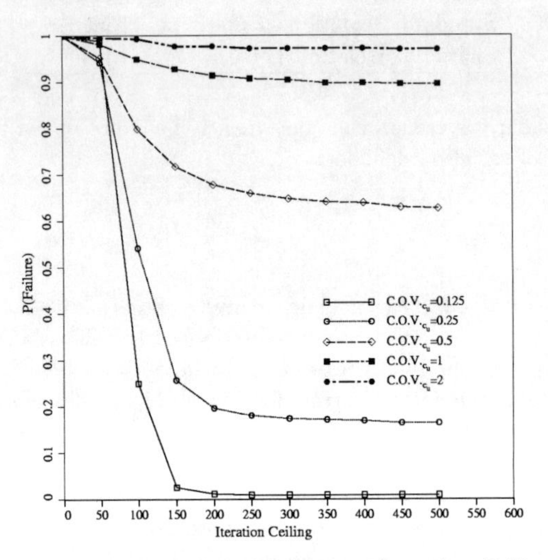

Figure 2: Probability of failure vs. Iteration Ceiling.

In brief, the analyses involve the application of gravity loading, and the monitoring of stresses at all the Gauss points. If the Tresca criterion is violated, the program attempts to redistribute those stresses to neighboring elements that still have reserves of strength. This is an iterative process which continues until the Tresca criterion and global equilibrium are satisfied at all points within the mesh under quite strict tolerances.

In this study,"failure" is said to have occurred if, for any given realization, the algorithm is unable to converge within 500 iterations. Following a set of 1000 realizations of the Monte-Carlo process the probability of failure is simply defined as the proportion of these realizations that required 500 or more iterations to converge.

While the choice of 500 as the iteration ceiling is subjective, Figure 2 confirms, for the case of $\theta_{\ln c_u}/H = 1$, that the probability of failure computed using this criterion is quite stable after about 200 iterations.

Parametric studies

In the parametric studies described in this section, the mean strength expressed in the form of a Stability Number

$$N_s = \mu_{c_u}/(\gamma H), \tag{2}$$

was given the values 0.15, 0.20, 0.25 and 0.30, and in each case, a range of $C.O.V._{c_u}$ and $\theta_{\ln c_u}/H$ values were investigated as follows:

$$\begin{aligned} \theta_{\ln c_u}/H &= 0,\ 0.5,\ 1,\ 2,\ 4,\ 8,\ \infty \\ C.O.V._{c_u} &= 0.125,\ 0.25,\ 0.5,\ 1,\ 2,\ 4 \end{aligned} \tag{3}$$

To put the probabilistic results in context, Table 1 shows the Factor of Safety F from conventional limit equilibrium analysis for the slope in Figure 1 assuming a homogeneous shear strength defined by the Stability Number N_s.

Table 1: Factors of Safety Assuming Homogeneous Soil

N_s	F
0.15	0.88
0.17	1.00
0.20	1.18
0.25	1.47
0.30	1.77

For each set of assumed statistical properties given by $C.O.V._{c_u}$ and $\theta_{\ln c_u}/H$, Monte-Carlo simulations were performed, typically involving 1000 repetitions or "realizations" of the shear strength random field and the subsequent finite element analysis. Each realization of the random field, while having the same underlying statistics, led to a quite different spatial pattern of shear strength values within the slope.

Figure 3 shows two typical random field realizations and associated failure mechanisms for slopes with $\theta_{\ln c_u}/H = 0.5$ and $\theta_{\ln c_u}/H = 2$. Notice how the higher $\theta_{\ln c_u}/H$ gives a more slowly varying shear strength over space and a smoother failure surface.

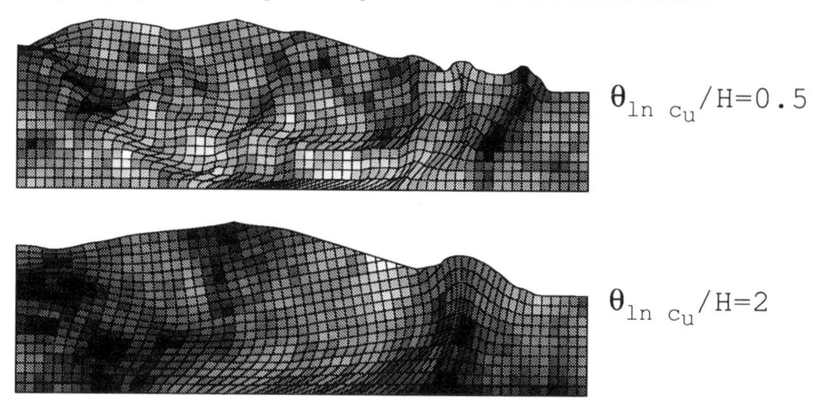

Fig 3. Typical random field realizations. Darker zones indicate weaker soil.

Single random variable approach

It is instructive to consider the special case of $\theta_{\ln c_u} = \infty$, which implies that each realization of the Monte-Carlo process gives a uniform strength, the same everywhere, but with the strength varying from one realization to the next according to the governing lognormal distribution. The probability of failure in such a case is simply equal to the probability that the Stability Number N_s will be below 0.17, the value that would give a Factor of Safety of unity.

For example, if $\mu_{c_u} = 0.25\gamma H$ and $\sigma_{c_u} = 0.125\gamma H$, corresponding to $C.O.V._{c_u} = 0.5$, the statistics of the Stability Number are therefore given by $\mu_{N_s} = 0.25$, $\sigma_{N_s} = 0.125$ and $C.O.V._{N_s} = 0.5$.

¿From standard relationships, the mean and standard deviation of the underlying *normal* distribution of the Stability Number are given by:

$$\sigma_{\ln N_s} = \sqrt{\ln\left\{1 + \left(\frac{\sigma_{N_s}}{\mu_{N_s}}\right)^2\right\}} \tag{4}$$

$$\mu_{\ln N_s} = \ln \mu_{N_s} - \frac{1}{2}\sigma^2_{\ln N_s} \tag{5}$$

hence $\mu_{\ln N_s} = -1.498$ and $\sigma_{\ln N_s} = 0.472$.

The probability of failure is therefore given by:

$$p(N_s < 0.17) = \Phi\left(\frac{\ln 0.17 - \mu_{\ln N_s}}{\sigma_{\ln N_s}}\right) \tag{6}$$

$$= 0.281 \tag{7}$$

where Φ is the cumulative normal distribution function. The relationship between the Factor of Safety (assuming a constant shear strength equal to μ_{c_u}) and the probability of failure assuming a Single Random Variable ($\theta_{\ln c_u} = \infty$) is summarized in Figure 4 for a range of $C.O.V._{c_u}$ values.

Apart from the rather obvious conclusion that the probability of failure goes up as the Factor of Safety goes down, it is also clear that for the majority of cases, the probability of failure also goes up as the $C.O.V._{c_u}$ of the shear strength increases. This result is not necessarily intuitive, since soil with a higher $C.O.V._{c_u}$ contains elements that are much weaker *and* much stronger than the mean. The result indicates however, that the weaker elements dominate the stability calculation.

The only exception to this trend occurs when the mean strength indicates a Factor of Safety of *less* than unity. As shown in Figure 4, the probability of failure in such cases is understandably high, however the role of $C.O.V._{c_u}$ has the opposite effect to that described above, with lowest values of $C.O.V._{c_u}$ tending to give the highest values of the probability of failure.

It is interesting to note that using this approach, a slope with a Factor of Safety of 1.50, based on the mean strength, would have a probability of failure as high as 27% if $C.O.V._{c_u} = 0.5$, the upper limit of the recommended range of Lee *et al* 1983 and others.

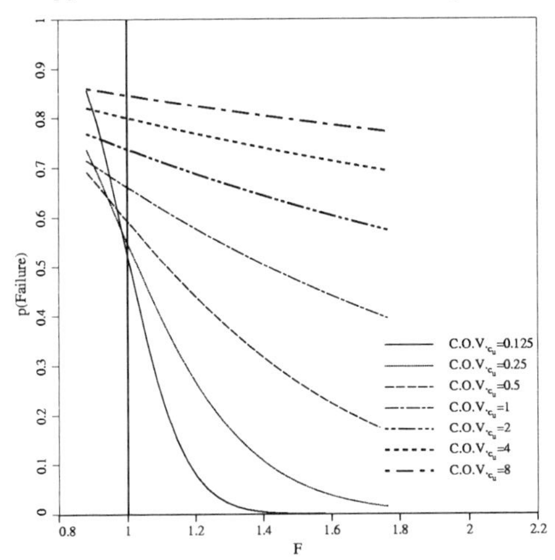

Fig 4. Factor of Safety vs. Probability of Failure. Single random variable approach, $\theta_{\ln c_u} = \infty$

Random field approach

The code developed by the authors enables a random field of shear strength values to be generated and subsequently mapped onto the finite element mesh. In a random field, the value assigned to each cell (or finite element in this case) is itself a random variable, thus the mesh of Figure 1 which has 910 finite elements consists of 910 random variables. The random variables can be correlated to one another by controlling the spatial correlation length $\theta_{\ln c_u}$ as described previously, hence the single random variable approach discussed in the previous section can now be viewed as just a special case of a much more powerful analytical tool.

Figures 5 and 6 show the effect of the spatial correlation length $\theta_{\ln c_u}/H$ on a soil with a Factor of Safety of 1.47 (based on the mean strength) for a range of $C.O.V._{c_u}$ values. Figure 5 clearly indicates two branches relating to the value of $C.O.V._{c_u}$. For low values of $0 < C.O.V._{c_u} < 0.5$, the probability of failure increases as $\theta_{\ln c_u}/H$ increases, indicating that the Single Random Variable approach in which $\theta_{\ln c_u} = \infty$ is conservative. For high values of $1 < C.O.V._{c_u}$ quite the reverse trend is apparent, with the higher values of $\theta_{\ln c_u}$ tending to *underestimate* the probability of failure.

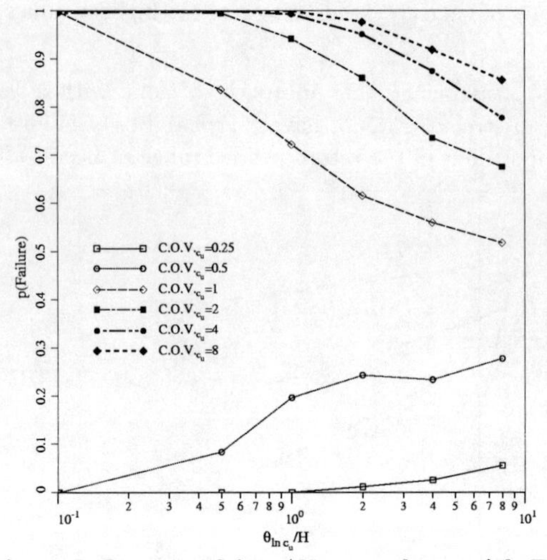

Fig 5. Influence of $\theta_{\ln c_u}/H$ on a slope with F=1.47.

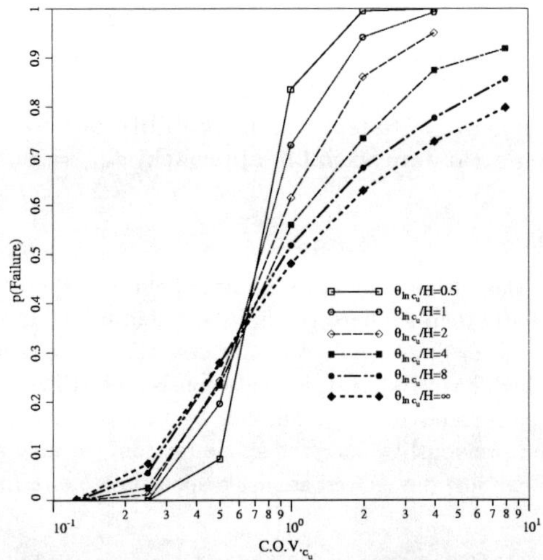

Fig 6. Influence of $C.O.V._{c_u}$ on a slope with F=1.47.

Figure 6 shows an alternative representation of the same data with $C.O.V._{c_u}$ plotted alon[g] the abscissa. This figure shows more clearly how $\theta_{\ln c_u} = \infty$ will tend to overestimate th[e] probability of failure for low $C.O.V._{c_u}$ values and underestimate it for high values. It i[s] also of interest to note the sensitivity of the probability of failure to the value of $C.O.V._{c_u}$

for low levels of correlation. For example, the line corresponding to $\theta_{\ln c_u}/H = 0.5$ rises steeply from zero to 100% probability of failure within the relatively narrow band of $0.25 < C.O.V._{c_u} < 2$. For even smaller values of $\theta_{\ln c_u}/H$ the rise was observed to be even more dramatic, although these results are not presented here. A further point of interest from Figure 6 is that all the lines appear to coincide at approximately the same value of $C.O.V._{c_u} \approx 0.65$, implying that at this level of shear strength variance, the probability of failure is independent of $\theta_{\ln c_u}/H$. This result and others are currently under further investigation by the authors.

The observations made with respect to Figures 5 and 6 were for the particular case of a mean shear strength that would have given a Factor of Safety of 1.47. The results from further analyses of a range of mean shear strength values corresponding to the Stability Numbers in Table 1 are shown in Figure 7. In order to reduce the number of variables, only the results assuming $C.O.V._{c_u} = 0.5$ are shown.

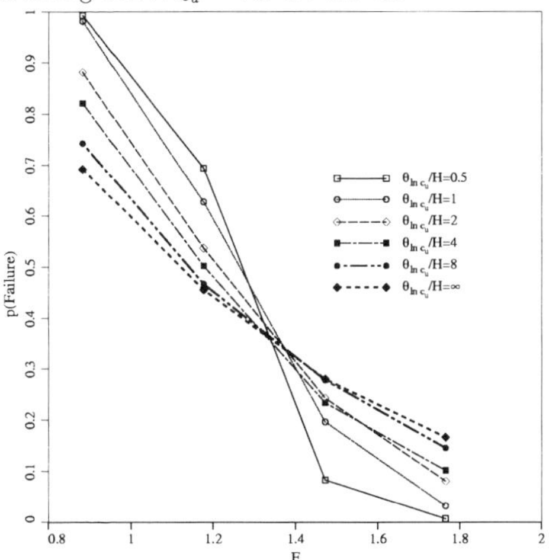

Fig 7. Influence of $\theta_{\ln c_u}/H$ on the Probability of Failure for a range of deterministic Factors of Safety ($C.O.V._{c_u} = 0.5$).

Figure 7 indicates another type of "cross over" with respect to the Factor of Safety. For the given value of $C.O.V._{c_u} = 0.5$, the Single Random Variable approach corresponding to $\theta_{\ln c_u} = \infty$ appears to overestimate the probability of failure for slopes with relatively high deterministic Factors of Safety ($F > 1.4$) and underestimate it for lower Factors of Safety ($F < 1.4$).

Concluding remarks

The paper has shown that soil strength heterogeneity in the form of a spatially varying lognormal distribution can significantly affect the stability of a slope of undrained clay when viewed in a probabilistic context. In this paper, particular attention was paid to the validity of treating the heterogeneity as a Single Random Variable which was shown to be a special case of the authors' formulation corresponding to an infinite correlation length of $\theta_{\ln c_u} = \infty$.

The following more specific observations can be made from the results presented in this paper:

1. For the slope considered in this study with a Factor of Safety of 1.47 based on the mean strength, the Single Random Variable approach gave conservative estimates of the probability of failure for Coefficient of Variation values in the "typical" range of $0 < C.O.V._{c_u} < 0.5$. For higher values of $C.O.V._{c_u}$ however, the Single Random Variable approach gave unconservative estimates.

2. For the slope considered in this study with $C.O.V._{c_u} = 0.5$, the Single Random Variable approach gave conservative estimates of the probability of failure for higher Factors of Safety in the approximate range $F > 1.4$ and unconservative estimates for lower Factors of Safety when $F < 1.4$.

More work remains to be done in this area, but the implications of this study are that the Single Random Variable approach is an acceptable guide to probabilistic slope stability providing the mean strength indicates a relatively high Factor of Safety. For more critical cases, in which the mean strength indicates a Factor of Safety closer to unity, the Single Random Variable approach can give an unconservative estimate of the probability of failure, i.e. *lower* than the "true" value.

A final comment relates to the influence of the Coefficient of Variation of the soil shear strength. While increasing the value of $C.O.V._{c_u}$ introduces both stronger and weaker zones of soil into the slope, the weaker soil always dominates the overall performance leading to a *less* stable slope.

ACKNOWLEDGEMENT

The results shown in this paper form part of a much broader study into the influence of soil heterogeneity on stability problems in geotechnical engineering. The writers wish to acknowledge the support of NSF Grant No. CMS-9877189

References

[1] J.M. Duncan. Factors of safety and reliability in geotechnical engineering. *J Geotech Geoenv Eng, ASCE*, 126(4):307–316, 2000.

[2] G.A. Fenton. *Simulation and analysis of random fields.* PhD thesis, Department of Civil Engineering and Operations Research, Princeton University, 1990.

[3] D.V. Griffiths and P.A. Lane. Slope stability analysis by finite elements. *Géotechnique*, 49(3):387–403, 1999.

[4] M.E. Harr. *Reliability based design in civil engineering*. McGraw Hill, London, New York, 1987.

[5] F.H. Kulhawy, M.J.S. Roth, and M.D. Grigoriu. Some statistical evaluations of geotechnical properties. In *Proc. ICASP6, 6th Int. Conf. Appl. Stats. Prob. Civ. Eng.* 1991.

[6] I.K. Lee, W. White, and O.G. Ingles. *Geotechnical Engineering*. Pitman, London, 1983.

[7] G.M. Paice and D.V. Griffiths. Reliability of an undrained clay slope formed from spatially random soil. In J-X. Yuan, editor, *IACMAG 97*, pages 1205–1209. A.A. Balkema, Rotterdam, 1997.

[8] I.M. Smith and D.V. Griffiths. *Programming the Finite Element Method*. John Wiley and Sons, Chichester, New York, 2nd edition, 1988.

[9] E.H. VanMarcke. *Random fields: Analysis and synthesis*. The MIT Press, Cambridge, Mass., 1984.

Effect of Deterministic and Probabilistic Models on Slope Reliability Index

Ahmed M. Hassan[1] and Thomas F. Wolff[2], M. ASCE

Abstract

Many deterministic and probabilistic models can be applied to reliability analysis of earth slopes. This paper investigates the effect of commonly used models by a set of comparative studies based on three prototype embankments. The deterministic models considered are the simplified Bishop method, the modified Swedish method, and Spencer's method. For Spencer's method, both circular and non-circular failure surfaces are considered. The probabilistic models considered are the mean-value first order second moment method (MFOSM), the point estimate method (PEM), and the advanced first order second moment method (AFOSM). The results indicate that the effect of these different models on the calculated slope reliability index varies with slope geometry and soil strength parameters. The effect of deterministic slope stability method is not significant; but the shape of the failure surface can significantly affect the slope reliability, especially for layered embankments. No practical difference was observed between the results of reliability analyses done by the mean-value First Order Second Moment method (MFOSM) and Point Estimate Method (PEM). For the cases studied the AFOSM gave a more conservative reliability index, but it requires more computational effort due to the iterative nature of slope stability problems and numerical problems may occur.

Introduction

When performing a reliability analysis of an earth slope, a number of deterministic and probabilistic models may be selected. Regarding the deterministic model, most published studies consider only one such model and most commonly consider only circular slip surfaces. In this study, several deterministic methods are compared in

---

[1] Lecturer, Civil Engineering Department, Faculty of Engineering, Minia University, Minia,Egypt
[2] Associate Professor and Associate Dean, College of Engineering, Michigan State University, East Lansing, MI, 48824, 517-355-5128, wolff@msu.edu

terms of their possible effects on the calculated reliability index and the location of the critical failure surface. Four combinations of deterministic slope stability model and failure surface shape were considered:

- The simplified Bishop method.
- The modified Swedish method.
- Spencer's method for circular failure surfaces.
- Spencer's method for non-circular failure surfaces.

The simplified Bishop method is probably the most commonly used method for slope stability analysis. It assumes horizontal interslice forces and satisfies vertical force equilibrium for each slice and moment equilibrium for the entire mass (Bishop and Morgenstern, 1960). It is restricted to circular failure surfaces. The modified Swedish method (Lowe, 1967) satisfies both vertical and horizontal force equilibrium, but not moment equilibrium. Both circular and non-circular failure surfaces may be used. Interslice forces are assumed parallel and their inclination is specified by the user. The U.S. Army Corps of Engineers (1970) version of the method specifies that this inclination be equal to the average slope of the embankment. This gives a higher factor of safety than the assumption of horizontal interslice forces, or some inclination flatter then the slope. However, lesser inclinations are frequently assumed in practice in the interest of conservatism. Spencer's method (Spencer, 1967) is a rigorous method that satisfies both force and moment equilibrium and solves for interslice force inclination assuming all side forces are parallel. The slip surface may be circular or non-circular. The computer program UTEXAS3 (Edris and Wright, 1993) was used for the analyses as it permits one to obtain a solution from all four methods at the same time and the results can be easily compared.

Regarding the probabilistic model, the following models were applied in this study:

- Mean-value first order second moment method (MFOSM)
- The point estimate method (PEM)
- Advanced first order second moment method (AFOSM).

A brief description of each of these models follows.

The Mean-Value First Order Second Moment Method (MFOSM)

In this method the variance of the factor of safety is determined by expanding it in a Taylor series at the point represented by the mean values of shear parameters (Cornell, 1971; Ang and Cornell, 1979; Ditlevsen, 1979; and others). The mean and variance of the factor of safety are given by:

$$E[FS] = \mu_{FS} = FS(\mu_{c_i}, \mu_{\phi_i}) \tag{1}$$

$$\mathrm{Var}[FS] = \sigma_{FS}^2 = \sum\left[\left(\frac{\partial FS}{\partial c_i}\right)^2 \sigma_{c_i}^2 + \left(\frac{\partial FS}{\partial \phi_i}\right)^2 \sigma_{\phi_i}^2\right] + 2\sum\left[\left(\frac{\partial FS}{\partial c_i}\frac{\partial FS}{\partial \phi_i}\right)\sigma_{c_i}\sigma_{\phi_i}\rho_{c_i,\phi_i}\right] \tag{2}$$

where i = 1, 2, …n and $\rho_{ci,\phi i}$ is the correlation coefficient between c_i and ϕ_i. (i is the soil number and n is the total number of soils). In the studies reported herein, no correlation was assumed between strength parameters for different types of soils; however, correlation was permitted between c and ϕ for a common soil.

In this study, the partial derivatives were calculated numerically by evaluating the change in the factor of safety over an increment of plus to minus one standard deviation following Corps of Engineers' practice (U.S. Army, 1992):

$$\frac{\partial FS}{\partial c_1} = \frac{FS_{c_{1+}} - FS_{c_{1-}}}{2\sigma_{c_1}} \tag{3}$$

where FS_{c1+} is the factor of safety obtained by taking c_1 at ($\mu_{c1} + \sigma_{c1}$), and FS_{c1-} is the value obtained by taking c_1 at ($\mu_{c1} - \sigma_{c1}$), with all other variables kept at their mean values. The factor of safety was assumed lognormally distributed, as it is bounded by zero and inifinity, and the reliability index is given by:

$$\beta = \frac{E[\ln FS]}{\sigma_{\ln FS}} \tag{4}$$

The Point Estimate Method (PEM)

To estimate moments of functions of random variables; the point estimate method was developed by Rosenblueth (1975, 1981). A random variable X is represented by two point estimates X_+ and X_- with two probability concentrations P_+ and P_-. These two estimates simulate the probability distribution of the random variable. For independent random variables with symmetrical distributions, the two point estimates are chosen as the expected value plus or minus one standard deviation:

$$X_+ = E[X] + \sigma_X \tag{5}$$

$$X_- = E[X] - \sigma_X \tag{6}$$

and the probability concentrations are:

$$P_+ = P_- = 0.50 \tag{7}$$

To obtain moments of functions of random variables, the function is evaluated for all possible combinations of the point estimates (2^N for N variables). The moments are calculated as the summation of these combinations each weighted by the product of the associated probability concentrations. For example, for two random variables,

$$E[Y^M] = P_{++}Y(X_{1+},X_{2+})^M + P_{+-}Y(X_{1+},X_{2-})^M + P_{-+}Y(X_{1-},X_{2+})^M + P_{--}Y(X_{1-},X_{2-})^M \quad (8)$$

The expected value is determined by substituting M=1 and the variance is determined by substituting M=2 to obtain $E[Y^2]$ and then applying the following identity:

$$Var[Y] = E[Y^2] - (E[Y])^2 \quad (9)$$

For correlated random variables, the joint probability concentrations are related to the correlation coefficient. For two correlated random variables:

$$P_{++} = P_{--} = \frac{1+\rho}{4} \quad (10)$$

$$P_{+-} = P_{-+} = \frac{1-\rho}{4} \quad (11)$$

Rosenblueth (1975) provides a solution for three or more random variables.

The Advanced First Order Second Moment Method (AFOSM)

The advanced first-order second-moment method (AFOSM) was introduced by Hasofer and Lind (1974) and is summarized as follows:

First the reduced variable is introduced:

$$X_i' = \frac{X_i - \mu_{X_i}}{\sigma_{X_i}} \quad (12)$$

Note that the variables, X_i, are assumed uncorrelated; however, if not, Hasofer and Lind (1974) presented an orthogonal transformation of correlated variables to uncorrelated ones. The safe, failure, and limit states may be represented in the space of the reduced variables. Hasofer and Lind (1974) showed that the distance from the origin to the limit state (failure surface) must be greater than the reliability index β. Shinozuka (1983) showed that the point on the failure surface with minimum distance to the origin is the *most probable failure point* and this distance could be a measure of the reliability index.

Expanding the performance function, $g(X_i)$ in a Taylor series at the most probable failure point X^* and keeping only the first-order terms gives the so-called *Hasofer-Lind reliability index*, β_{HL}, as:

$$\beta_{HL} = \frac{\mu_g}{\sigma_g} = \frac{-\sum_{i=1}^{N} X_i^{'*}\left(\frac{\partial g}{\partial X_i^{'}}\right)_*}{\sqrt{\sum_{i=1}^{N}\left(\frac{\partial g}{\partial X_i^{'}}\right)_*^2}} \tag{13}$$

where $\left(\frac{\partial g}{\partial X_i^{'}}\right)_*$ indicates that the partial derivatives are performed at most probable failure point (X^*).

An iterative procedure is required to locate the most probable failure point (X^*) and β_{HL} (see Hasofer and Lind, 1974; Ang and Tang, 1984; Chowdhury and Xu, 1994). The Hasofer-Lind reliability has the advantage of being invariant, i.e. independent of the form of performance function. Hence, β_{HL} values may be directly compared regardless of the format of performance functions they are derived from. The AFOSM is more commonly applied in structural engineering than in geotechnical engineering. However, it has been applied to probabilistic slope stability analysis by Li and Lumb (1987), and Chowdhury and Xu (1994).

Case Studies

Cannon Dam. The structure is located on the Salt River in northeastern Missouri, U.S.A. (Wolff, 1985). The dam has a 1000 ft (305 m) long earth embankment, a gated concrete spillway section and a concrete powerhouse. The cross-section of the earth-fill embankment is 139 ft (42.4 m) high (Figure 1). The embankment consists of two compacted clay zones, Phase I and Phase II, over a 15 ft thick sand foundation which overlays a thick limestone stratum. Phase I includes a cutoff trench through the sand down to limestone. A vertical sand chimney drain is located in the Phase II clay. The Phase I fill was compacted more wet of optimum than had been foreseen in the design, leading to an adjustment in the placement moisture conditions of Phase II. The structure was previously analyzed by Wolff (1985), and Wolff, et. al. (1995).

Shelbyville Dam. The structure is located on the Kaskaskia River in central Illinois, USA, about 100 miles (160 km) northeast of St. Louis, Missouri (Humphrey and Leonards, 1985). The dam is a combined earth-fill and concrete gravity structure with a total length of 3,392 ft (1,034 m). The main earth embankment cross-section is shown in Figure 2. Its maximum height is 108 ft (33 m) and consists of a homogeneous section with upstream and downstream berms, an inclined chimney drain and a horizontal drainage blanket. The embankment is constructed over a thin sand layer resting on a firm rock foundation. A cutoff trench extends through sand to the rock foundation. The structure was of interest due to a slide during construction and was previously analyzed by Humphery and Leonards (1984, 1986), Wolff (1985), and Hassan (1998).

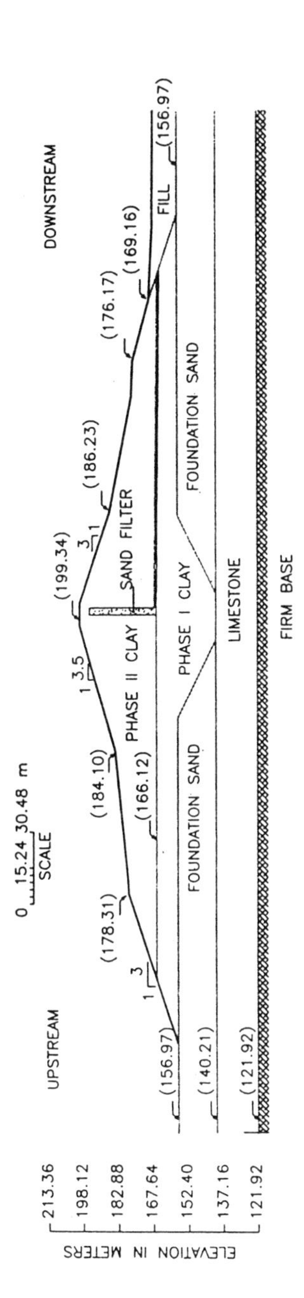

Figure 1 Cross-Section of Cannon Dam

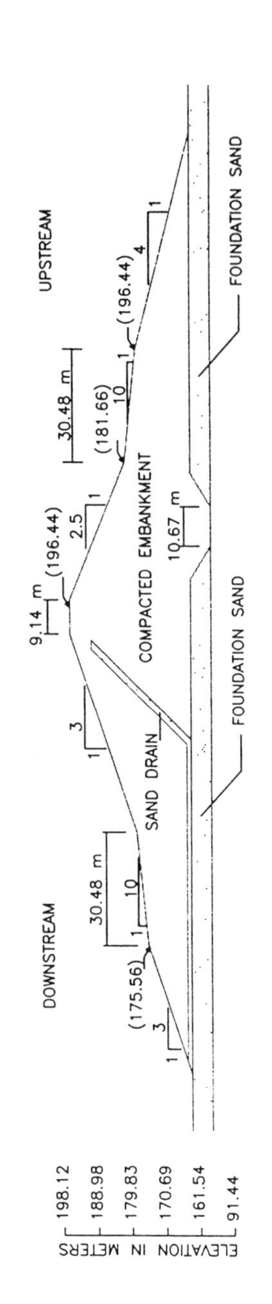

Figure 2 Cross-Section of Shelbyville Dam

Bois Brule Levee. The Bois Brule Levee is located along the west bank of the middle Mississippi River in Perry County, Missouri, USA (Wolff et. al., 1995). It was analyzed as representative of a typical hydraulic structure for which reliability assessment may be required in Corps of Engineers economic studies. The levee encloses a rectangular area approximately 15 miles long and 2 miles wide. A cross-section adjacent to Bois Brule pumping station was analyzed. This cross-section is shown in Figure 3.

Strength Parameters

Probabilistic moments of the strength parameters for the three structures were taken from previously published values summarized in the report by Wolff, et. al. (1995), and are listed in Tables 1 through 3.

Table 1 Cannon Dam, Moments of Soil Strength Parameters (Total Stress Characterization)

Material	*Parameter*	*Mean*	*Standard Deviation*	*Coefficient of Variation*	*Correlation Coefficient*
Phase I clay fill	c_1	117.79 kN/m^2	58.89 kN/m^2	50 %	+0.10
	ϕ_1	8.5°	8.5°	100 %	
Phase II clay fill	c_2	143.64 kN/m^2	79.00 kN/m^2	55 %	-0.55
	ϕ_2	15°	9°	60 %	

Table 2 Shelbyville Dam, Moments of Soil Strength Parameters (Total Stress Characterization)

Material	*Parameter*	*Mean*	*Standard Deviation*	*Coefficient of Variation*
Embankment clay	c_1	113.00 kN/m^2	37.25 kN/m^2	33 %
	ϕ_1	9.6°	7.7°	80 %
Sand Foundation	ϕ_2	32°	2°	6.25 %

(correlation coefficient was taken as zero)

Table 3 Bois Brule Levee, Moments of Soil Strength Parameters (Total Stress Characterization)

Material	*Parameter*	*Mean*	*Standard Deviation*	*Coefficient of Variation*	*Correlation Coefficient*
Embankment clay	c_1	41.18 kN/m^2	28.73 kN/m^2	69.8 %	+ 0.50
	ϕ_1	18°	4.5°	25 %	
Foundation clay	c_2	47.88 kN/m^2	38.31 kN/m^2	80 %	+0.70
	ϕ_2	18°	6°	33.3 %	
Foundation sand	ϕ_3	32°	2°	6.25 %	

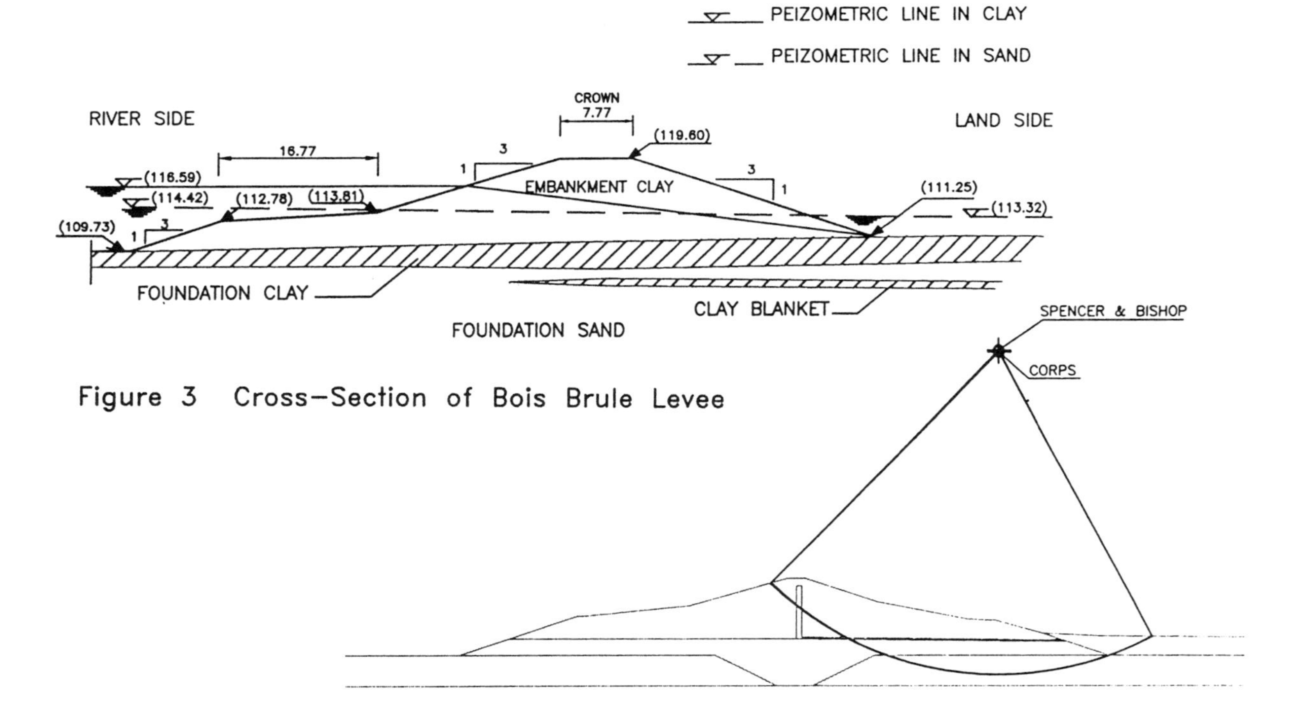

Figure 3 Cross-Section of Bois Brule Levee

Figure 4 Cannon Dam, Circular Slip Surfaces

The probabilistic moments of the strength parameters for Cannon Dam and Shelbyville Dam were obtained from UU tests on record samples cut from the embankments during construction; those for the Bois Brule levee were obtained from undisturbed piston samples on the completed embankment many years after construction. The presence of both c and ϕ components reflects the linear approximation of the curved strength envelope for partially saturated soils.

Slope Stability Analyses

Slope stability analyses were performed using the computer program UTEXAS3 (Edris and Wright, 1993). Before performing probabilistic analyses, a deterministic analysis was performed to locate the surface of minimum factor of safety. This surface will be referred to as the *critical deterministic surface* throughout the present paper. This surface was located with the soil parameters set at their mean values. As shown previously, probabilistic methods require using different combinations of parameter values to numerically calculate the required partial derivatives. It may be noted that Hassan and Wolff (1999) developed an algorithm using parameter combinations weaker than the mean to estimate the location of the critical probabilistic surface (of minimum β).

For circular surfaces, UTEXAS3 has its own search algorithm which starts with a user-defined initial surface, proceeds to change either the center or the radius of the surface in a systematic manner, and each time calculates a factor of safety. The process ends when a minimum factor of safety is found, or a certain number of iterations are reached. A similar search process is used by the program for non-circular surfaces which also begins with a user-defined surface. The program then systematically changes the coordinates of the failure surface segments and calculates a factor of safety each time until a minimum value is reached. The process for both circular and non-circular surfaces depends on the specified initial surface and hence several initial surfaces should be examined to ensure than a minimum factor of safety is actually achieved. For the three embankments considered, different initial slip surfaces were specified and the search process was repeated until a surface of minimum factor of safety was located corresponding to each deterministic model analyzed using mean strength values.

Figure 4 shows critical circular surfaces for Cannon Dam corresponding to Spencer, Bishop, and Modified Swedish methods. The non-circular critical surface is shown in Figure 5. It may be noted that the three circular slip surfaces almost coincide, although they correspond to different slope stability methods. For Shelbyville Dam, Figure 6 shows three circular critical surfaces corresponding to Spencer, Bishop and Modified Swedish methods, and the non-circular critical surface is shown in Figure 7. Again, the three circular surfaces almost coincide, which suggests that the critical deterministic surface is not significantly sensitive to the deterministic model.

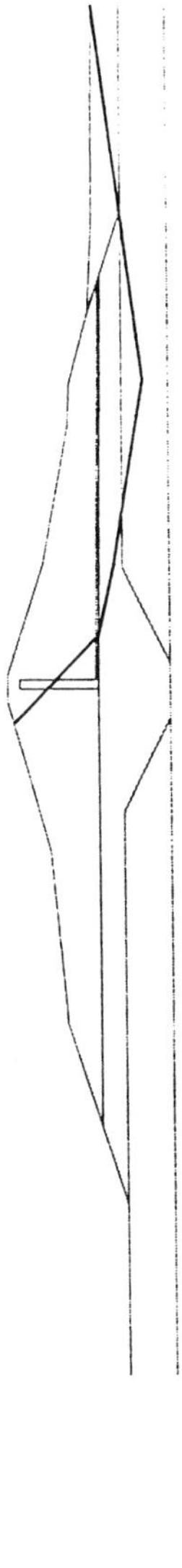

Figure 5 Cannon Dam, Non-Circular Failure Surface

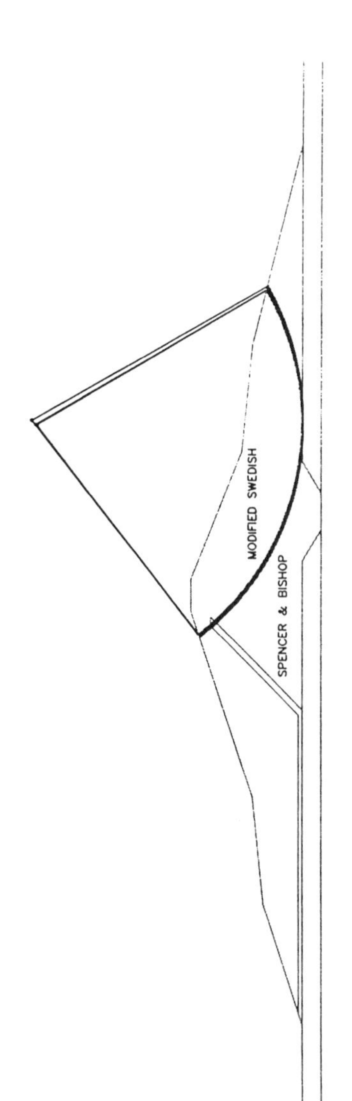

Figure 6 Shelbyville Dam, Circular Failure Surfaces

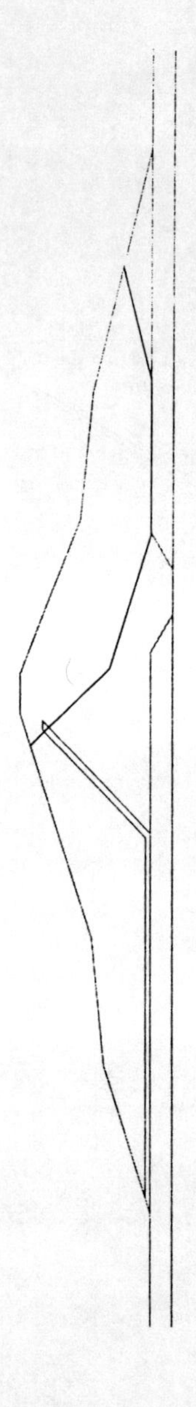

Figure 7 Shelbyville Dam, Critical Non-Circular Surface

The values of factor of safety for each case are given in the following Table 4.

Table 4 Factor of Safety Corresponding to Different Slope Stability Methods

Structure	*Spencer circular*	*Simplified Bishop*	*Modified Swedish*	*Spencer non-circular*
Cannon Dam	2.775	2.753	2.789	2.647
Shelbyville Dam	3.115	3.118	3.117	3.071
Bois Brule Levee	2.552	2.538	2.577	2.718

It can be observed that, for each embankment, factors of safety are similar and the influence of the deterministic model on the factor of safety and the location of the critical deterministic surface is not significant. The non-circular slip surface gives a somewhat smaller factor of safety for the two dams, and a somewhat greater value for the levee.

Reliability Analysis

Each of the four critical deterministic surfaces for each of the three structures was analyzed using the MFOSM. In addition, the four surfaces for the two dam structures were analyzed using the PEM. Finally, selected analyses were performed using the AFOSM. Results are shown in Table 5.

Table 5 Comparison Between β values from MFOSM, PEM, and AFOSM Analyses

		Deterministic Model			
Structure	*Probabilistic Model*	*Spencer Circular*	*Bishop Circular*	*Corps Circular*	*Spencer Non-Circular*
Cannon Dam	MFOSM	10.85	10.36	10.28	7.03
	PEM	11.34	10.83	10.74	7.39
	AFOSM	--	--	--	did not converge
Shelbyville Dam	MFOSM	3.40	3.40	3.43	3.35
	PEM	3.39	3.40	3.43	3.35
	AFOSM	1.67	--	--	--
Bois Brule Levee	MFOSM	3.78	3.73	2.97	3.81
	PEM	--	--	--	--
	AFOSM	did not converge	--	--	--

It may be noted that the AFOSM yielded a β value significantly lower than MFOSM and PEM for Shelbyville Dam. The method did not converge for Cannon Dam because the failure surface passes through the sand layer for which ϕ was taken as a deterministic value. The failure point could not be located using the other four parameters (c_1, ϕ_1, c_2, ϕ_2) because the strength ϕ for the sand alone was sufficient to provide a factor of safety greater than one. In other words, no values of the random variables correspond to the limit state. Hence, it was learned that the practice of taking certain random variables with inherently low variability as deterministic variables may not be appropriate in the AFOSM.

Conclusions

Four deterministic models and three probabilistic models were used to calculate the reliability index of three actual embankments. For the cases studied, the selection of deterministic model did not have a significant effect on the results. The non-circular slip surface provided a more critical β than the circular failure surfaces for Cannon and Shelbyville Dams. If extensive and consistent reliability analyses are to be performed; several deterministic models should examined. Generally, a non-circular slip surface should be considered, especially in case of layered earth embankments.

For the structures studied, no practical difference was observed between the results of reliability analyses done by the mean-value First Order Second Moment method (MFOSM) and Point Estimate Method (PEM). The MFOSM may be preferred due to its simplicity.

The AFOSM has the advantages of being invariant and independent of the form of the performance function and providing a more conservative β. However, due to the iterative nature of slope stability problems, use in practice (the thrust of this work) is cumbersome. Furthermore, numerical problems may occur as shown in case of Cannon Dam.

A quite large difference in β may be obtained depending on the shape of the slip surface considered. The analyses presented in this section gave a β ranging from 11.34 to 7.03 calculated for the same section of Cannon Dam. Further analyses in Hassan and Wolff (1999) showed that an even smaller value of β may be obtained for the same section when searching for the surface of minimum β. Hence, these findings suggest that intensive reliability analysis may be necessary especially if the results are related to economical decisions.

Shelbyville Dam was less sensitive to failure surface shape than was Cannon Dam, apparently because the critical surface occurs entirely within one embankment zone. This suggests that the sensitivity to deterministic and probabilistic models varies from structure to another.

Acknowledgments

Data and some computer models used in this study were from several research efforts sponsored by the U.S. Army Corps of Engineers. Nothing herein should be considered an endorsement by the Corps of Engineers.

Appendix I. References

Ang, A.H.-S., and Cornell, C.A. (1979). "Reliability Bases of Structural Safety and Design." *J. of the Struct. Div.*, ASCE, 100, 10777-1769.

Ang, A.H.-S., and Tang W.H. (1984). *Probability Concepts in Engineering Planning and Design, Volume II : Decision, Risk, and Reliability*, John Wiley and Sons, New York.

Bergado, D.T., Anderson, L. R. (1985). "Stochastic Analysis of Pore Pressure Uncertainty for the Probabilistic Assessment of The Safety of Earth Slopes." *Soils and Foundations*, 25(2), 87-105.

Bishop, A.W., and N. Morgenstern (1960). "Stability Coefficients for Earth Slopes." *Geotechnique*, London, England, 10(4), 129-150.

Chowdhury, R.N., and Xu, D.W. (1994). "Slope System Reliability with General Slip Surfaces." *Soils and Foundations*, Japanese Society of Soil Mech. and Found. Engrg, 34 (3),99-105.

Cornell, C.A. (1971). "First-Order Uncertainty Analysis of Soils Deformation and Stability." *Proc. of The Int. Conf. on Applications of Statistics and Probability to Soil and Struct. Engrg.*

Ditlevsen, O. (1979). "Generalized Second Moment Reliability Index." *J. of Struct. Mech.*, ASCE, 7(4), 435-451.

Edris, E.V., Jr., and Wright, S.G. (1993). "User's Guide: UTEXAS3 Slope Stability Package: Volume 4- User's Manual." *Instruction Report GL-xx-x*, U.S. Army Engineer Waterways Experiment Station, Vicksburg, MS.

Hassan, A.M. (1998). "A Practical Approach to Combined Probabilistic Analysis of Slope Suability and Seepage Problems," Ph.D. Dissertation, Michigan State University, East Lansing MI.

Hassan, A.M. and Wolff, T.F. (1999) "Search Algorithm For Minimum Reliability Index of Earth Slopes" *J. of Geotechnical and Geoenvironmental Engineering*, ASCE 125(4), 301-308.

Hasofer, A.A., and Lind, A.M. (1974). "Exact and Invariant Second- Moment Code Format." *J. Engrg Mech. Div.*, ASCE, 100(EM1), 111-121.

Humphery, D.N., and G.A. Leonards (1984). "A Slide in The Upstream Slope of Lake Shelbyville Dam," Report to St. Louis District, U.S. Army Corps of Engineers.

Humphery D.N., and G.A. Leonards (1986). "Slide in Upstream Slope of Lake Shelbyville Dam." *J. Geotech. Engrg*, ASCE, 112(5), 564-577.

Li, K.S., and Lumb, P. (1987). "Probabilistic Design of Slopes." *Can. Geotech J.*, Ottawa, Canada, 24(4), 520-535.

Lowe, J. (1967). "Stability Analysis of Embankments." *J. Soil Mech. and Found. Div.*, ASCE, 93(SM4), 1-34.

Spencer, E. (1967). "A Method of Analysis of The Stability of Embankments Assuming Parallel Inter-Slice Forces.", *Geotechnique*, London, England, 17(1), 11-26.

Shinozuka, M. (1983). "Basic Analysis of Structural Safety." *J. of Struct. Engrg*, ASCE, 109(3), 721-740.

U.S. Army (1970). "Engineering and Design, Stability of Earth and Rockfill Dams." *Engineer Manual 1110-2-1902*, Department of the Army, Corps of Engineers, Office of the Chief of Engineers, Washington, DC.

U.S. Army (1992). "Reliability Assessment of Navigation Structures." *Engineer Technical Letter No. 1110-2-532*, Department of the Army, US Army Corps of Engineers, Washington, DC.

Wolff, T.F. (1985). "Analysis and Design of Embankment Dam Slopes: A Probabilistic Approach." Ph.D. Dissertation, Purdue University, West Lafayette, IN.

Wolff, T. F., Hassan, A., Khan, R., Ur-Rasul, I, and Miller, M. (1995). "Geotechnical Reliability of Dam and Levee Embankments," technical report prepared to U.S. Army Engineer Waterways Experiment Station Geotechnical Laboratory, Vicksburg, MS.

Appendix II. Notation

The following symbols are used in this paper:

c cohesion
FS factor of safety
i soil number
n number of soils
V_X coefficient of variation of X
X random variable
β reliability index
ϕ friction angle
μ mean value
ρ correlation coefficient
σ standard deviation

SHEAR STRENGTH EVALUATION OF CLAY-ROCK MIXTURES

Anthony T. Iannacchione[1] and Luis E. Vallejo[2]

ABSTRACT

At present, there is little knowledge concerning the shear strength of clays containing floating rock particles with concentrations from 0 to 30%. In practice, the effect of rock particles is typically disregarded in shear strength analysis. The two primary reasons for this are: 1) a lack of agreement concerning the influence of rock particles on material strength, and 2) the expense and difficulty of testing clay-rock mixtures with existing laboratory procedures. These factors have impeded the development of successful slope remediation design techniques for colluvium derived from resistant sedimentary rocks or spoil produced from surface mining. This study reviewed 31 technical papers which contain analysis of shear strengths for clay and sands with varying mixtures of rock particles. These technical papers, published over the last 40 years, are comprised of field case studies, laboratory evaluations, and theoretical analysis. Evaluation of this body of knowledge has shown that the shear strength gradually increases with increasing percentages of floating particles in unsaturated clays.

1.0 INTRODUCTION

One of the most important design input parameter needed for geotechnical design is soil's shear strength. The shear strength is commonly defined by the Mohr-Coulomb failure envelope:

$$\tau = \sigma_n \tan\phi + C \tag{1}$$

where τ = Shear strength,
σ_n = Normal stress,

[1]Deputy Director, National Institute for Occupational Safety and Health, P.O. Box 18070, Pittsburgh, PA.

[2]Assistant Professor, University of Pittsburgh, 949 Benedum Hall, Pittsburgh, PA.

ϕ = Angle of shearing resistance,
C = Cohesion.

The characteristic soil shear strength is defined by the angle of shearing resistance (ϕ) and the cohesion (C). For soils with varying percentages of rock particles, standard procedures for developing shear strength parameters are complicated and are not universally accepted in practice. In fact, examination of common design practices indicates that most slope design efforts disregard the rock particles when determining site-specific shear strength parameters for the soil (Iannacchione et al. 1994).

Soils derived from steep slopes in the central Appalachian Basin are often a clayey sand or a sandy clay. The percentage concentration of rock particles found in these slopes typically ranges from 10 to 30% (Iannacchione and Vallejo, 1995). This range produces rock particle arrangements which reduce contact to such a degree as to allow the oversized particles to "float" in the clay matrix.

Generally, a rock particle is considered oversized if it is normally discarded in a laboratory shear strength test. Numerous authors have discussed appropriate ratios among testing vessels and maximum particle size. Rathee (1981) examined several of these studies and found that recommended ranges of maximum particle size to testing vessel dimension varied from 1/5 to 1/40. Typical triaxial tests reported in most geotechnical investigations are 7.1 cm in diameter, the maximum particle size accommodated ranges from 0.8 cm to 1.4 cm. Therefore, particles within the gravel and above category are typically classified as oversized and discarded during standard laboratory testing, with smaller particles making up the soil matrix.

2.0 ROCK PARTICLES IN CONTACT

The strengths of cohesionless (granular) soils, whether wet or dry, are most dependent upon the frictional properties of the material (ϕ). Granular frictional properties are affected by the surface roughness and interlocking characteristics and the size, shape, and strength of the particles. Leps (1970) sought to explain these factors from the evaluations of a large database of triaxial tests on rock fill dam materials, and found a linear relationship between effective angle of shearing resistance (ϕ') and effective normal stress (σ_n'). These data showed that materials at low confining stresses have more strength than at high confining stresses. This was due to the dilation of the material at low effective normal stress and significant crushing of contact points with reduced dilation at high stress.

The effects of surface roughness were evaluated by Vallerga et al. (1957), with glass beads sheared under equal compactive effort. Beads with etched surfaces showed a considerable increase in the internal angle of friction. Conversely, low shear strength cohesionless soils are loose, with grains of round shape and a smooth surface. Density of these materials is affected by many factors, including gradation of the soil and confining stress.

The strength characteristics of rock particles in contact with their neighbors have been extensively studied in conjunction with the widespread use of rock fills in dams and embankments in the 1960's and 70's (Marachi et al., 1972). Leslie (1963) reviewed a significant volume of artificially generated gradation relationships from gravelly soils and found that the highest values of the angle of shearing resistance were obtained from the densest sample with the largest maximum size particles. The authors also found that for any given porosity, the more uniform samples with smaller maximum sizes had higher values of the internal angle of friction. Marachi et al. (1972) observed that several factors influenced particle crushing, including: 1) increased water content, 2) increased uniformity, 3) increased angularity, 4) reduced particle strength, 5) increased effective confining pressure, 6) increased shear stress under a given confining pressure, 7) testing in a triaxial cell as compared to plane strain testing, and 8) increased particle sizes.

Marsal (1967) also observed that increased particle sizes reduced shear strength. However, Leussink (1965) disputed Marsal's testing approach and research conclusions, indicating that his studies had found a linear relationship between strength and porosity. In another study, Morgan and Harris (1967) concluded that there were no significant strength increases due to increased maximum particle size.

3.0 COHESIVE SOIL WITH ROCK PARTICLES

The strength of clayey soil is influenced by void ratio, composition, and angle of shearing resistance. The degree of saturation also plays a significant role in strength determination. The composition characteristics of cohesive soils are defined in terms of plasticity, where higher Plasticity generally yield lower angles of shearing resistance.

The earliest reference to laboratory-generated shear strength data from clay-rock mixtures was by Hall (1951). This study focused on the development of a triaxial apparatus for testing large soil specimens of at least 30.5 cm. Several specimens were tested which ranged from clayey sandy gravel to gravel. No conclusions were made concerning the influence of rock particles, but examination of the data clearly shows an increase in strength with increasing gravel content.

The influence of varying concentrations of rock particles on the shear strength of cohesive soil-rock mixtures was first investigated by Miller and Sowers (1957). These tests were carried out on consolidated, undrained triaxial specimens of remolded river sand and sandy clay from a decomposed gneiss. Sand versus clay mixtures ranged from 0 to 100%. Each specimen was compacted to its maximum dry density and optimum moisture contents. The experiments showed that increases in cohesionless material up to 67% had no effect upon the angle of shearing resistance but there was a gradual decrease in the cohesion of the sample (Figure 1). Between 67 and 74% cohesionless soil, the internal angle of friction increased and the cohesion decreased significantly. Beyond 74%, the internal angle of friction rose at a gradual rate. Miller and Sowers (1957) concluded that

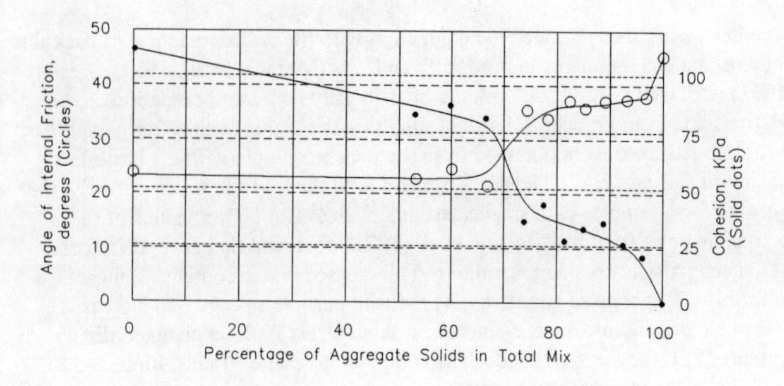

Figure 1. Relationship between aggregate percentage and angle of shearing resistance and cohesion for a cohesive soil (Miller and Sowers, 1957).

the dramatic changes in shear strength between 67 and 74% cohesionless material were a result of the granular structure controlling strength at the expense of a clayey matrix. Unfortunately, these tests were performed at relatively high effective normal stress (>200 kPa) and under optimum compaction conditions. The Mohr-Coulomb failure envelopes shown in Figure 2 indicate a reversal of trends at low confinements. However, this phenomenon seems unlikely and is probably more a result of the authors using a linear failure envelope than a physical reality.

Holtz and Ellis (1961) formulated a testing program to evaluate the effect of gravel content on shear strength for partially saturated materials containing particles up to 7.6 cm in size. This research found that shear strength did not

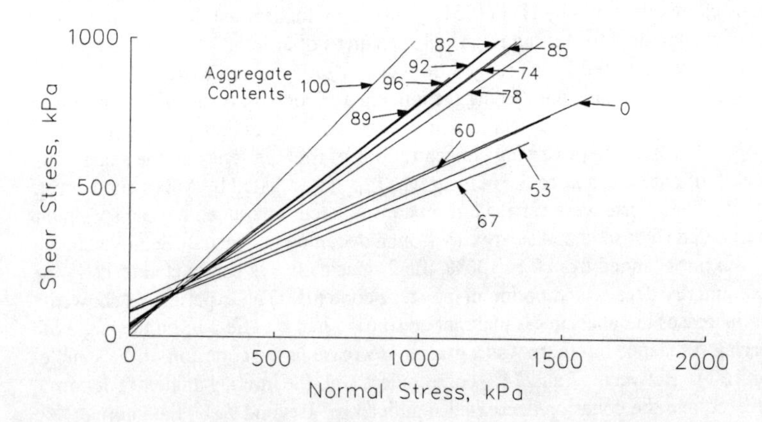

Figure 2. Shear strength data from Miller and Sowers (1957) showing the increase in angle of shearing resistance with increasing gravel content.

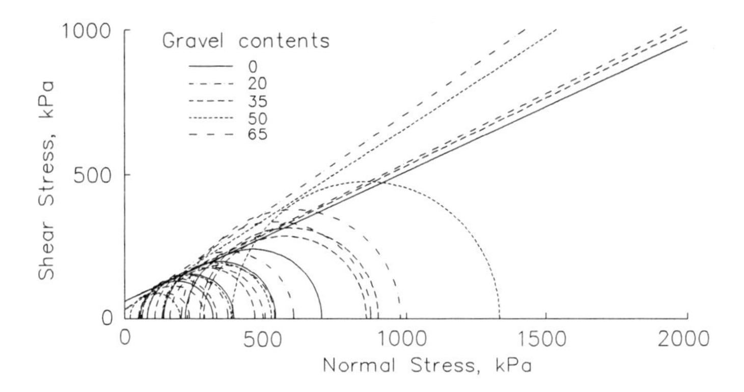

Figure 3. Shear strength data from Holtz and Ellis (1961) for cohesive soil with increasing gravel content.

significantly change for gravel contents up to about 35% (by weight). Beyond 35%, shear strengths increased significantly to about the 50% range (Figure 3). However, these data reveal an unrealistic reversal of the angle of shearing resistance at low normal stress conditions (<100 kPa).

Dobbiah et al. (1969) expanded on earlier work by examining the influence of maximum particle size on shear strength. The soil used in these tests contained mixtures of clay, silt, and sand. The authors found that increases in gravel sizes produced increases in shear strength. The density of the experimental clay-rock mixtures reached a maximum at approximately 50% gravel content, then decreased rapidly with increasing gravel concentrations. At this point, particle contact must have dominated due to the limited availability of clay within the available void spaces.

Patwardhan et al. (1970) found that considerable shearing

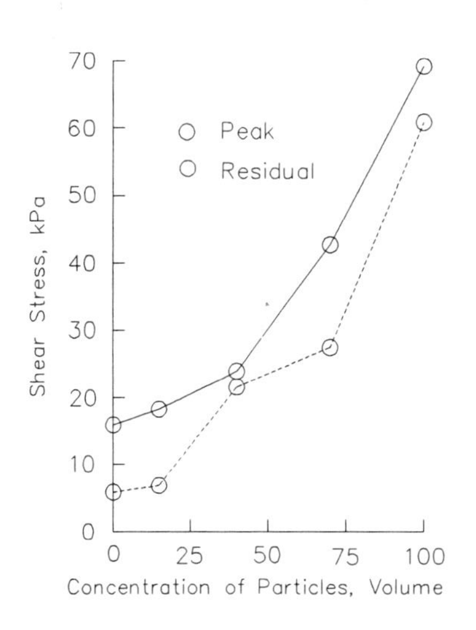

Figure 4. Shear stress values versus rock particle concentration for a clay-boulder/cobble mixture report by Patwardhan, et. al (1970).

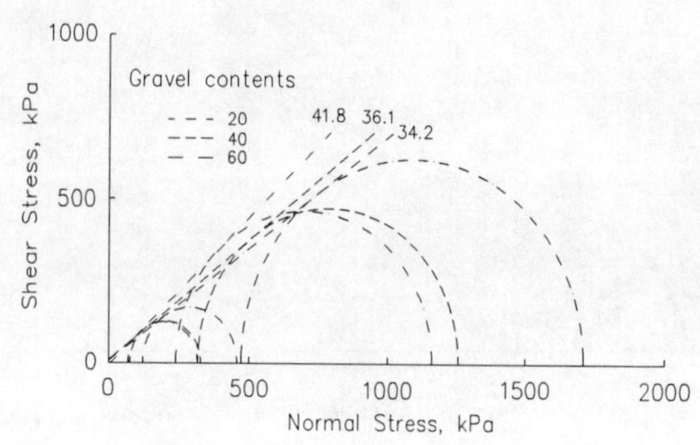

Figure 5. Shear strength data from a sand-clay soil with gravel size particles from Donaghe and Torrey (1979).

resistance was mobilized in assemblages of boulders/cobble-clay mixtures. Tests were performed in a 91-by-91-by-153 cm direct shear box using boulders with an average size of 15 cm. These test conditions produced a ratio (d_{max}/d) of about 1/6 on average, certainly close to the lower acceptable range for a direct shear test. Both the boulders and the clay came from a weathered basalt formation. Samples were saturated prior to testing and were loosely compacted, achieving pre-testing void ratios of 0.7 to 0.8. No vertical confinement was applied. The shearing resistances measured in this research were in the range of 8 to 70 kPa (Figure 4) and are much more representative of low consolidation stress conditions than the results reported above.

In another study, Rico and Orozco (1975) tested the reaction of varying concentrations of fines added to a sandy gravel material of mixed granitic and volcanic rocks. The fines material was taken from a commercial kaolinite and bentonite source and was classified as CL-ML material. Each sample was dynamically compacted at optimum water content and tested in the undrained state. In general, the undrained strength increased with increases in the fines up to about 5 to 10% depending on the type of matrix material, but then decreased sharply to a value below that of the aggregate-only soil.

The Army Corps of Engineers (Donaghe and Torrey, 1979) assessed the effects of both scalping and replacement methods (to be described later) on the shear strength of soil/rock mixtures. The tests were carried out as consolidated undrained triaxial tests on 38.1 cm specimens of gravel-sand-clay mixtures. The specimens were compacted to 95% of their standard compaction maximum dry density. Gravel sizes ranged from 0.4 to 7.6 cm and were tested at concentrations of 20, 40, and 60%. Here again, the effective angle of shearing resistance (ϕ') increased with increasing gravel contents (Figure 5).

Vallejo (1989) discussed the occurrence of large particles in rock fill dams, glacial tills, mud flows, debris flows, solifluction sheets, and residual soil deposits. Vallejo and Zhou (1994) examined consolidation and stability characteristics of simulated soil-rock mixtures. From testing mixtures of kaolinite clay with glass beads or sand, the author found that the percentage of the granular phase in the mixtures has a marked influence on the compression index, coefficient of permeability, and shear strength of the mixtures. It was further found that when the volume of concentration of the sand varied between 80 and 100%, the shear strength of the mixtures was governed mainly by frictional resistance between the sand (Figure 6). When the concentration of sand varied between 50 and 80%, the shear strength of the mixture was provided in part by the shear strength of the kaolinite clay and in part by the frictional resistance between the sand grains. When the sand concentration was less than 50%, the shear strength of the mixture was entirely dictated by the strength of the clay.

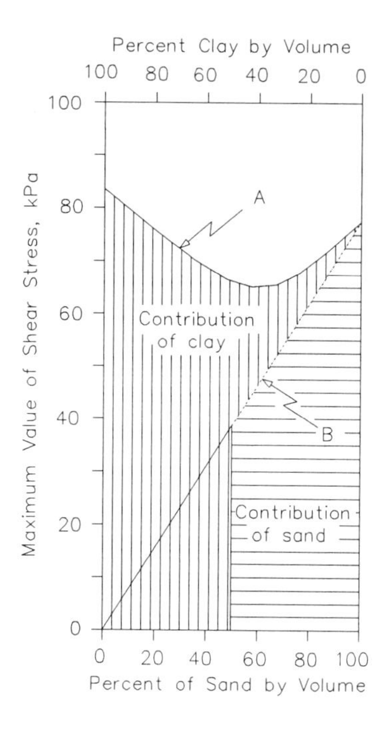

Figure 6. Shear strength of mixtures of clay and sand under a normal stress of 150 kPa (Vallejo and Zhou, 1994).

4.0 METHODS FOR DETERMINING SHEAR STRENGTH OF SOIL-ROCK MIXTURES

Several methods have been developed for determining the shear strengths of soil-rock mixtures. These methods can be grouped into four general categories: 1) Back-analysis, 2) Physical properties alteration, 3) Empirical, and 4) Analytical. The back-analysis method uses actual geometric properties of failed slopes to identify ranges of material properties which could produce these failures. The physical properties alteration method relies on adjustments to test samples to account for missing oversized rock particles. Analytical strength methods adjust the strength formulas, while empirical methods rely on past experience or large databases to assign shear strength parameters. A more detailed analysis of each method follows.

4.1 Back-analysis methods

The back-analysis method has the distinct advantage of being used with in situ conditions. Geometric conditions--such as 1) slope angle, 2) material type, thickness, and density, and 3) location of failure plane--are known inputs into standard slope stability programs. Parametric studies are then performed by estimating the shear strength parameters and checking for a slope safety factor equal to one. This technique was demonstrated earlier to prove the inadequacy of laboratory-determined shear strength values in this study. The disadvantage of this method is the researcher's inability to 1) know the location of phreatic surfaces and, 2) deal with localized phenomena (changes in density, percentage of water, etc.).

4.2 Physical properties alteration methods

Three models of soil strength determination with the use of large particles have been proposed: 1) the parallel method; 2) the replacement method; and 3) the matrix method. The parallel modeling method of estimating the field properties of rock fill material was first suggested by Lowe (1964). In this method, specimens with parallel gradation are constructed with maximum particle sizes of 3.8 cm (1.5 in), adding fines to make up for the removal of oversized particles (Figure 7). Unfortunately, this method has proven to be unsatisfactory because of its failure to consider the shape, crushing, and surface roughness properties of the oversized material.

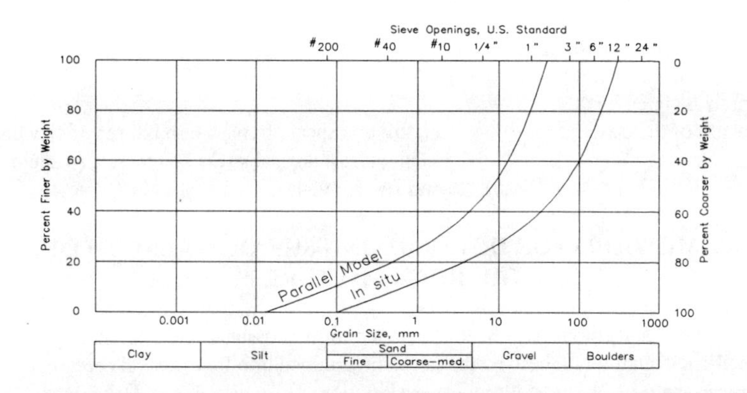

Figure 7. Parallel method illustrated with gradational analysis.

The replacement modeling method was introduced by the Army Corps of Engineers (1970) and suggests that particles larger than 1/6 of the triaxial test chamber's size be removed. If these particles compose more than 10% dry weight of sample, then an equal percentage of material retained on the #4 sieve but less than the maximum allowable sieve size should be introduced into the specimen (Figure 8). Donaghe and Torrey (1979) studied this process and found that the replacement procedures generally provided conservative strength parameters for earth-rock mixtures based on effective stresses.

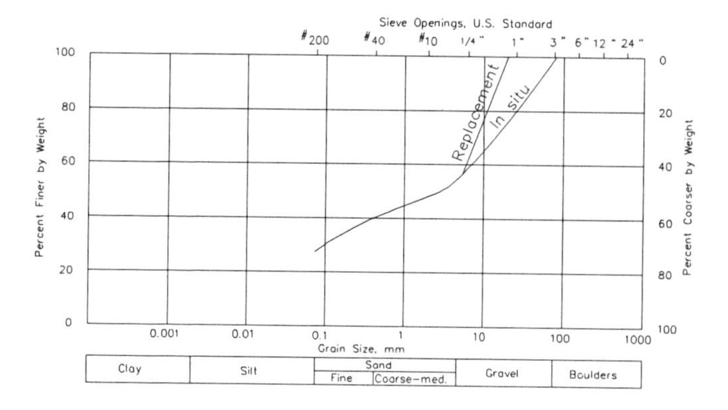

Figure 8. Replacement method illustrated with gradational analysis.

The matrix modeling method, introduced by Siddiqi (1984), removes the oversized particles from the specimen and examines the far-field soil matrix away from the particle. It is based upon the assumption that rock particles in a matrix of cohesionless material do not significantly affect the strength and deformation characteristics of the mixture. When less than 40% of the sample is composed of rock particles, there should be little contact among the particles. In this case, the far-field matrix contains a greater volume of material than the near-field; therefore it is the dominant strength member. Conversely, when the rock particles compose greater than say 65% of the sample, particle contact dominates. The soil matrix material simply fills voids created by the bridging action between non-floating particles.

More recent work by Su (1989) and Fragaszy et al. (1990) has shown that rock particles affect the density of the near-field matrix material. The authors determined that for state conditions in cohesionless material, the void ratio should increase around rock particles. This is a result of rock particles promoting void development based upon packing arrangements. The studies showed that average matrix density measurements lead to strengths that are too low. A method was proposed to determine the density of the far-field soil matrix, which according to the authors' findings controls the static strength of the material.

4.3 Empirical method

An interesting modeling technique has been proposed by Barton and Kjaernsli (1981) for estimating the shear strength of rockfill dams. This procedure requires the following input data: 1) the uniaxial compressive strength of the rock material, 2) the particle size (D_{50}), 3) the degree of particle roughness, 4) the porosity following compaction, and 5) normal stress of interest. This method serves to obtain preliminary estimates of the peak drained friction angle of rock fill, whether it consists of angular quarried rock or well-rounded gravel. Although this example is restricted to soils with high concentrations of rock particles in contact and is specifically useful for dam design, it illustrates the character and practical nature of this approach.

4.4 Analytical method

As with the empirical method, few examples exist for the analytical method. Hencher et al. (1984) proposed the following formula to calculate the shear strength of boulder colluvium:

$$\tau' = \sigma' \tan (\phi' + i_m + i_f) \tag{2}$$

where τ' = Effective shear stress,
σ' = Effective normal stress,
ϕ' = Effective angle of shearing resistance for the matrix (corrected for dilation),
i_m = Stress-dependent dilation angle for the matrix,
i_f = Dilation angle for the overall shear plane, taking boulder interference into account.

The values of the effective angle of shearing resistance (ϕ') and the stress-dependent dilation angle for the matrix (i_m) were obtained from the direct shear test results. The value of the dilation angle for the shear plane (i_f) was taken to be the averaged value of deviations of slip surface from the main direction of failure. Irfan and Tang (1993) indicated that this method gave an estimate of the upper-bound shear strength for the boulder colluvium in the Hong Kong area.

In a second example of the analytical methods, Vallejo (1979, 1989) examined the Skempton and DeLory (1957) approach for analysis of infinite slopes and found it inadequate for mud flows and debris flows that had a mixture of clay and rock lumps in a soft mud matrix. Vallejo examined the ratio of the volume occupied by the large particles and the volume of the whole mass, C, and determined the following relationships:

- C greater than 0.8 => frictional shear resistance between the large particles dominates,
- C less than 0.55 => shear strength for the soil dominates,
- C between 0.55 and 0.8 => shear strength of clayey matrix and frictional shear resistance of the large particles interact.

When the intermediate condition exists, Vallejo recommended that the following formula be used:

$$\tau = C\ \sigma\ \tan \phi + [1 - C]\ c \tag{3}$$

where ϕ' is the effective angle of shearing resistance between the large particles and c_u is the undrained shear strength of the matrix (mud).

5.0 CURRENT THEORY EXPLAINING THE BEHAVIOR OF A COHESIVE SOIL-ROCK MIXTURE

The effect rock particles have on cohesive soil matrix systems is significantly different in relation to previously explained theories for cohesionless soil matrix systems. The major expression of this difference is found in the way stiff particles attract stress, alter strain patterns, and affect density. In a cohesionless soil matrix, a high concentration of rock particles (>40%) produces grain-to-grain contact and high angles of internal friction with little cohesion (Figure 9a). The addition of small amounts of cohesive soil matrix produces a shape drop in the angle of shearing resistance (Miller and Sowers, 1957) and a rapid increase in cohesion. This indicates that some of the clay is trapped between rock particles, preventing particle-to-particle contact (Figure 9b). The soil matrix between the rock particle contact points is highly compacted, while in other areas some open voids and grain-to-grain contact persists. As the clay content increases,

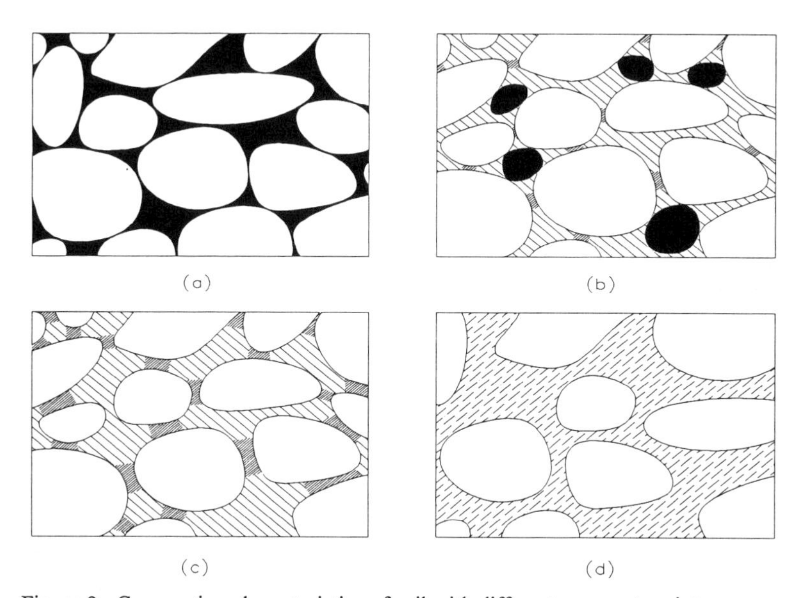

Figure 9. Compaction characteristics of soil with different aggregate mixtures (Miller and Sowers, 1957).

the cohesion increases, but at a decreasing rate. This reflects increasing clay compaction and a greater degree of void-filling by the clay (Figure 9c). At this point, there is a sufficient soil matrix to fill the voids loosely. At some point, enough clay exists in the system to cause the particles to float in the matrix of compacted clay (Figure 9d).

6.0 VARIATION IN CLAY SHEAR STRENGTH WITH CHANGING ROCK PARTICLE CONCENTRATIONS

Previously presented studies by Hall (1951), Miller and Sowers (1957), Holtz and Ellis (1961), Dobbiah et al. (1969), Patwardhan et al. (1970), and Donaghe and Torrey (1979) provided shear strength parameter data for varying concentrations of rock particles (Table 1). Most of the soil matrix material consisted of sandy clay with varying plasticity characteristics. Generally these test samples were wet but unsaturated material compacted close to maximum dry density. In many cases only limited tests were performed on particle concentrations where floating rocks would dominate (<40%). Confining pressures in the form of normal or lateral stresses ranged from 29 to 1379 kPa.

An evaluation of the effects rock particles have on cohesive soils in the unsaturated state provides some relevant data. Data contained within the above reports were utilized to evaluate the shear strength characteristics of material confined at approximately 200 kPa. This pressure simulates approximately 10 m of overburden. Examination of eastern Kentucky colluvium landslides showed that most failures occur between 5 and 15 m of overburden. However the Patwardhan

Table 1. Characteristics of previous laboratory test performed on cohesive soils with varying concentrations of rock particles.

Author(s), date	Soil matrix type	Distribution, C_c	Plastic Index, PI	Max. Particle size, cm	Compaction characteristic	Type of test	Moisture condition	Confining stress, kPa	Concentration of rock particles, %
Hall, 1951	Clayey (27) sand (73)	0.01	12	7.6	Close to MDD	Triaxial, CU	Unsaturated	104 to 414	53,85
Miller and Sowers, 1957	Sandy clay	0.01	6	0.5	Close to MDD	Triaxial, NA	Unsaturated	35 to 207	0,53,60, 67,74,78, 82,89,92, 96,100
Holtz and Ellis, 1961	SC-CL	0.8	28	7.6	Close to MDD	Triaxial, CU	Unsaturated	29 to 215	0,20,35, 50,65
Dobbiah et al. 1969	Clayey (36) sand (64)	0.6	17	2.5	Close to MDD	Triaxial, NA	Unsaturated	69 to 414	10,20,30, 40,50,60, 70,80
Patwardhan, et al. 1970	Clay	NA	7.3	15 (avg.)	Initial VR = 0.8	Direct Shear	Unsaturated	0	0,15,40, 70,100
Donaghe and Torrey, 1979	Clay and sand	0.6	21	7.6	Close to MDD	Triaxial, CU	Unsaturated	414 to 1379	20,40,60

Max. = Maximum dry density and optimum moisture content
UU = Unconsolidated-undrained
CU = Consolidated-undrained
CD = Consolidated-drained
MDD = Maximum dry density
VR = Void ratio
NA = Not Available

et al. (1970) data were not normalized because these tests were performed without confining pressure.

Figure 10 shows how shear strength of clay-rock mixtures is affected by varying rock particle concentration. In general, a gradual increase in strength is recognized as particle concentrations increase. In several of the tests, there is a marked increase in shear strength at a particle content of approximately 50%. This is undoubtably in response to significant particle-to- particle interaction occurring at and above this concentration. In general, high concentrations of rock particles (rock fills) have higher shear strengths. Leps (1970) examined 18 laboratory tests where the strengths of rock fills were determined. He found that the average rock fill had an internal angle of about 45° at 200 kPa normal pressure, which yields a shear strength of approximately 200 kPa.

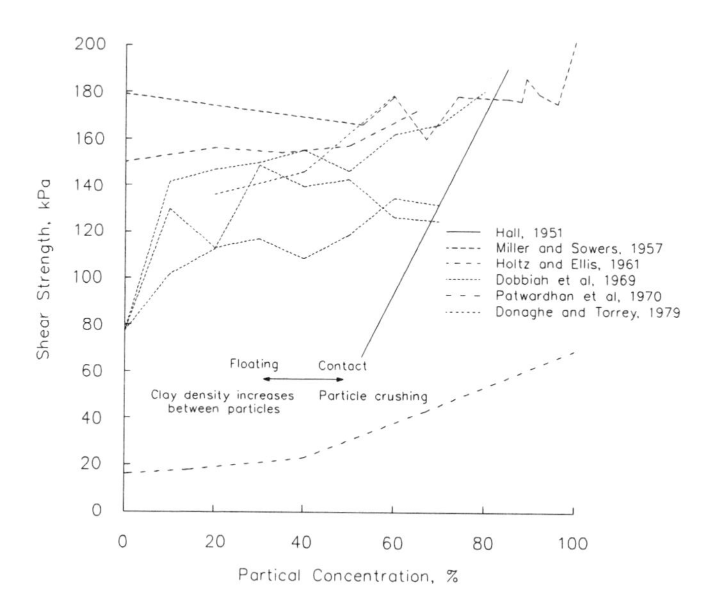

Figure 10. Relationship between shear strength and particle concentration for six past studies where the matrix material contained unsaturated clay.

7.0 SUMMARY AND CONCLUSION

This study has established that the shear strength of unsaturated clay can be affected by floating, oversized rock particles. This has significant practical implications because many colluvium soils have oversized particles within their matrix. Historically, design of slope remediation projects has been hampered by two factors: 1) an inability to test material with oversized particles, and 2) a lack of knowledge about the relationship between particle concentration and shear strength. Miller and Sowers (1957) first proposed a theory to explain the changes in behavior of cohesive material as floating particle concentrations increased. In this theory, floating particles compacted the soil matrix between the rock particle contact points and also changed void ratios. Miller and Sowers (1957) did not indicate that a relationship between shear strength and floating rock particle concentration exists; however, this study found six laboratory cases where shear strength was shown to gradually increase with increasing floating particle concentrations. This implies that slope remediation efforts based solely upon shear strengths developed from standard laboratory tests, with all oversized particles removed, could produce conservative designs.

REFERENCES

Barton, N. and B. Kjaernsli, "Shear Strength of Rockfill", Norges Geotekniske Institutt, No. 136, 1981, 19 pp.

Dobbiah, D., H.S. Bhat, P.V. Somasekhar, H.B. Sosalegowda, and K.N. Ranganath, "Shear Characteristics of Soil-Gravel Mixtures", J. of the Indian Nat. Soc. of Soil Mech. and Found. Eng., No. 8, 1969, pp. 57-66.

Donaghe, R.T. and V.H. Torrey, III., "Scalping and Replacement Effects on Strength Parameters of Earth-Rock Mixtures", Proc. of the European Conf. on Soil Mech. and Found. Eng., Vol. 2, Brighton, England, September 1979, pp. 27-34.

Fragaszy, R.J., W. Su, and F.H. Siddiqi, "Effects of Oversize Particles on Density of Clean Granular Soils", Geot. Testing J., Vol. 12, No. 2, 1990, pp. 106-114.

Hall, E.B., "A Triaxial Apparatus for Testing Large Soil Specimens", Special Tech. Publ. No. 106 on Triaxial Testing of Soil Bituminous Mixtures, (ASTM, 1951), pp. 152-161.

Hencher, S.R., J.B. Massey, and E.W. Brand, "Application of Back Analysis to Some Hong Kong Landslides", Proc. of the 4th Intern. Symp. on Landslides, Toronto, Vol. 1, 1984, pp. 631-638.

Holtz, W.G. and W. Ellis, "Triaxial Shear Characteristics of Clayey Gravel Soils," Proc. of the 5th Intern. Conf. On Soil Mech. and Found. Eng., Vol. 1, 1961, pp. 143.

Iannacchione, A.T., S.K. Bhatt, and J. Sefton, "Geotechnical Properties of Kentucky's AML Landslides and Slope Failure Evaluation", Proc. of the Assoc. of Aband. Land Progr., 16th Ann. Conf. Reclaiming for the Future, Park City, UT, Sept. 18-22, 1994, pp. 129-145.

Iannacchione, A.T. and L.E. Vallejo, "Factors Affecting the Slope Stability of Kentucky's Abandoned Mine Lands", Proc. of the 35th U.S. Symp. on Rock Mech., Tahoe, NV, June 5-7, 1995, pp. 837-842.

Irfan, T.Y. and K.Y. Tang, Effect of the Coarse Fractions on the Shear Strength of Colluvium, Geot. Eng. Office, Report No.23, Civil Eng. Dept., Hong Kong Government, Hong Kong, June 1993, 232 pp.

Laboratory Testing Manual, Appendix X (U.S. Army Corps of Eng., EM 1110-2-1906, 1970).

Leps, T.M., "Review of Shearing Strength of Rockfill", J. of the Soil Mech. and Found. Div., Am. Soc. of Civil Eng., Vol. 96, No. SM4, Proc. Paper 7394, July, 1970, pp. 1159-1170.

Leslie, D.D., "Large-Scale Triaxial Tests on Gravelly Soils", Proc. of the Pan American Conf. on Soil Mech. and Found. Eng., Vol. 1, 1963, pp. 181-202.

Leussink, H., "Discussion", Proc., 6th Intern. Conf. on Soil Mech. and Found. Eng., Vol. 3, 1965, pp. 310-316.

Lowe, J. III, "Shear Strength of Coarse Embankment Dam Materials", Proc. of the 8th Congress on Large Dams, 1964, pp. 745-761.

Marachi, N.D., C.K. Chan, and H.B. Seed, "Evaluation of Properties of Rockfill Materials", J. of Soil Mech. and Found. Div., Am. Soc. Of Civil Eng., Vol. 98, Proc. Paper 8672, SM1, January 1972, pp. 95-114.

Marsal, R.J., "Large Scale Testing of Rockfill Materials", J. of Soil Mech. and Foundation Eng., ASCE, Vol. 93, No, SM2, March 1967, pp. 27-43.

Miller, E.A. and G.F. Sowers, "The Strength Characteristics of Soil-Aggregate Mixtures", Highway Research Board Bull., No. 183, 1957, pp. 16-23.

Morgan, G.C. and M.C. Harris, "Portage Mountain Dam - II Materials", Canadian Geot. J., Vol. 4, No. 2, 1967, pp. 142-166.

Patwardhan, A.S., J. Shivaji Rao, and R.B. Gaidhane, "Interlocking Effects and Shearing Resistance of Boulders and Large Size Particles in a Matrix of Fines on the Basis of Large Scale Direct Shear Test", Proc. of the 2nd Southeast Asian Conf. on Soil Eng., Singapore, June 11-15, 1970, pp. 265-273.

Rathee, R. K., "Shear Strength of Granular Soils and Its Prediction by Modelling Techniques", Indian Geot. J., Vol. 5, No. 2, 1981, pp. 98-113.

Rico, A. and J.M. Orozco, "Effect of Fines on the Mechanical Behaviour of Granular Roadbase Materials",. Proc. of the 5th Panamerican Conf. on Soil Mech. and Found. Eng., Buenos Aires, Vol. 1, 1975, pp. 31-41.

Siddiqi, F.H., "Strength Evaluation of Cohesionless Soils with Oversize Particles" (unpublished Ph.D. Dissertation, School of Eng., Univ. of California, Davis, CA., Nov., 1984), pp. 179.

Skempton, A.W. and DeLory, F.A., "Stability of Natural Slopes in London Clay", Proc. of the 4th Intern. Conf. on Soil Mech. and Found. Eng., London, Vol. 2, 1975, pp. 378-381.

Su. W., "Static Strength Evaluation of Cohesionless Soil with Oversize Particles", (unpublished Ph.D. Dissertation, Washington State Univ., Pullman, WA, May 1989), pp. 247.

Vallejo, L.E., "An Explanation for Mudflows", Geotechnique, Vol. 29, 1979, pp. 351-354.

Vallejo, L.E., "An Extension of the Particulate Model of Stability Analysis for Mudflows", Soil and Found., Vol. 29, No. 3, Sept. 1989, pp. 1-13.

Vallejo, L.E. and Y. Zhou., "The Mechanical Properties of Simulated Soil-Rock Mixtures", Proc. of the 13th Intern. Conf. on Soil Mech. and Found. Eng., New Delhi, India, Vol. 1, January 10, 1994, pp. 365-368.

Vallerga, B.A.; H.B. Seed; C.L. Monismith; and R.S. Cooper, "Effect of Shape, Size and Surface Roughness of Aggregate Particles on the Strength of Granular Material", Special Tech. Publ., Am. Soc. for Testing and Materials, No. 212, 1957, pp. 63-74.

Finite Element Methods for 3D Slope Stability Analysis

Boris Jeremić
Associate Member, Department of Civil and Environmental Engineering,
University of California, Davis, CA 95616,

Abstract

In this paper, a new approach for modeling of three dimensional slope stability problems is presented. To this end, p-version of the finite element method together with large deformation hyperelastic-plastic formulation is utilized to model localized, continuous deformation that has been observed in failure mechanisms of slopes. In particular, it is shown how the new method can be used with a rather small number of finite elements to model sharp deformation gradients resulting from shear localization during slope failures. Method is well suited for both 2D and 3D computations. In 3D, we show clear advantages of our formulation over classical approach which rely on large number of elements or special elements with enhanced shape functions. In addition to that, solution advancement strategies that allow for accurate following of the equilibrium path in hardening, softening and snap–back regimes are presented.

Proposed method is used to simulate behavior of three dimensional slopes. Reduction of safety factors (obtained for plain strain conditions) is observed as the out the of plane curvature of the slope is increased.

1 Introduction

The failure of soil slopes come in a wide variety of conceivable manners. Qualitative definition is given in a book by Terzhagi, Peck and Mesri ([16]): *“ The failure of a mass of soil located beneath a slope is called a slide. It involves a downward and outward movement of the entire mass of soil participating in the failure."*

It is important to observe that in general all slope failures are occurring under fully three dimensional conditions. Moreover, failure is usually accompanied by the occurrence of large deformations (large strains, large rotations and large translations) of the soil mass. Large strains are concentrated in narrow, continuous bands, as shown in Figure 1. Figure 1(a) (cf. Nemat–Nasser and Okada [12]) presents such an observed shear band in a granular material while Figure 1(b) (cf. Zornberg et al. [19]) shows failed slope with highly localized, but again, continuous shear deformation. In addition to that, accurate following of the equilibrium

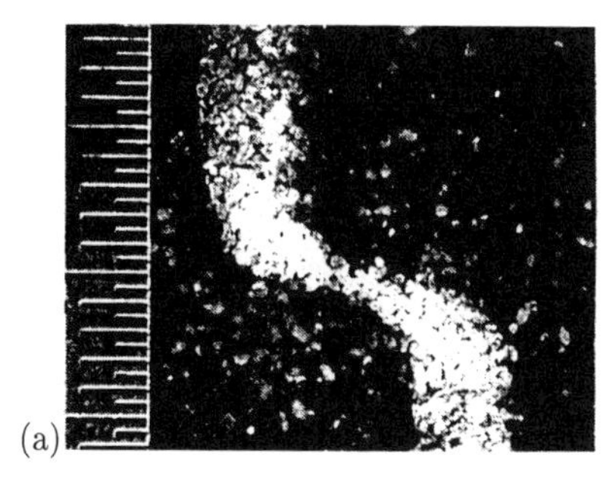

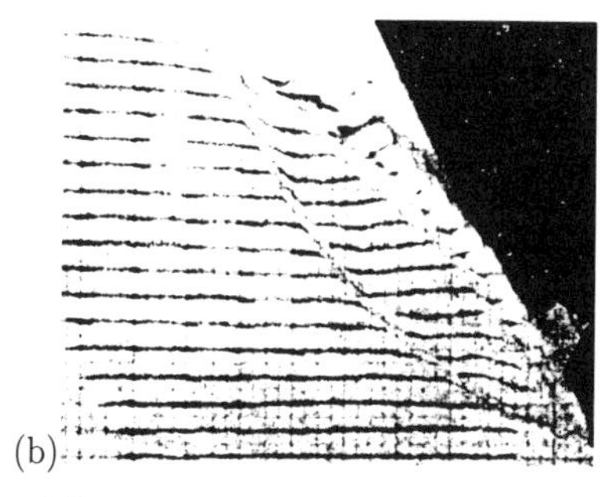

Figure 1: (a) Shear Band in soil material, (b) Centrifuge modeling of a slope failure with continuously sheared zone.

path under hardening, softening and snap–back behavior has to be provided for.

Therefore, modeling of slope failure by the finite element method must address the following issues: (a) occurrence of large deformations; (b) the effect of three dimensional conditions; (c) accurate following of the equilibrium path under hardening, softening and snap–back behavior; (d) occurrence of narrow, continuous localized zone on which the slope slides. In this paper we attempt to describe finite element methodology which takes into the account above issues. Section 2 describes the large deformation hyperelastic–plastic finite element format used. The p–version of the finite element method is briefly described in section 3. Section 4 describes numerical strategies and techniques used to follow the equilibrium path in hardening, softening and snap back regimes. Selected numerical examples are presented in section 5. Summary and a brief description of future work on the simulations of failure for three dimensional slopes is given in section 6.

2 Large Deformation Hyperelastic–Plastic Format

In this Section we elaborate on the Total Lagrangian finite element formulations for solving material and geometric nonlinear problems in geomechanics (see also Jeremić et al. [9] and [10]). We choose the Newton type procedure for satisfying equilibrium, i.e. virtual work for a given computational domain. The local form of equilibrium equations in material format (Lagrangian) for static case can be written as:

$$P_{iJ,J} - \rho_0 b_i = 0 \tag{1}$$

where $P_{iJ} = S_{IJ}(F_{iI})^t$ and S_{IJ} are first and second Piola–Kirchhoff stress tensors, respectively and b_I are body forces. Weak form of equilibrium equations is obtained by premultiplying Eq. (1) with virtual displacements δu_i and integrating by parts on the initial configuration B_0 (initial volume V_0):

$$\int_{V_0} \delta u_{i,j} P_{ij} dV = \int_{V_0} \rho_0 \delta u_i b_i dV - \int_{S_0} \delta u_i \bar{t}_i dS \tag{2}$$

It proves beneficial to rewrite Lagrangian format of weak form of equilibrium equations by using symmetric second Piola–Kirchhoff stress tensor S_{ij}:

$$\int_{V_0} \delta u_{i,j} F_{jl} S_{il} dV = \int_{V_0} \frac{1}{2} \left((\delta u_{i,l} + \delta u_{l,i}) + (\delta u_{i,j} u_{j,l} + u_{l,j} \delta u_{j,i}) \right) S_{il} dV \tag{3}$$

where we have used the symmetry of S_{il}, definition for deformation gradient $F_{ki} = \delta_{ki} + u_{k,i}$. We have also conveniently defined differential operator $\hat{E}_{il}(\delta u_i, u_i)$ as

$$\hat{E}_{il}(\delta u_i, u_i) = \frac{1}{2} (\delta u_{l,i} + \delta u_{i,l}) + \frac{1}{2} (u_{l,j} \delta u_{j,i} + \delta u_{i,j} u_{j,l}) \tag{4}$$

Given the displacement field $u_i^{(k)}(X_j)$, in iteration k, the iterative change Δu_i

$$u_i^{(k+1)} = u_i^{(k)} + \Delta u_i \tag{5}$$

is obtained from the linearized virtual work expression

$$\int_{\Omega_c} \hat{E}_{ij}(\delta u_i, {}^{n+1}_{\ 0}u_i^{(k)})\ {}^{n+1}_{\ 0}S_{ij}^{(k)}\ dV + \int_{\Omega_c} \rho_0\ \delta u_i\ {}^{n+1}_{\ 0}b_i\ dV + \int_{\partial\Omega_c} \delta u_i\ {}^{n+1}_{\ 0}t_i\ dS \simeq$$
$$W(\delta u_i, u_i^{(k)}) + \int_{\Omega_c} \hat{E}_{ij}(\delta u_i, u_i)\ \mathcal{L}_{ijkl}\ \hat{E}_{kl}(\Delta u_i, u_i)\ dV + \int_{\Omega_c} d\hat{E}_{ij}(\delta u_i, \Delta u_i)\ S_{ij}\ dV \quad (6)$$

The vectors of external and internal forces are defined as

$$\mathbf{f}_{int} = \frac{\partial(W^{int}(\delta u_i, {}^{n+1}_{\ 0}u_i^{(k)}))}{\partial(\delta u_i)} \quad (7)$$

$$\mathbf{f}_{ext} = \frac{\partial(W^{ext}(\delta u_i))}{\partial(\delta u_i)} \quad (8)$$

with:

$$W^{int}(\delta u_i, {}^{n+1}_{\ 0}u_i^{(k)}) = \int_{\Omega_c} \hat{E}_{ij}(\delta u_i, {}^{n+1}_{\ 0}u_i^{(k)})\ {}^{n+1}_{\ 0}S_{ij}^{(k)}\ dV \quad (9)$$

$$W^{ext}(\delta u_i) = -\int_{\Omega_c} \rho_0\ \delta u_i\ {}^{n+1}_{\ 0}b_i\ dV - \int_{\partial\Omega_c} \delta u_i\ {}^{n+1}_{\ 0}t_i\ dS \quad (10)$$

The Algorithmic Tangent Stiffness (ATS) tensor $\mathcal{L}_{ijkl}^{ATS}$ is defined as a linearization of second Piola–Kirchhoff stress tensor S_{ij} with respect to the right deformation tensor C_{kl} through $dS_{ij} = 1/2\ \mathcal{L}_{ijkl}\ dC_{kl}$ with $dC_{kl} = 2\ \hat{E}_{kl}(du_i, u_i)$.

On the constitutive level we use the free energy density ψ, defined as

$$\rho_0\psi(\bar{C}_{ij}^e, \kappa_\alpha) = \rho_0\psi^e(\bar{C}_{ij}^e) + \rho_0\psi^p(\kappa_\alpha) \quad (11)$$

where $\psi^e(\bar{C}_{ij}^e)$ represents a suitable hyperelastic model in terms of the elastic right deformation tensor $\bar{C}_{ij}^e$, whereas $\psi^p(\kappa_\alpha)$ represents the hardening. We now define the elastic domain $\mathcal{B}$ as

$$\mathcal{B} = \{\bar{T}_{ij}, K_\alpha \mid \Phi(\bar{T}_{ij}, K_\alpha) \leq 0\} \quad (12)$$

When Φ is isotropic in Mandel stress $\bar{T}_{ij}$ (which is the case here) in conjunction with elastic isotropy, we can conclude that $\bar{T}_{ij}$ is symmetrical and we may replace $\bar{T}_{ij}$ by τ_{ij} in Φ. The constitutive relations can now be written as

$$\bar{L}_{ij}^p := \dot{F}_{ik}^p \left(F_{jk}^p\right)^{-1} = \dot{\mu}\frac{\partial\Phi^*}{\partial\bar{T}_{ij}} = \dot{\mu}\bar{M}_{ij} \quad (13)$$

$$K_\alpha = K_\alpha(\kappa_\beta) \quad , \quad \dot{\kappa}_\beta = \dot{\mu}\ \frac{\partial\Phi^*}{\partial K_\beta} \quad , \quad \kappa_\beta(0) = 0 \quad (14)$$

where $F^p_{ik} = (\bar{F}^e_{li})^{-1} F_{lk}$ is the plastic part of the deformation gradient.

The flow rule from equation (13) can be integrated to give

$$ {}^{n+1}F^p_{ij} = \exp\left(\Delta\mu\, {}^{n+1}\bar{M}_{ik}\right) {}^{n}F^p_{kj} \tag{15} $$

and by using multiplicative decomposition of deformation gradient we obtain

$$ {}^{n+1}\bar{F}^e_{ij} = {}^{n+1}\bar{F}^{e,tr}_{ik} \exp\left(-\Delta\mu\, {}^{n+1}\bar{M}_{kj}\right) \quad \text{with} \quad {}^{n+1}\bar{F}^{e,tr}_{ik} = {}^{n+1}F_{im} \left({}^{n}F^p_{mk}\right)^{-1} \tag{16} $$

The elastic deformation is then

$$ {}^{n+1}\bar{C}^e_{ij} \stackrel{\text{def}}{=} \left({}^{n+1}\bar{F}^e_{im}\right)^T {}^{n+1}\bar{F}^e_{mj} = \exp\left(-\Delta\mu\, {}^{n+1}\bar{M}^T_{ir}\right) {}^{n+1}\bar{C}^{e,tr}_{rl} \exp\left(-\Delta\mu\, {}^{n+1}\bar{M}_{lj}\right) \tag{17} $$

By recognizing that the exponent of a tensor can be expanded in Taylor's (MacLaurin's) series and after some algebraic manipulations we obtain

$$ {}^{n+1}\bar{C}^e_{ij} = {}^{n+1}\bar{C}^{e,tr}_{ij} - \Delta\mu\, {}^{n+1}\bar{M}_{ir}\, {}^{n+1}\bar{C}^{e,tr}_{rj} - \Delta\mu\, {}^{n+1}\bar{C}^{e,tr}_{il}\, {}^{n+1}\bar{M}_{lj} \tag{18} $$

The hardening rule (13) can be integrated to give

$$ {}^{n+1}\kappa_\alpha = {}^{n}\kappa_\alpha + \Delta\mu \left.\frac{\partial \Phi^*}{\partial K_\alpha}\right|_{n+1} \tag{19} $$

The incremental problem is defined by equations (18), (19), the constitutive relations

$$ {}^{n+1}\bar{S}_{IJ} = 2 \left.\frac{\partial W}{\partial C_{IJ}}\right|_{n+1} \tag{20} $$

$$ {}^{n+1}K_\alpha = - \left.\frac{\partial W}{\partial \kappa_\alpha}\right|_{n+1} \tag{21} $$

and the Karush–Kuhn–Tucker (KKT) conditions

$$ \Delta\mu < 0 \quad ; \quad {}^{n+1}\Phi \le 0 \quad ; \quad \Delta\mu\, {}^{n+1}\Phi = 0 \tag{22} $$

This set of nonlinear equations will be solved with a Newton type procedure, described in some more details by Jeremić et al. [9] and Jeremić [8].

It is important to note that the above finite element formulation accounts for material and geometric nonlinear (large deformation, elasto–plastic) effects in a general geomaterial. Application to metallic materials (pressure insensitive, obeying J_2 yield criteria) can be derived as a special case of the above formulation.

3 p–version Finite Element Method

In this section we describe the set of hierarchical shape functions that we use to approximate continuous displacement field. Detailed description of the higher order p–version finite element formulation for linear elastic materials is given by Babuška and Szabó [3]. We choose the Legendre polynomials as shape functions that produce stiffness matrices with favorable properties (Babuška et al. [2]). For example, such shape functions produce sparse element stiffness matrices and element distortion has little effect on computation accuracy.

Shape function for the three–dimensional brick elements have general form $\phi_i(r)\phi_j(s)\phi_k(t)$. Here, $\phi_i(r)$ is a shape function of degree i and is defined in terms of the Legendre polynomial $P_{i-1}(\xi)$ as:

$$\phi_i(r) = \sqrt{\frac{2i-1}{2}} \int_{-1}^{r} P_{i-1}(\xi) d\xi \tag{23}$$

Figure 2(a) and Eq. (24) present the 1D set of hierarchical shape functions. N_{C1} and N_{C2} are called nodal (or linear) shape functions while the nonlinear polynomials N_{Si} are called internal shape functions.

$$N_{C1} = 0.5(1-r) \qquad N_{C2} = 0.5(1+r) \qquad N_{Si} = \phi_i(r) \tag{24}$$

Fig. 2(a) shows that the hierarchical shape functions N_{Si} exhibit spectral prop-

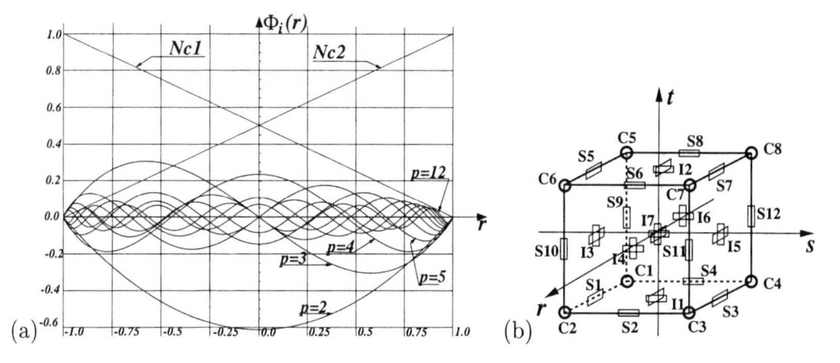

Figure 2: (a) Set of nodal and side hierarchical shape functions based on Legendre polynomials. (b) The standard higher order 3D brick elements.

erties. This property is extensively used to simulate continuous mode of deformation as found in slope failure mechanics. Over 3D elements, the shape functions are divided into three categories : nodal shape functions (Δ), side and face shape functions (Γ) and internal shape functions (Ω). In particular, as seen in Figure 2(b), the standard 3D brick element features 8 corner, 12 side, 6 face and 1 internal nodes. Our implementation uses element with polynomials of order $p = 8$. Such high order elements make it possible to follow initiation and propagation of the continuous localized zone within a single element.

4 Solution Advancement Strategy

In section 2 we have derived the basic equations of materially nonlinear analysis of solids. Discretization of such problems by finite element methods results in a set of nonlinear algebraic equations called residual force equations:

$$\mathbf{r}(\mathbf{u}, \lambda) = \mathbf{f}_{int}(\mathbf{u}) - \lambda \mathbf{f}_{ext} = 0 \tag{25}$$

where $\mathbf{f}_{int}(\mathbf{u})$ are the internal forces which are functions of the displacements, $\mathbf{u}$, the vector $\mathbf{f}_{ext}$ is a fixed external loading vector and the scalar λ is a load–level parameter that multiplies $\mathbf{f}_{ext}$. Numerical procedures used for solving Eq. (25) are rooted in the idea of "advancing the solution" by continuation. Except in very simple problems, the continuation process is multilevel and involves hierarchical breakdown into stages, incremental steps and iterative steps. Processing a complex nonlinear problem generally involves performing a series of analysis stages. Multiple control parameters are not varied independently in each stage and may therefore be characterized by a single stage control parameter λ. Stages are only weakly coupled in the sense that end solution of one may provide the starting point for another.

Various forms of path following methods with various methods of approximating the exact length of an arc, have stemmed from the original work of Riks [14], [15] and Wempner [17]. They aimed at finding the intersection of equation (25) with $s = constant$ where s is the arc-length , defined as[1] an incremental form:

$$a = (\Delta s)^2 - (\Delta l)^2 = \left(\frac{\psi_u^2}{u_{ref}^2} \Delta \mathbf{u}^T \mathbf{S} \Delta \mathbf{u} + \Delta \lambda^2 \psi_f^2 \mathbf{f}_{ext}^T \mathbf{f}_{ext} \right) - (\Delta l)^2 \tag{26}$$

where Δl is the radius of the desired intersection (see Figure (3) and represents an approximation to the incremental arc length. Scaling matrix $\mathbf{S}$ is usually diagonal

[1] A bit different form in that it is scaled with scaling matrix $\mathbf{S}$, introduced by Felippa [6].

nonnegative matrix that scales the state vector $\Delta\mathbf{u}$ and u_{ref} is a reference value with the dimension of $\sqrt{\Delta\mathbf{u}^T\mathbf{S}\Delta\mathbf{u}}$. It is important to note that the vector $\Delta\mathbf{u}$ and scalar $\Delta\lambda$ are incremental and not iterative values, and are starting from the last converged equilibrium state. The main essence of the arc-length methods is

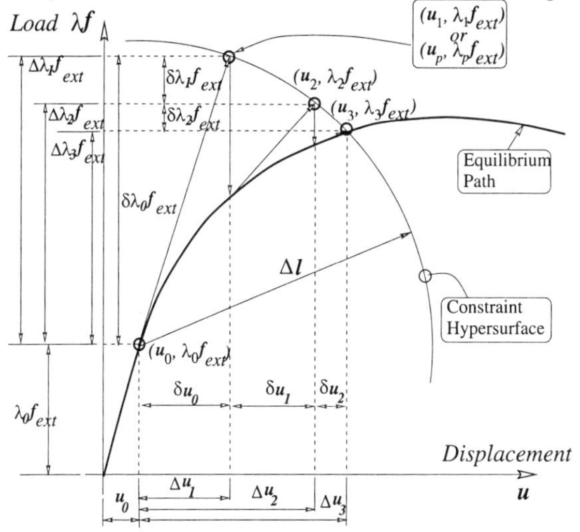

Figure 3: Spherical arc-length method and notation for one DOF system.

that the load parameter λ becomes a variable. With load parameter λ variable we are dealing with $n+1$ unknowns (n unknown displacement variables and on extra unknown in the form of load parameter λ. In order to solve this problem we have n equilibrium equations (25) and the one constraint equation (26). We can solve the augmented system of $n+1$ equations by using a truncated Taylor series expansion of equations (25) and (26) to obtain

$$\begin{aligned} \mathbf{r}^{new}(\mathbf{u},\lambda) &= \mathbf{r}^{old}(\mathbf{u},\lambda) + \frac{\partial \mathbf{r}(\mathbf{u},\lambda)}{\partial \mathbf{u}}\,\delta\mathbf{u} + \frac{\partial \mathbf{r}(\mathbf{u},\lambda)}{\partial \lambda}\,\delta\lambda = \\ &= \mathbf{r}^{old}(\mathbf{u},\lambda) + \mathbf{K}_t\,\delta\mathbf{u} - \mathbf{f}_{ext}\,\delta\lambda = 0 \end{aligned} \tag{27}$$

$$a^{new} = a^{old} + 2\frac{\psi_u^2}{u_{ref}^2}\Delta\mathbf{u}^T\mathbf{S}\delta\mathbf{u} + 2\Delta\lambda\delta\lambda\,\psi_f^2\mathbf{f}_{ext}^T\mathbf{f}_{ext} = 0 \tag{28}$$

where $\mathbf{K}_t = \partial\mathbf{r}(\mathbf{u},\lambda)/\partial\mathbf{u}$ is the tangent stiffness matrix. The aim is to have

$\mathbf{r}^{new}(\mathbf{u}, \lambda) = 0$ and $a^{new} = 0$ which can then be solved for $\delta\mathbf{u}$ and $\delta\lambda$ to get:

$$\begin{bmatrix} \delta\mathbf{u} \\ \delta\lambda \end{bmatrix} = - \begin{bmatrix} \mathbf{K}_t & -\mathbf{f}_{ext} \\ 2\frac{\psi_u^2}{u_{ref}^2}\Delta\mathbf{u}^T\mathbf{S} & 2\Delta\lambda\, \psi_f^2 \mathbf{f}_{ext}^T \mathbf{f}_{ext} \end{bmatrix}^{-1} \begin{bmatrix} \mathbf{r}^{old} \\ a^{old} \end{bmatrix} \tag{29}$$

System described in equation (29) is sufficient to solve for $\delta\mathbf{u}$ and $\delta\lambda$ in each incremental step.

The constraining equation (26) was given in a rather general form. Some further comments and observations are in order. By assigning various numbers to parameters ψ_u, ψ_f, $\mathbf{S}$ and u_{ref} one can obtain different constraining schemes from (26). Coefficients ψ_u and ψ_f may not be simultaneously zero. Useful choices for $\mathbf{S}$

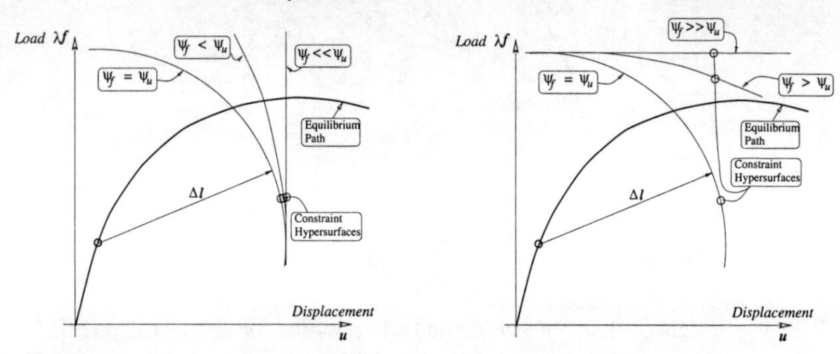

Figure 4: Influence of ψ_u and ψ_f on the constraint surface shape.

are $\mathbf{I}$, $\mathbf{K}_t$ and $diag\,(\mathbf{K}_t)$. If $\mathbf{S} = \mathbf{I}$ and $u_{ref} = 1$ the method is called the arclength method[2]. If we choose $\mathbf{S} = diag\,(\mathbf{K}_t)$ nice scaling is obtained but otherwise no physical meaning can be attributed to this scaling type. With $\mathbf{S} = \mathbf{K}_t$ and $\psi_f \equiv 0$ external work constraint of Bathe and Dvorkin [4] is obtained. A rather general equation (26) can be further specialized to load (λ) control with $\psi_u \equiv 0; \psi_f \equiv 1$ and state control with $\psi_u \equiv 1; \psi_f \equiv 0$ and $\mathbf{S} = \mathbf{I}$. In the finite element literature, the term displacement control has been traditionally associated with the case in which only one of the components of the displacement vector $\mathbf{u}$[3] is specified. This may be regarded either as a variant of state control, in which a norm that singles out the ith component is used, or as a variant of the λ control if the control parameter is taken as $\lambda\mathbf{u}_i$. It is, of course, possible to make the previous

[2]It actually reduces to the original work of Riks [14], [15] and Wempner [17].

[3]Say $\mathbf{u}_i$.

parameters variable, functions of different unknowns. For example if one defines $u_{ref} = \Delta\mathbf{u}^T\mathbf{S}\Delta\mathbf{u}$ then close to the limit point $\Delta\mathbf{u} \to 0 \Rightarrow \psi_u^2/u_{ref}^2 \gg \psi_f^2$ that makes our constraint from equation (26) behave like state control. One important aspect of scaling constraint equations by using $\mathbf{S} = diag(\mathbf{K}_t)$ or $\mathbf{S} = \mathbf{K}_t$ is the possibility of non–positive definite stiffness matrix $\mathbf{K}_t$. It usually happens that after the limit point is passed, at least one of the eigenvalues of $\mathbf{K}_t$ is non–positive, thus rendering the constraint hypersurface non–convex.

For slope stability analysis, the procedure starts with $\psi_f >> \psi_u$ and then when the limit point is approached, ψ_f gradually changes to $\psi_f = \psi_u$. This is particularly helpful if the softening regime (after limit point) is rather sharp or if snap–back is encountered (as observed in laboratory tests by Zietlow and Labuz [18].

Solution of the system of equations (29) will have two roots in general (constraint will intersect equilibrium path at two points). In order to be able to traverse the equilibrium path in positive sense and chose the correct root, we apply two rules. The first one requires that the external work expenditure over the predictor step be positive:

$$\Delta W = \mathbf{f}_{ext}^T\Delta\mathbf{u} = \mathbf{f}_{ext}^T\mathbf{K}_t^{-1}\mathbf{f}_{ext}\delta\lambda > 0 \tag{30}$$

The simple conclusion is that $\delta\lambda$ should have the sign of $\mathbf{f}_{ext}^T\mathbf{K}_t^{-1}\mathbf{f}_{ext}$. This condition is particularly effective at limit points. However, it fails when $\mathbf{f}_{ext}$ and $\mathbf{K}_t^{-1}\mathbf{f}_{ext}$ are orthogonal $\mathbf{f}_{ext}^T\mathbf{K}_t^{-1}\mathbf{f}_{ext} = 0$ which can happen at bifurcation and turning points (see Figure 5).

Near a turning point application of the positive work rule, given by equations (30) causes the path to double back upon itself. When it crosses the turning point it reverses so the turning point becomes impassable. Physically, a positive work rule is incorrect because in passing a turning point the geomechanics solid releases external work until another turning point is encountered.

To pass a turning point imposing a condition on the angle of the prediction vector proves more effective. The idea is to compute both solutions $\Delta\mathbf{u}_1^{new} = \Delta\mathbf{u}^{old} + \delta\bar{\mathbf{u}} + \delta\lambda_1\delta\mathbf{u}_t; \Delta\mathbf{u}_2^{new} = \Delta\mathbf{u}^{old} + \delta\bar{\mathbf{u}} + \delta\lambda_2\delta\mathbf{u}_t$. The one that lies closest to the old incremental step direction $\Delta\mathbf{u}^{old}$ is the one sought. This should prevent the solution from double backing. The procedure can be implemented by finding the solution with the minimum angle between $\Delta\mathbf{u}^{old}$ and $\Delta\mathbf{p}^{new}$, and hence the

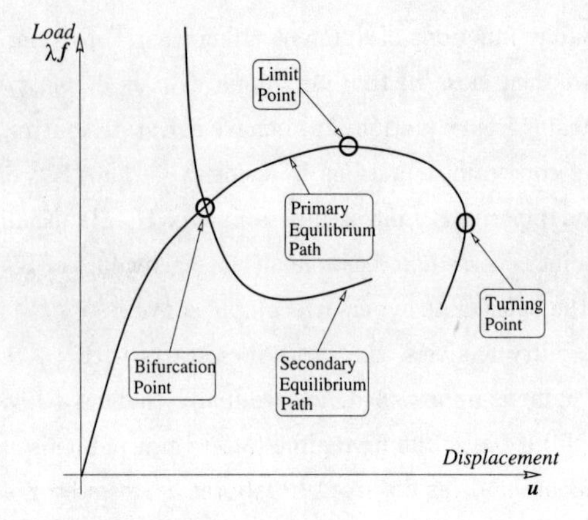

Figure 5: Simple illustration of Limit, Turning and Bifurcation points.

maximum cosine of the angle:

$$\cos\phi = \frac{\left(\Delta \mathbf{u}^{old}\right)^T \left(\Delta \mathbf{u}^{new}\right)}{\|\Delta \mathbf{u}^{old}\| \, \|\Delta \mathbf{u}^{new}\|} = \frac{\left(\Delta \mathbf{u}^{old}\right)^T \left(\Delta \mathbf{u}^{old} + \delta\bar{\mathbf{u}} + \delta\lambda\delta\mathbf{u}_t\right)}{\|\Delta \mathbf{u}^{old}\| \, \|\Delta \mathbf{u}^{old} + \delta\bar{\mathbf{u}} + \delta\lambda\delta\mathbf{u}_t\|} \tag{31}$$

where $\delta\bar{\mathbf{u}} = -\mathbf{K}_t^{-1}\mathbf{r}^{old}$ and $\delta\mathbf{u}_t = \mathbf{K}_t^{-1}\mathbf{f}_{ext}$. Once the turning point has been crossed, the work criterion should be reversed so the external work is negative.

5 Numerical Simulation Results

The described finite element formulation and techniques were applied to the problem of slope stability in three dimensions (3D). To this end we analyzed a set of three dimensional slopes with cross section as shown in Figure 6. The focus of the study was in assessing the reduction of safety factor when slopes are extending and curving into the third dimension (perpendicular to the plan in Figure 6) resulting in the violation of the plane strain assumption. For this exercise we set the density of soil to $\gamma = 19.61 kN/m^3$, cohesion to $c = 28 kN/m^2$ and friction angle to $\phi = 18°$. There was no water present. Method of slices (Fellenius [7] and Petterson [13]) for this particular 2D setup will yield a factor of safety of 2.0 (see for example Cernica [5] or Atkinson [1]).

In order to investigate the effect of slopes which are subject to a more general 3D conditions, different finite element models were constructed. Here we present two such 3D slope models. Figure 7(a) depicts finite element model for a slope

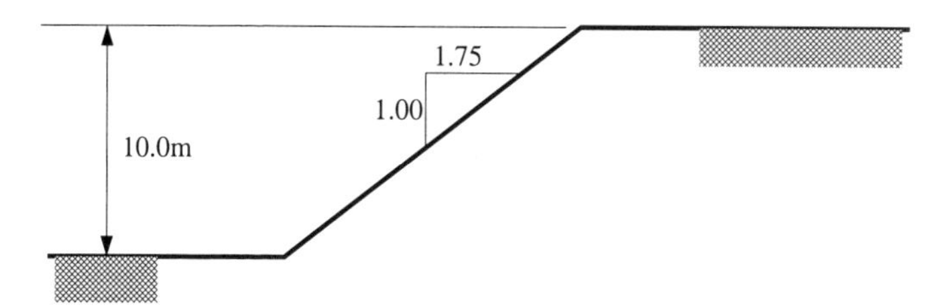

Figure 6: Plan of the soil slope cross section.

which is curving by 45° and features 7050 higher order brick elements. Figure 7(b) presents finite element model for a slope which is curving by full 90° and features 8460 higher order brick elements. Boundary conditions for both models are full support on the lower plane (corresponding to an underlying rock layer) and vertical sliders on sides. Slope curves in the central section. The radius of the curve is 32.0m to the top of the slope (2.0m from the end of the FE model). Model extends horizontally 30m from the top and bottom of the slope. It should be noted that both models are symmetric with vertical plane of symmetry dividing the center of curved parts of slopes. Even-though only one half of the model could have been used for this deterministic study, we retain the full model for comparison with future simulations with random soil material properties.

The elasto–plastic material model used was a simple version the B material model described in more details by Jeremić et al. [9]. B material model features hierarchical set of yield surface, potential surfaces and hardening–softening laws. In this simple instance B model is described by a rounded Mohr–Coulomb like yield surface with deviatoric shape based on recent work of Krenk [11]. Model follows elastic perfectly plastic flow and the non–associativity is controlled by potential surface similar to the yield surface.

For this exercise we define slope failure as the first limit point on the equilibrium path for the node at the top of the slope. Slope behavior is followed until initial failure while the material friction angle is reduced. The analysis shows that the factor of safety for the 45° curved slope is $F_s = 1.65$. In the case of 90° curved slope the safety factor (first occurrence of the limit point or beginning of softening behavior) is at $F_s = 1.38$. This represents a significant reduction from a value of 2.0 obtained by the method of slices (widely used in practice).

Although we are able to follow the slope behavior in post-peak regime, for this

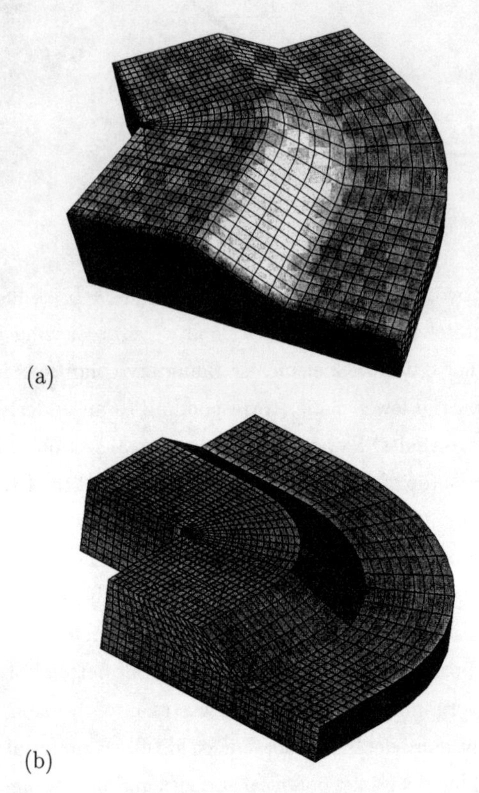

Figure 7: Finite element models for the 3D slope problem (a) 45° slope and (b) 90° slope.

study we only observe the limit point and record the material parameters (friction angle reduction). Further simulation of the slope failure mechanism reveals that the initial failure can be followed with a stabilization of failure, as the mass of the slope is redistributed and the slope attains next equilibrium, before complete collapse. This is more evident for slopes that are closer the plain strain conditions (straight slope) since they have additional support in the out of plane direction. In all cases large strains and large rotations of material points were observed even before the first limit point.

6 Summary

In this paper finite element formulation and technology for assessing failure of soil slopes in 3D has been presented. Large deformation hyperelasto–plastic finite element format in spatial (Lagrangian) frame of reference has been formulated. Particular attention was focused on the continuous localization of deformation which is commonly observed in failed geomechanics solids. Higher order finite element shape functions, based on Legendre polynomials have been used in conjunction with hyperelastic–plastic format. Techniques for following the equilibrium path of geomechanics solids during failure process have been described.

Three dimensional slope stability problem was analyzed using previously described formulation and techniques. It has been shown that factor of safety can be significantly reduced if the slope is curved out of plane in a convex fashion.

This initial study has focused on a limited number of slope designs. Further work on analyzing various concave and convex slopes designs, the influence of varying water table and randomness of soil material parameters is currently under way.

References

[1] ATKINSON, J. *An Introduction to the Mechanics of Soils and Foundations.* Series in Civil Engineering. McGraww–Hill, 1993. ISBN 0-07-707713-X.

[2] BABUŠKA, I., GRIEBEL, M., AND PITKÅRANTA, J. The problem of selecting the shape functions for a p–type finite element. *International Journal for Numerical Methods in Engineering 28* (1989), 1891–1908.

[3] BABUŠKA, I., AND SZABÓ, B. *Finite element analysis.* John Wiley & Sons Inc., 1991.

[4] BATHE, K.-J., AND DVORKIN, E. On the automatic solution of nonlinear finite element equations. *Computers & Structures 17*, 5-6 (1983), 871–879.

[5] CERNICA, J. N. *Geotechnical Engineering Soil Mechanics.* John Wiley & Sons, Inc., 1995.

[6] FELIPPA, C. A. Dynamic relaxation under general incremental control. In *Innovative Methods for Nonlinear Problems*, W. K. Liu, T. Belytschko, and K. C. Park, Eds. Pineridge Press, Swansea U.K., 1984, pp. 103–133.

[7] FELLENIUS, W. Calculation of the stability of earth dams. In *Trans. 2nd Congr. Large Dams* (Washington D.C., 1936), vol. 4.

[8] Jeremić, B. *Finite Deformation Hyperelasto–Plasticity of Geomaterials.* PhD thesis, University of Colorado at Boulder, July 1997.

[9] Jeremić, B., Runesson, K., and Sture, S. A model for elastic–plastic pressure sensitive materials subjected to large deformations. *International Journal of Solids and Structures 36*, 31/32 (1999), 4901–4918.

[10] Jeremić, B., Runesson, K., and Sture, S. Finite deformation analysis of geomaterials. *International Journal for Mechanics of Cohesive-Frictional Materials* (2000). Submitted for publication.

[11] Krenk, S. Family of invariant stress surfaces. *ASCE Journal of Engineering Mechanics 122*, 3 (1996), 201–208.

[12] Nemat-Nasser, S., and Okada, N. Strain localization in particulate media. In *Proceedings of the 12th Engineering Mechanics Conference* (May 17-20 1998), H. Murakami and J. E. Luco, Eds., ASCE.

[13] Petterson, K. E. The early history of circular sliding surfaces. *Geotechnique 5* (1955).

[14] Riks, E. The application of Newton's method to the problems of elastic stability. *Journal of Applied Mechanics 39* (December 1972), 1060–1066.

[15] Riks, E. An incremental approach to the solution of snapping and buckling problems. *International Journal for Solids and Structures 15* (1979), 529–551.

[16] Terzaghi, K., Peck, R. B., and Mesri, G. *Soil Mechanics in Engineering Practice*, third ed. John Wiley & Sons, Inc., 1996.

[17] Wempner, G. A. Discrete approximations related to nonlinear theories of solids. *International Journal for Solids and Structures 7* (1971), 1581–1599.

[18] Zietlow, W. K., and Labuz, J. F. Measurement of the intrinsinic process zone in rock using acoustic emission. *International Journal of Rock Mechanics and Mining Sciences 35*, 3 (1998), 291–299.

[19] Zornberg, J. G., Sitar, N., and Mitchell, J. K. Performance of geosynthetic reinforced slopes at failure. *ASCE Journal of Geotechnical and Geoenvironmental Engineering 124*, 8 (1998), 670–683.

Slope Stability Calculations Using Limit Analysis

Jyant Kumar[1]

Abstract

By using the upper bound theorem of limit analysis, and assuming the yielding of soil with its partial shear strength, stability calculations have been performed for determining the factor of safety of soil slopes. During failure, the slope material was assumed to rotate as a single rigid mass bounded with logarithmic spiral rupture surface. A detailed stepwise solution procedure has been provided. The results have been compared with those obtained using the Bishop simplified method. In all the cases except for very steep slopes, the upper bound theorem of limit analysis provides almost the same answers as predicted by the Bishop method. For very steep slopes, the Bishop method gives a conservative estimate of the factor of safety as compared to the upper bound limit analysis approach.

Introduction

The solution of any stability problem with the use of limit analysis requires: (i) complete mobilisation of the soil shear strength along rupture/velocity discontinuity surfaces; (ii) satisfaction of the kinematics of the problem; and (iii) applicability of the associated flow rule in soil material. From these considerations with the application of this method, stability numbers for slopes, as defined by Taylor (1948), have already been developed by Chen (1975). With the use of these stability numbers, the limiting height for a slope with given slope inclinations and soil parameters can be directly established. However, the stability factor which is

---

[1] Asistant Professor, Department of Civil Engineering
Indian Institute of Science, Bangalore 560012, India
Email: jkumar@civil.iisc.ernet.in

recommended in most of the design codes for slopes and embankments is based on the factor of safety with respect to shear strength of soil mass. The determination of such factor of safety requires the consideration of limiting equilibrium of soil with respect to its partial shear strength. Although Karal (1977) has already demonstrated the use of upper bound theorem of limit analysis to find the factor of safety of slopes with the soil mass assumed to yield with its partial shear strength. However, this concept has hardly been extended to actual practice perhaps on account of associated computational difficulties. All the existing literature on the application of upper bound theorem of limit analysis for slope stability problems deals with the determination of stability numbers on the basis of complete mobilisation of shear strength of the soil mass [Chen & Giger (1971), Chen (1975), Michalowski (1995), etc.]. In the present paper, using the upper bound theorem of limit analysis and on the basis of assumption of occurrence of soil yield with partial shear strength, a detailed stepwise solution procedure has been given for obtaining the factor of safety of homogenous soil slopes underlain by a hard stratum, in the absence of pore water pressures. The results have been compared with those obtained on the basis of the Bishop method [Bishop (1955) and Bishop & Morgenstern (1960)].

Factor of Safety

The factor of safety (F) as often recommended in practice is the ratio of available shear strength of soil to that required to maintain equilibrium. Under equilibrium, the mobilised shear strength (τ_m) will therefore be equal to $\tau_m = 1/F\ (c + \sigma_n'.\tan\phi)$; where σ_n' = effective normal stress and, c & ϕ are the given shear strength parameters of soil.

Upper Bound Theorem of Limit Analysis

By extending the use of the upper bound theorem of limit analysis to a soil mass which is assumed to yield with its partial shear strength (τ_m), having mobilised strength parameters c_m and ϕ_m , where $c_m = c/F$ and $\phi_m = \tan^{-1}(\tan\phi/F)$, the factor of safety of a slope can be determined from the condition that the rate of dissipation of total internal energy in any kinematic admissible collapse mechanism should be equal to the rate of total work done by various external and body forces. For two dimensional problems, mathematically this equality can be expressed as :

$$\int_A \sigma_{ij}\, \delta\varepsilon_{ij}\, dA + \int_L t_i\, [V]_i\, dL = \int_S t_i\, V_i\, dS + \int_A \gamma_i\, V_i\, dA \qquad (1)$$

where the first two terms represent the rate of internal work done (dissipation of internal energy) by the stresses σ_{ij} over the strain rates $\delta\varepsilon_{ij}$ within region A, and by the tractions t_i over the velocity jump $[V]_i$ along the velocity discontinuity line L. The last two terms define the rate of external work of tractions t_i over velocities V_i along boundary line S, and of body forces γ_i over velocities V_i in region A.

The stresses satisfy the yield condition (corresponding to partial shear strength parameters of soil) and are related to the strain rates by the associated flow rule (normality condition) :

$$\delta\varepsilon_{ij} = \delta\lambda \; \partial f(\sigma_{ij}) / \partial\sigma_{ij} \qquad (2)$$

Where $\delta\varepsilon_{ij}$ is the strain rate tensor, σ_{ij} is the corresponding stress tensor satisfying the yield condition $f(\sigma_{ij}) = 0$ with its partial shear strength parameters c_m and ϕ_m, and $\delta\lambda$ is a non-negative plastic multiplier.

If the material bounded by rupture and boundary surfaces is subdivided into different rigid regions undergoing either translation or rotation, the strain rates $\delta\varepsilon_{ij}$ within region A will become equal to zero, and equation (1) will become

$$\int_L t_i [V]_i \, dL = \int_S t_i V_i \, dS + \int_A \gamma_i V_i \, dA \qquad (3)$$

For a material obeying the Mohr-Coulomb yield condition and the associated flow rule, the velocity jump vector $[V]_i$ must incline at an angle ϕ_m with the velocity discontinuity line; the magnitude of product $t_i[V]_i$ then equals $c_m[V]_i \cos\phi_m$.

Kinematically Admissible Rupture Surface

If the soil mass bounded by the periphery of the slope and the associated rupture surface is assumed to rotate as a single rigid body, the shape of the rupture surface must become an arc of logarithmic spiral, as shown in Fig.1, with radius, $r = r_0 \exp(\theta \tan\phi_m)$. The angle between the direction of velocity and tangent to the spiral arc at any point will become ϕ_m.

Expressions for the Rate of Work Done and Energy

For any value of ϕ_m, the rupture surface can be drawn if the locations are assumed for the pole (P) of log-spiral arc and a point A where the rupture surface joins near the crest of the slope (refer Fig.1). The rupture surface shown in Fig.1 is an example of the base failure of the slope, where the rupture surface touches the hard stratum below the toe of the slope. The rate of total work done by the body forces due to soil mass $ABCDC'B'A$ can be found by summing algebraically the rates of work done due to log spiral sectors PAB', $PB'C'$ & $PC'D$ and the triangular sectors PAB, PBC & PCD,

$$W_{TOTAL} = W_{ABCDC'B'A} = W_{PAB'} + W_{PB'C'} + W_{PC'D} - W_{PAB} - W_{PBC} - W_{PCD} \qquad (4)$$

Here, any symbol W_i represents the rate of external work done by body forces due to the soil mass contained within area A_i. The rates of work done by various log-spiral sectors and triangular sectors in the above equation were determined by deriving first the basic expressions for the rate of work done due to log-spiral sector opq and triangular sector ost as shown in Figs.1(b) & 1(c). The final expressions are given below:

The rate of work done by body forces for the log-spiral sector opq :

$$W_{opq} = \frac{\gamma \omega r_0^3}{3(1+9\tan^2\phi_m)} \{ \exp[3\tan\phi_m(\theta_1-\theta_0)](3\tan\phi_m\cos\theta_1+\sin\theta_1)- 3\tan\phi_m\cos\theta_0-\sin\theta_0\} \quad (5)$$

The rate of work done by body forces due to triangular sector ost:

$$W_{ost} = \frac{\gamma \omega R_0^3 \sin^3(\psi_0+\alpha)}{3} [0.5\cos\alpha/\sin^2(\psi_0+\alpha) + \sin\alpha\cot(\psi_0+\alpha) - 0.5\cos\alpha/\sin^2(\psi_1+\alpha) - \sin\alpha\cot(\psi_1+\alpha)] \quad (6)$$

Fig.1 can be referred to for the definitions of various symbols in the above expressions; ω is the angular velocity of the rigid mass.

While using the above expressions for determining the rate of work done by various sectors for given values of γ, ω & ϕ_m, the appropriate values were substituted for basic parameters: (i) θ_0, θ_1, r_0 & r_1 in case of log spiral sectors; and (ii) ψ_0, ψ_1, α, R_0 & R_1 in case of triangular sectors. The rate of dissipation of total internal energy along the periphery of log-spiral arc pq (refer Fig.1b) is given by the following expression:

$$E_{pq} = \frac{\omega c_m r_0^2}{2\tan\phi_m} \{ \exp[2(\theta_1-\theta_0)\tan\phi_m] -1 \} \quad \text{for } \phi \# 0 \quad (7\text{ a})$$

$$E_{pq} = \omega c_m r_0^2(\theta_1-\theta_0) \quad \text{for } \phi=0 \quad (7\text{ b})$$

For any trial rupture surface, the rate of dissipation of internal energy in the slope mass for given values of c_m and ϕ_m can be determined by substituting the appropriate values of parameters θ_1, θ_0 & r_0 in the above expressions.
The rate of dissipation of total energy, $E_{TOTAL} = E_{AB}/c/_D$

Likewise, for the slope and toe failures, the corresponding final expressions for the rate of work done by body forces and the rate of dissipation of internal energy can be determined by using basic equations 5, 6 & 7.

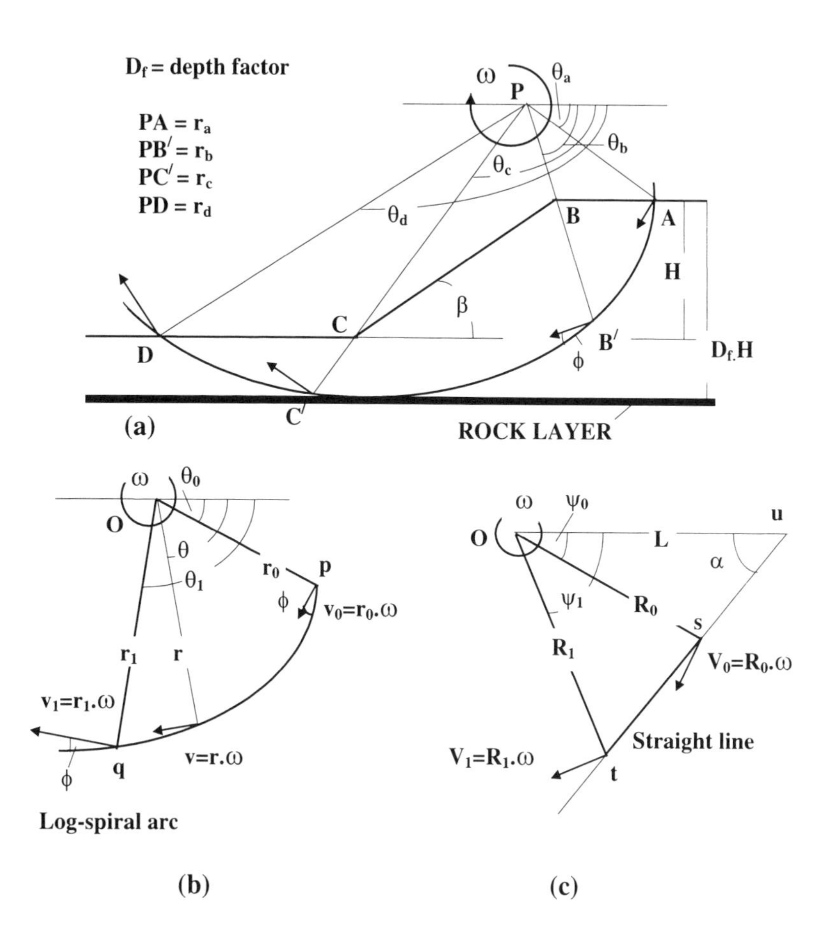

Figure 1. Slope Geometry and Collapse Mechanism.

Solution Procedure

Case 1: ϕ=0. For ϕ=0, the equation of log-spiral ($r = r_o e^{\theta \tan\phi}$) becomes same the as that of a circle. The circular rupture surface was therefore drawn by selecting suitable locations of points P and A. The magnitude of W_{TOTAL} was then determined for ϕ_m=0 by making use of Eqs. 4, 5 & 6. By making use of the upper bound theorem of limit analysis, E_{TOTAL} was equated to W_{TOTAL}. From the known value of E_{TOTAL}, the equation 7(b) was then used to determine unknown c_m, from which the factor of safety (F) becomes, $F = c/c_m$.

Case 2: ϕ # 0. In this case, the values of both E_{TOTAL} and W_{TOTAL} are functions of unknown parameter ϕ_m. Even with the same locations of points P and A, the path of rupture surface varies with changes in ϕ_m. The magnitude of ϕ_m is correct if it gives exactly the same values for both E_{TOTAL} and W_{TOTAL}. Let $\phi_{m,i}$ be the trial value of ϕ_m at any stage, and $\phi_{m,min}$ & $\phi_{m,max}$ are corresponding minimum and maximum possible values of $\phi_{m,i}$ during that stage. The following solution technique was adopted in order to determine the correct value of ϕ_m :

(i) Locations of points P and A were assumed.

(ii) The initial values of $\phi_{m,min}$ & $\phi_{m,max}$ were set equal to zero & $\pi/2$ respectively. With trial value of $\phi_{m,i} = \phi_{m,min}$, the rupture surface was drawn.

(iii) The values of E_{TOTAL} and W_{TOTAL} were determined. If the value E_{TOTAL} becomes nearly equal to W_{TOTAL}, the upper bound theorem of limit analysis is satisfied and $\phi_{m,i}$ is the correct value of ϕ_m.

(iv) If the value of E_{TOTAL} is found to be less than W_{TOTAL}, $\phi_{m,min}$ was equated to $\phi_{m,i}$. On the other hand, if the value of E_{TOTAL} was found to be greater than W_{TOTAL}, the value of $\phi_{m,max}$ was set equal to $\phi_{m,i}$.

(v) The value of $\phi_{m,i}$ was then replaced by ($\phi_{m,min} + \phi_{m,max})/2$. The rupture surface was redrawn using the new value of $\phi_{m,i}$ keeping the same locations of points P & A , and steps (iii) & (iv) were followed again. This procedure continued till the convergence was attained.

(vi) The factor of safety was finally determined using $F = \tan\phi/\tan\phi_{m,i}$.

While doing the computations at any stage if the difference of E_{TOTAL} and W_{TOTAL} becomes less than 0.1% of the maximum of the absolute values of E_{TOTAL} and W_{TOTAL}, the convergence was assumed to be achieved. After obtaining the factor of safety for the assumed locations of P and A, it was also checked that the associated rupture surface does not go below the hard stratum.

After having obtained the magnitude of factor of safety for each assumed location of points P & A, a number of different possible combinations for the locations of points P & A were tried as to find the absolute minimum value of the factor of safety. The location of point P was varied in a square grid with spacing interval 0.02 H, and the location of point A was varied with an increment of distance of 0.02H along the horizontal ground starting from the crest of the slope. With these selections, the convergence up to second decimal places was found to be attained in the magnitude of factor of safety. Also, all different possible modes of failure mechanisms, viz. slope, toe and base failures, were considered so as to find the minimum factor of safety.

Results and Comparison

The results were obtained with respect to the variation of slope inclination (β) from 10 to 90 degrees, cohesion (c) from 0.025 to $0.1\gamma H$, friction angle ϕ from 0 to 50 degrees, and depth factor (D_f) from 1.0 to 2.0. For the purpose of comparison, the results were also determined in all the cases by using the Bishop simplified method on the basis of circular slip surface. A separate computer program was written for this purpose. The results obtained with the developed program on the basis of the Bishop method were further verified by using commercially available computer software SLOPE (1997).

The comparison of results between the upper bound limit analysis and the Bishop method is shown in (i) Figs. 2 & 3 for D_f=1.5 ; and (ii) Table 1 for D_f =1.0 & 2.0. It can be seen that the factor of safety of the slope decreases with (i) increase in slope inclination (β) ; (ii) increase in depth factor (D_f); (iii) decrease in non-dimensional soil cohesion factor, $c/(\gamma H)$; and (iv) decrease in friction angle (ϕ). The uppermost curves in the figures are corresponding to $\phi=50^o$, and corresponding lowermost curves are for $\phi=0$. The results using the upper bound theorem of limit analysis compare well with the Bishop method in most of the cases except for values of slope inclinations(β) approximately greater than around 70 degrees. For slope inclinations lesser than 70 degrees the maximum difference between the two methods remains less than about 7%. However, for slopes with too steep inclinations the difference between the two methods increases. It can be seen from Table 1 that the maximum difference between the two methods for D_f=1.0 is (i) 36.86 % for β=90 degrees ; (ii) 6.71 % for β=10 degrees; and (iii) 2.93 % for β=40 degrees. The corresponding maximum difference between the two methods for D_f =2.0 is (i) 22.63% for β=90 degrees ; (ii) 2.71% for β=10 degrees ; and 1.33% for β=40 degrees. The present findings are similar to those of Michalowski(1995), where he presented a comparison between the two methods, in terms of non-dimensional stability numbers, for homogeneous slopes without any hard stratum at its bottom (i.e. D_f is infinite theoretically). The slope inclination was varied from 20 to 90 degrees. It was shown in his comparisons that the upper bound limit analysis, using rigid rotational collapse mechanism, compares well with that of the Bishop method in most of the cases except for very steep slope inclinations where the Bishop method provides conservative estimates of the factor of safety as compared to that of upper bound theorem of limit analysis.

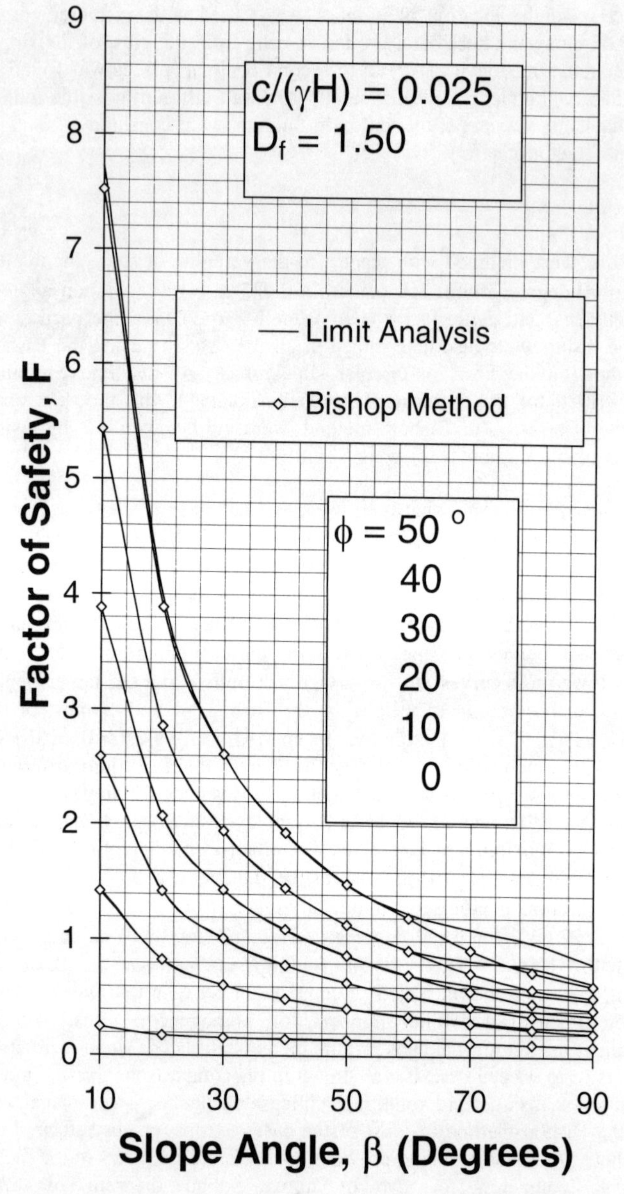

Figure 2. Factor of Safety for D_f=1.50 and $c/(\gamma H)$=0.025.

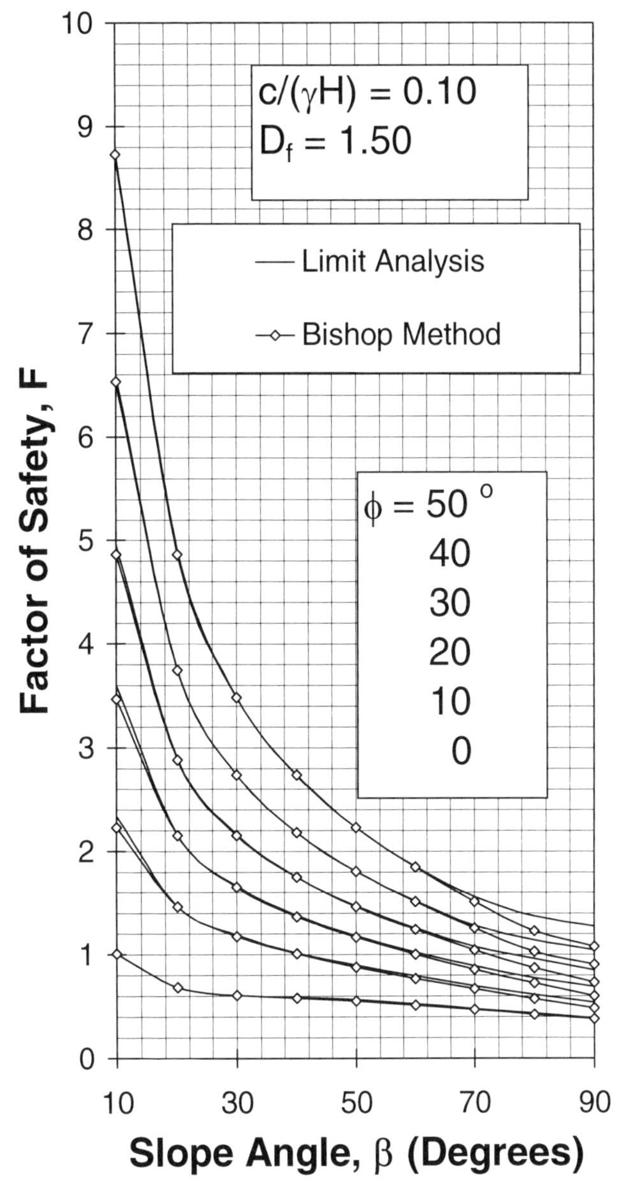

Figure 3. Factor of Safety for D_f=1.50 and c/(γH)=0.10.

Table 1. Difference between the Bishop method and upper bound limit analysis.

$c/\gamma H$	D_f	β	$\phi=10^o$			$\phi=50^o$		
			Factor of Safety		Difference (%)	Factor of Safety		Difference (%)
			Limit Analysis	Bishop Method		Limit Analysis	Bishop Method	
0.025	1.0	10	1.63	1.53	6.71	7.72	7.56	2.18
		40	0.50	0.49	1.74	1.99	1.97	1.26
		90	0.29	0.25	14.4	0.92	0.67	36.86
	2.0	10	1.42	1.42	0.00	7.72	7.52	2.71
		40	0.49	0.49	0.00	1.93	1.93	0.00
		90	0.22	0.20	8.78	0.76	0.62	22.63
0.10	1.0	10	2.89	2.81	2.98	9.49	9.08	4.47
		40	1.06	1.03	2.93	2.78	2.75	1.19
		90	0.65	0.57	14.02	1.51	1.28	17.67
	2.0	10	2.15	2.15	0.00	8.73	8.73	0.00
		40	1.02	1.01	1.33	2.73	2.73	0.00
		90	0.53	0.48	10.38	1.15	1.08	6.76

The present study also reveals that in all the cases that the computational efforts involved in finding the factor of safety using the upper bound theorem of limit analysis is much more extensive than that of the Bishop method. For any given locations of P and A, in order to achieve convergence in the value of the factor of safety, the upper bound theorem of limit analysis requires at least about 50 to 100 iterations (changes in $\phi_{m,i}$ during each iteration), whereas the Bishop method requires hardly 10 to 20 iterations (changes in FOS during each iteration) in most of the cases.

Soil Dilatancy

It should be mentioned that the results have been obtained with the assumption that soil material obeys the associated flow rule. The dilatancy (increase in volume during soil shear) which is predicted on the basis of this assumption is usually more than is actually observed in most soils. It has already been shown by Drescher & Detournay (1993) and Rowe & Davis (1982) that the assumption of associated flow rule material gives higher values of stability numbers for slopes and greater collapse loads for foundations/anchors. As a result, the true factor of safety values for slopes with non-associated flow rule material will be a little lower than those given in this article.

Concluding Remarks

Step wise procedure to compute factor of safety using upper bound theorem of limit analysis has been given. The results have been compared with those using the Bishop simplified method. The difference between the two methods remains small in most of the cases except for very steep slopes in which cases the Bishop method provides a little lower factor of safety. Because of the consideration of the kinematics of the problem, the upper bound theorem of limit analysis has an advantage over the limit equilibrium method. However, its application to the problem of the factor of safety determination of slopes is computationally more difficult. The ways to include the effect of pore water pressures and earthquake forces in carrying out the upper bound limit analysis are reported in literature [Michalowski (1995) and Soubra (1997)]. However, still hardly any information exists where this method has been extended to a general slope stability problem with the existence of a number of different soil layers.

References

Bishop, A. W. (1955). "The use of slip circle in the stability analysis of slopes." *Geotechnique*, 5(1), 7-17.

Bishop, A. W., and Morgenstern, N. R. (1960). "Stability coefficients for earth slopes." *Geotechnique*, 10(4), 129-150.

Chen, W.F., and Giger, M. W. (1971). "Limit analysis of stability of slopes." *Journal Soil Mech. Found. Engrg. Div., ASCE*, 97 (1), 19-26.

Chen, W.F. (1975). "Limit analysis and soil plasticity." Amsterdam: *Elsevier*.

Drescher, A. & Detournay, E.(1993). "Limit load in translational failure mechanisms for associated and non-associated materials." *Geotechnique*, 43(3), 443-456.

Karal, K. (1977). "Energy method for soil stability analyses." *Journal Geotech. Engrg. Div., ASCE*, 103(5), 431-447.

Karal, K. (1977). "Application of energy method." *Journal Geotech. Engrg. Div., ASCE*, 103(5), 381-399.

Michalowski, R. L. (1995). "Slope stability analysis: a kinematical approach." *Geotechnique*, 45(2), 283-293.

Morgenstern, N. R. and Price, V. E.(1965). "The analysis of the stability of general slip surfaces." *Geotechnique*, 15(1), 79-93.

SLOPE (1997). "Slope stability analysis program by D.L. Borin." Version: 8, *Geosolve*, London.

Soubra, A.H.(1997). "Seismic bearing capacity of shallow strip footings in seismic conditions." *Proc. Inst. of Civil Engineers, Geotechnical Engineering*, 125(4), 231-241.

Taylor, D. W. (1948). "Fundamental of soil mechanics." New York. *John Wiley*, 1948.

Analysis of the Progression of Failure of Earth Slopes by Finite Elements

J. B. Lechman[1]
D. V. Griffiths[2]
Member A.S.C.E.

Abstract

A significant advantage of the finite element method over traditional limit equilibrium methods for analyzing slope stability is its ability to monitor the progress of the failure zone up to and including overall shear failure. This advantage results from the finite element's ability to calculate stresses in the slope that result from the application of gravity, or other applied loads, and subsequently compare these stresses to some failure criterion. Within the method, various construction procedures (i.e. loading paths) such as the embankment or excavation can be modeled. In the present paper, finite element stability analysis of a slope with a relatively simple elasto-plastic, Mohr-Coulomb soil model is used to investigate the spread of yield within the slope under differing loading strategies. In particular, loading is accomplished in three ways: gravity applied instantaneously to an initially weightless slope, gravity applied incrementally to an initially weightless slope, and through building the slope up one row of elements at a time (i.e. embankment construction). The influence of loading strategies on the spread of failure is highlighted through the use of contour plots, which indicate the spread of yield within the slope and hence the location and shape of the potential failure surface. Furthermore, the affects of the flow rule used on the yielding zones is shown.

1.0 *Introduction*

Prior to the advent of modern computing techniques, methods for determining the stability of slopes were, by necessity, a matter of making various assumptions to allow for the solving of equations of static equilibrium. The results of these limited, simplified analyses are presented in the form of charts of stability numbers (e.g. Taylor 1937, Bishop and Morgenstern 1960, Spencer 1967, Janbu

[1]Geomechanics Research Center, Colorado School of Mines, Golden, CO 80401
[2]Division of Engineering, Colorado School of Mines, Golden, CO 80401

1968, Cousins 1978) from which the factor of safety can be determined based on the soil's strength properties and slope geometry. Common to all methods from which the charts are derived is the assumption that the slope can be divided into slices. After this initial step, further assumptions are made in order to solve the problem of static indeterminacy created by the side forces acting on the failing mass slice. While the variety of assumptions for dealing with these side forces are in generally good agreement with regards to overall safety factor, they may produce large discrepancies in the distribution of stresses throughout the failure mass (Whitman and Bailey 1967, Wright, et al. 1973). Cousins (1978) also lists several restrictions to the use of these charts which deal with matters ranging from pore pressure, to limited slope angle range, to limited slope geometry, to limited or the lack of information about the failure surfaces. These charts are expedient and relatively easy to use, giving conservative values for the safety factor; however the aforementioned problems with limit equilibrium methods coupled with advances in computational ability beg the question of the possibility of more robust methods for analyzing slope stability.

One such method is the finite element method. The advantages of the method over limit equilibrium methods are stated by Griffiths and Lane (1999) as:

1. No assumption needs to be made in advance about the shape or location of the failure surface. Failure occurs "naturally" through the zones within the soil mass in which the soil shear strength is unable to sustain the applied shear stresses.

2. Since there is no concept of "slices" in the finite element approach there is no need for assumptions about slice side forces.

3. If realistic soil compressibility data is available, the finite element solutions will give information about deformations at working stress levels.

4. The finite element method is able to monitor progressive failure up to and including overall shear failure.

This method has shown itself to be in good agreement with the various charts and proves to be very robust in that complications arising from the geometry of the slope and material property variations can be easily managed by it (Zienkiewicz et al. 1975, Griffiths 1980, 1989, Matsui and San 1992, Griffiths and Lane 1999). Advances made in refining the method and its applications may well prove to be a defining step in the maturation of soil mechanics.

One such area that the method could be refined is that of understanding how yield develops and spreads through slopes as a function of loading path, yield criteria, and flow rule used. This understanding has potential benefits in the area of progressive failure of soils that exhibit shear strength reduction (i.e. where the back-calculated average shear strength after failure is less then the measured peak strength). Lo and Lee list the current state of knowledge, as of 1973, of conditions germane to the development of progressive failure as: 1)brittleness of soil; 2) non-uniformity in the distribution of shear stresses; and 3) a deterioration of strength with time because of such mechanisms as softening and change in ground water conditions. The interrelationships of these three factors quickly overtake traditional limit equilibrium methods, and their complexity suggests the use of numerical methods for their analysis (Bishop 1971). The fourth item listed by Griffiths and Lane (1999) is a result of the finite element method's ability to track the non-uniform distribution of stresses that result from loading the slope. This paper examines the effect of non-uniform distributions of stress, which result due to the nature of the slope stability problem (e.g. Clough and Woodward 1967), on the state of yielding in the slope via contour plots of the failure criterion. Traditional progressive failure with its implications in strain softening soil is not in view here. In fact, shear strength reduction is not incorporated into the soil model used here. The focus is strictly on the development of yield zones in an elasto-plastic, Mohr-Coulomb soil as a function of loading and flow rule.

2.0 *Method of Analysis*

2.1 *Finite element method used*

Two programs are used in the analysis. The first is based closely on Program 6.2 in the text by Smith and Griffiths (1998)–the main differences being the ability to model more realistic geometries, better graphical output facilities, and the ability to apply gravity in percentage increments. The second program is a similarly modified version of Program 6.9 from the same text. However, it has the added capability of adding one row of elements at a time to the mesh, which starts from a rectangular foundation layer with initial stresses due to gravity and at rest lateral earth pressure. Stresses accumulate in this program due to the weight of each successive row of elements that are added.

Both programs are for 2-d (plane strain) analysis of elastic-perfectly plastic soils with a Mohr-Coulomb failure criterion. Both use 8-node quadrilateral elements with reduced integration (4 Gauss-points per element) in both the stiffness and stress redistribution phases of the algorithm. A gravity 'turn-on' procedure generates nodal forces which act in the vertical direction at all nodes. These loads are applied in a single increment, or in percentage increments which generate normal and shear stresses at all the Gauss-points within the mesh. These stresses are then compared with the Mohr-Coulomb failure criterion. If the stresses at a particular Gauss-point lie within the Mohr-Coulomb failure envelope then that location is assumed to remain elastic. If the stresses lie on or outside the failure envelope, then that location is assumed to be yielding. Global shear failure occurs when a sufficient number of Gauss-points have yielded to allow a global failure mechanism to develop; causing the displacements in the slope to increase dramatically.

The analysis is based on an iterative Modified Newton-Raphson method called the Viscoplastic algorithm (Zienkiewicz *et al* 1975). The algorithm forms the global stiffness matrix once (except for the embankment program where the stiffness matrix is reformed with each new lift added, but otherwise the analysis is the same) with all nonlinearity being transferred to the right hand side. If a particular zone within the soil mass is yielding, the algorithm attempts to redistribute those excess stresses by sharing them with neighboring regions that still have reserves of strength. The redistribution process is achieved by the algorithm generating self-equilibrating nodal forces which act on each element that contains stresses that are violating the failure criterion. These forces, being self-equilibrating, do not alter the overall gravity loading on the finite element mesh, but do influence the stresses in the regions where they are applied. In reducing excess stresses in one part of the mesh however, other parts of the mesh that were initially 'safe' may now start to violate the failure criterion themselves necessitating another iteration of the redistribution process. The algorithm will continue to iterate until both equilibrium and the failure criterion at all points within the soil mass are satisfied within quite strict tolerances. Convergence is achieved in a global sense by observing the change in nodal displacements from one iteration to the next. Convergence is satisfied when this change is less than 0.1%.

If the algorithm is unable to satisfy these criteria at all yielding points within the soil mass, 'failure' is said to have occurred. Failure of the slope and numerical non-convergence occur together, and are usually accompanied by a dramatic increase in the nodal displacements. Within the data, the user is asked to provide an iteration ceiling beyond which the algorithm will stop trying to redistribute the stresses. Failure to converge implies that a mechanism has developed and the algorithm is unable to simultaneously satisfy both the failure criterion (Mohr-Coulomb) and global equilibrium.

2.2 *Soil model*

An elastic-perfectly plastic (Mohr-Coulomb) model has been used in this work consisting of a linear (elastic) section followed by a horizontal (plastic) failure section.

The soil model used in this study consists of six parameters as shown in Table 1.

The dilation angle ψ affects the volume change of the soil during yielding. In this simple model ψ is assumed to be constant which is unrealistic, but will serve to show the affect of dilation on the final failure region. ψ becomes important during failure due to it use in calculating the visco-plastic strains. The assump-

Table 1: Six–parameter model

ϕ'	Friction angle
c'	Cohesion
ψ	Dilation angle
E	Young's modulus
ν	Poisson's ratio
γ	Unit weight

tion commonly used is that the plastic potential function has the same form as the failure function, but with ϕ replaced by ψ.

The parameters c' and ϕ' refer to the cohesion and friction angle of the soil. Although a number of failure criteria have been suggested for use in representing the strength of soil as an engineering material, the one most widely used in geotechnical practice is due to Mohr-Coulomb. In terms of principal stresses and assuming a compression-negative sign convention, the criterion can be written as follows:

$$F = \frac{\sigma_1' + \sigma_3'}{2} \sin\phi' - \frac{\sigma_1' - \sigma_3'}{2} - c' \cos\phi' \tag{1}$$

where c' and ϕ' represent the shear strength parameters of the soil and σ_1' and σ_3' the major and minor principal effective stresses at the point under consideration. The failure function F can be interpreted as follows:

$F < 0$	inside M-C envelope (elastic)
$F = 0$	on M-C envelope (yielding)
$F > 0$	outside M-C envelope (yielding) and stresses must be redistributed

The unit weight γ assigned to the soil is important because it is proportional to the nodal loads generated by the gravity turn-on procedure.

In summary, the most important parameters in a finite element slope stability analysis are the unit weight γ which is directly related to the nodal forces trying to cause failure of the slope, and the shear strength parameters c' and ϕ' which measure the ability of the soil to resist failure.

2.3 *Determination of the factor of safety*

The Factor of Safety (FOS) of a soil slope is defined here as the factor by which the original shear strength parameters must be reduced to bring the slope to the point of failure. The factored shear strength parameters c_f' and ϕ_f', are therefore given by:

$$c'_f = c'/SRF \tag{2}$$

$$\phi'_f = \arctan(\frac{\tan\phi'}{SRF}) \tag{3}$$

where SRF is the strength reduction factor, and at failure $SRF = FOS$.

This method has been referred to as the 'shear strength reduction technique' (e.g. Matsui and San 1992) and allows for the interesting option of applying different factors of safety to the c' and $\tan\phi'$ terms. In this paper however, the same factor is always applied to both terms. To find the 'true' factor of safety, it is necessary to initiate a systematic search for the value of SRF that will just cause the slope to fail. This is achieved by the program solving the problem repeatedly using a sequence of user-specified SRF values.

However, when gravity is applied incrementally, the factored strength parameters that will allow the slope to fail at full gravity are held constant as loading is increased; therefore only the case where $SRF = FOS$ will be considered during this loading condition. A similar approach was used during the embankment procedure, where the strength parameters are factored to cause failure at the same embankment height as the slope considered with other loading strategies.

2.4 *Visualization technique*

For each successive trial safety factor, gravity load increment, or embankment lift the programs described above give values of the failure criterion (F) at each Gauss-point at convergence or failure. The coordinates of each Gauss-point and the corresponding value of F are then written to an output file which can be used in a commercial software package to produce contour plots which show the yielding regions of the slope. Of interest in the plots is the affect of an associated flow rule ($\phi = \psi$) on the spread of yield through the slope.

Furthermore, since contour plots described also give the location and shape of the final controlling mechanism, it is interesting to compare the results with traditional methods, namely those reported by Cousins (1978) which were arrived at by a modified Taylor method. The charts produced by Cousins are of particular interest because they not only give the stability factor but also the depth that the failure surface extends into the foundation of the slope. The finite element results can be compared to these results on the basis of overall factor of safety. Or, when the failure criterion is contoured, the two methods can be compared on the basis of the location and shape of the global failure surface to determine whether the final mechanism that results from the finite element analysis is in agreement with mechanisms from traditional methods (e.g. circular in shape, passes through the toe of the slope, etc.).

3.0 *Results*

For the purposes of this paper, a simple, dry slope composed of a soil exhibiting friction and cohesion was analyzed. The slope was taken from Griffiths and Lane (1999) example 2. In general, this simplification is not necessary in the finite element method. However, when comparing the results to traditional methods, simplification is necessary on the grounds that limit equilibrium methods typically require this in order that the problems may be solved with charted solutions. From the geometry and strength parameters, specific points on Cousins' charts at which the depth factor (D, the amount by which the height of the slope should be multiplied by to give the maximum height of the failure surface) can be determined explicitly. It should be noted that the definition of D in the context of Cousins' paper differs from other definitions of D. Cousins' paper uses D as a multiplier of the height of the slope (H) to give the maximum depth through which the failure surface passes and not as a multiplier of the height to give the depth to a solid base. Defined this way, a depth factor equal to 1 (D=1) implies a toe failure, where D=1.5 implies a failure surface that extends into the foundation material a distance equal to half the height and does not necessarily pass through the toe of the slope. Furthermore, gravity loading methods are examined as they relate to the spread of yield and the shape and location of the yielding zone at failure.

Also, an examination of the influence of dilation on the spread of yield and the final state of the yielding zone at failure is contained here. The analyses in this paper have, by virtue of the variability of the amount of actual volume change, been limited to two simple cases. The cases considered are where $\psi = 0$ and $\psi = \phi'$. When the angle of dilation equals the internal angle of friction ($\psi = \phi'$) the flow rule is "associated"; thus direct comparisons to classical plasticity can be made.

Some results of the analysis follow. It should be noted, however, that Figures 1 and 2 show the state of yield in a slope with various strength parameters, given by Equations 2 and 3, for a slope with gravity applied to the full end of construction slope instantaneously. This is useful to show the affects of the non-uniform distribution of stresses on the state of yield in the slope, but Figures 1 and 2 are not intended to imply a connection from one slope stability analysis (i.e., SRF value) to the next. Furthermore, some figures imply regions of the slope which yield (e.g. the deep failure in Figure 4). This may be due to 'round-off' error in choosing the value of the contour interval and slight fluctuations in the stress redistribution. Otherwise, these regions can be interpreted as indicating zones of unloading where some plastic deformation has occurred, but the current stress state due to redistribution of excess stresses is slightly below one that would cause failure.

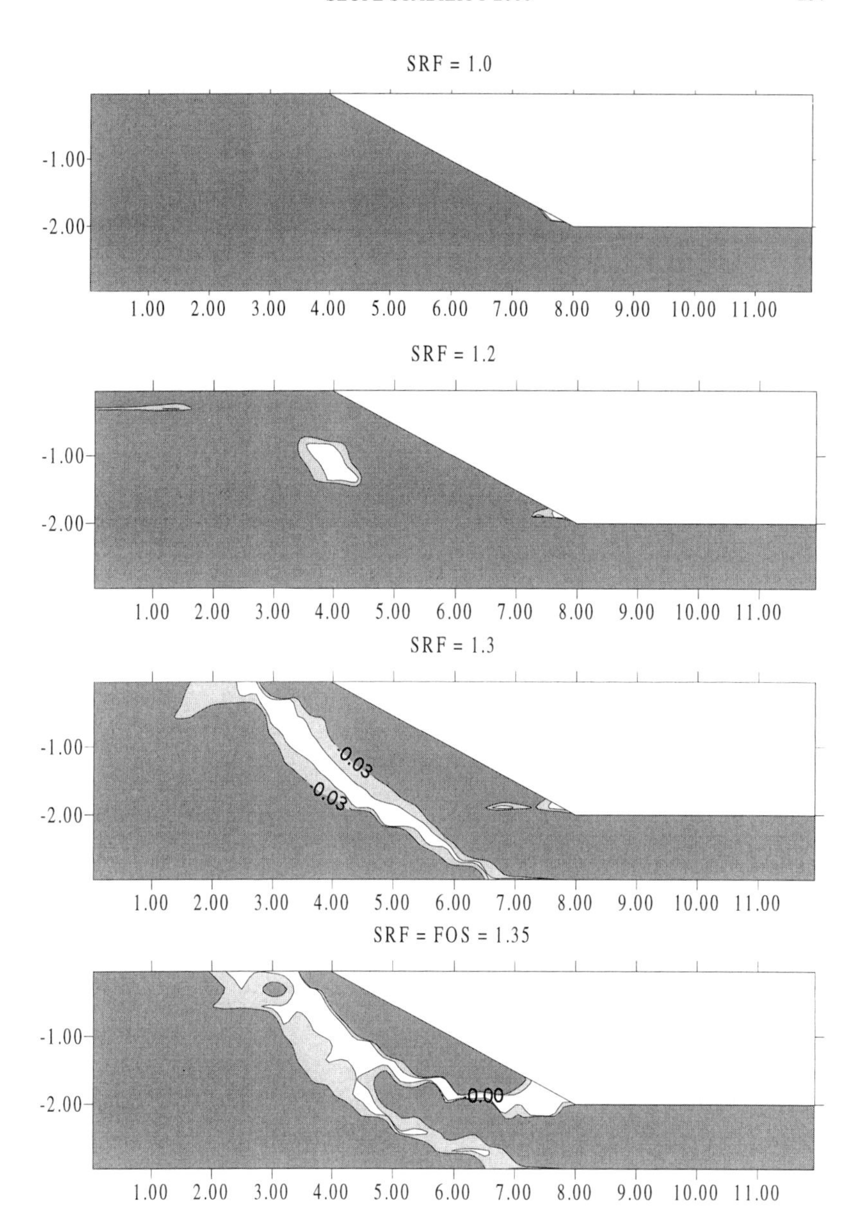

Figure 1: Figure 1. Embankment with a 2:1 slope, $c'/(\gamma H) = 0.05$, $\phi' = 20.0°$, $\psi = 0°$, and gravity applied instantaneously. Plots show contours of F at various SRF values. White regions are yielding ($F = 0$).

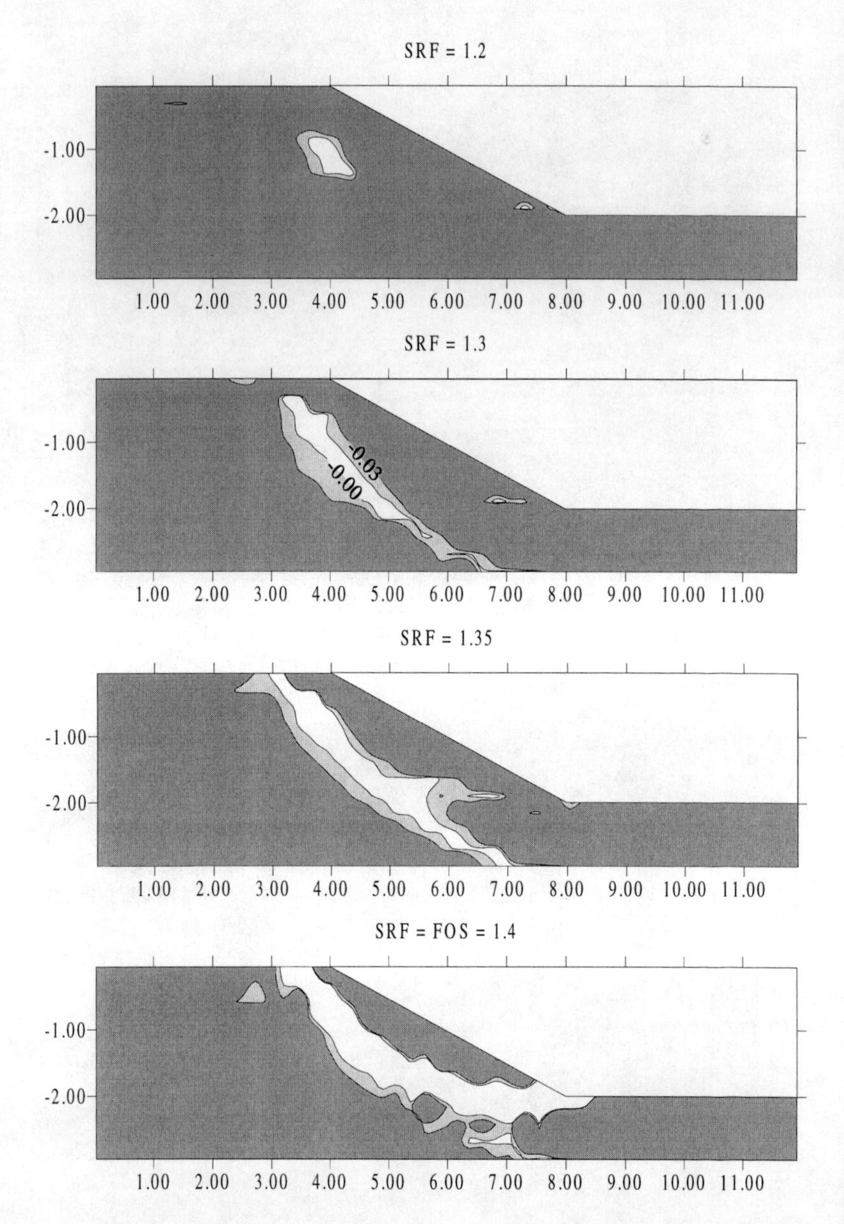

Figure 2: Figure 2. Embankment with a 2:1 slope, $c'/(\gamma H) = 0.05$, $\phi' = \psi = 20.0°$, gravity applied instantaneously. Plots show contours of F at various SRF values. White regions are yielding ($F = 0$).

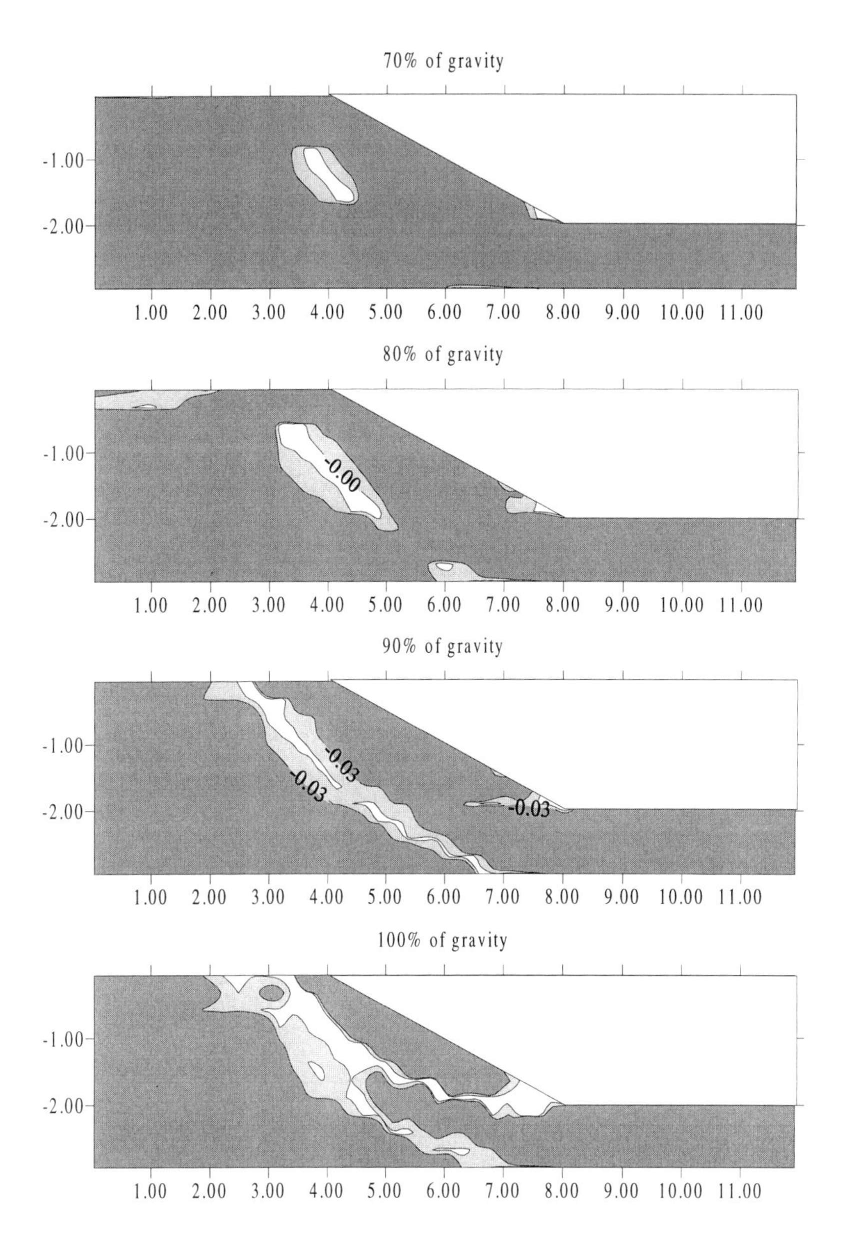

Figure 3: Figure 3. Embankment from Figure 1. Plots show contours of F at $SRF = FOS = 1.35$ with gravity applied incrementally. White regions are yielding ($F = 0$).

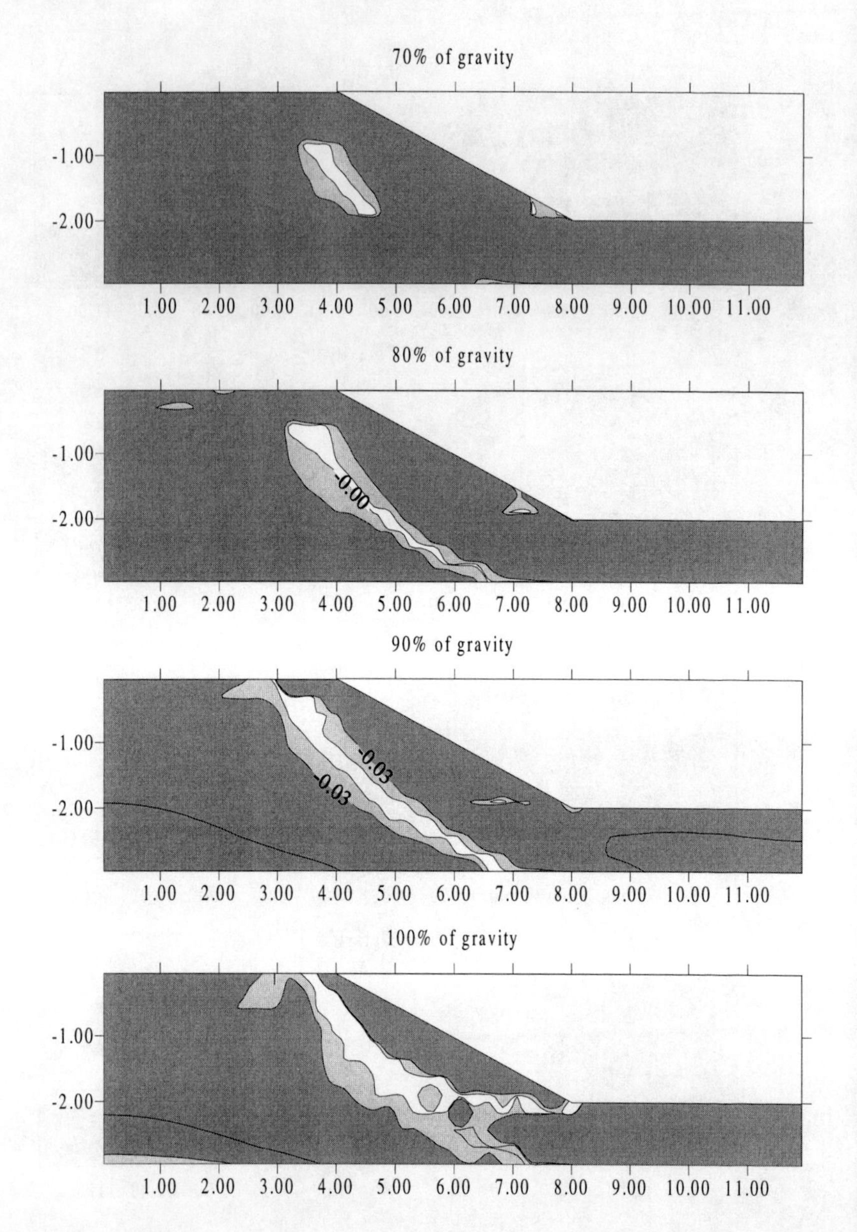

Figure 4: Figure 4. Embankment from Figure 2. Plots show contours of F at $SRF = FOS = 1.35$ with gravity applied incrementally. White regions are yielding ($F = 0$).

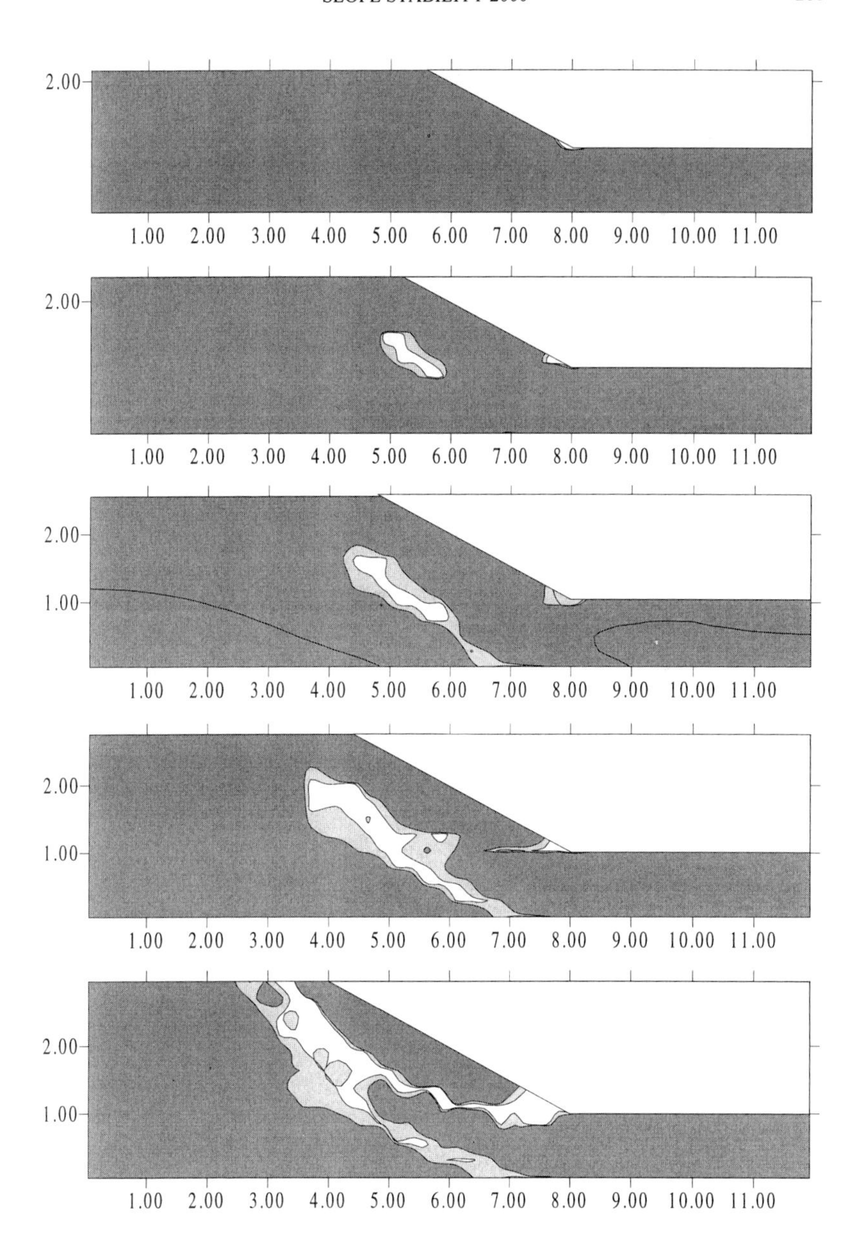

Figure 5: Figure 5. "Built-up" embankment with a 2:1 slope, $c'/(\gamma H) = 0.05$, $\phi' = 20.0°$, $\psi = 0°$. Plots show contours of F at various SRF values. White regions are yielding ($F = 0$).

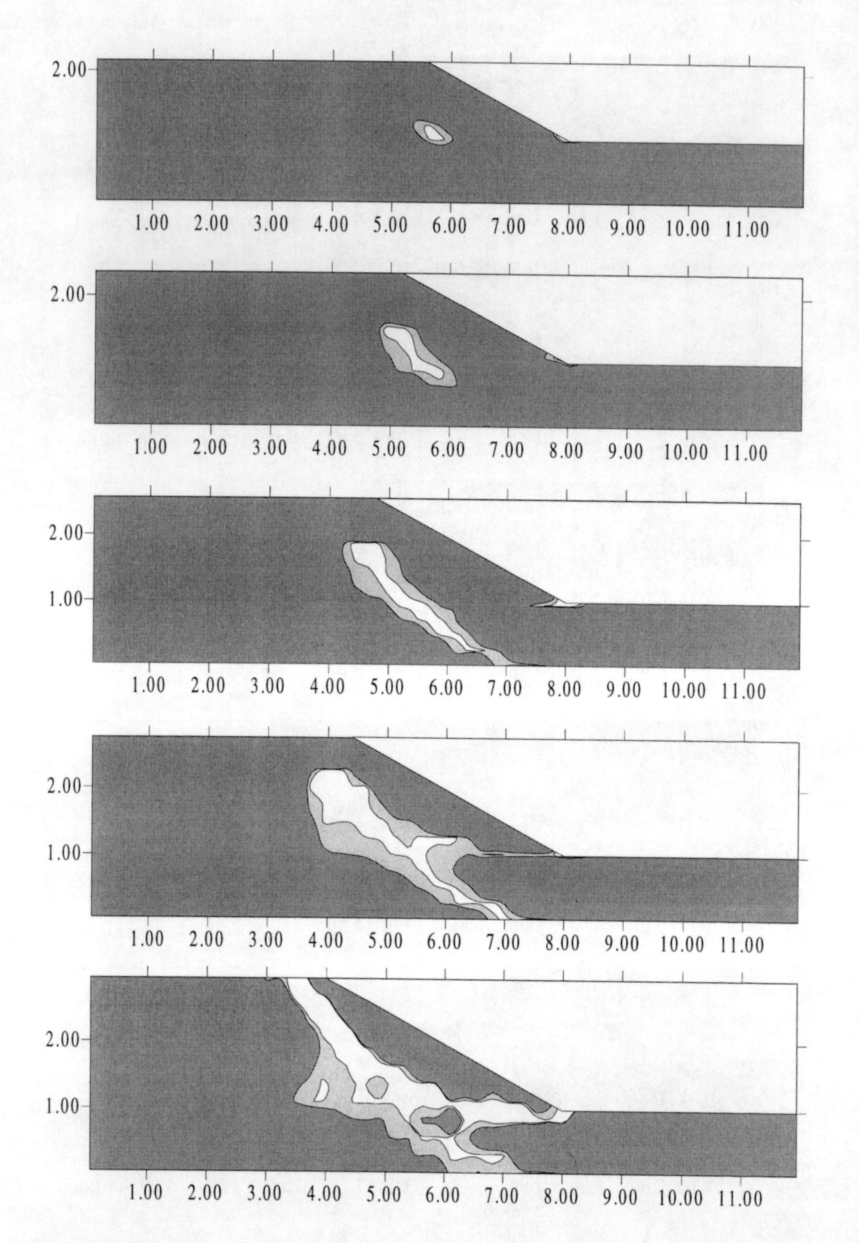

Figure 6: Figure 6. "Built-up" embankment with a 2:1 slope, $c'/(\gamma H) = 0.05$, $\phi' = \psi = 20.0°$. Plots show contours of F at various SRF values. White regions are yielding ($F = 0$).

4.0 *Conclusions*

Three methods for loading a soil slope have been discussed, and contour plots of the failure function have been made and compared for different cases of the flow rule. Although not all the results of the analyses are shown above, some general conclusions drawn from the investigation can be summarized as follows:

4.1 *Comparisons to Cousin's results*

1. Agreement was good for overall factor of safety. For the slope in the previous figures, Cousins' charts give a factor of safety of approximately 1.38 compared to a factor of safety of 1.35 to 1.38 for the finite element analysis.

2. Cousins' charts also give a depth factor (D) of 1.0 which is approximately equal to that indicated by the finite element analysis. As stated earlier, a priori knowledge of the failure surface is not required in finite element analysis as it is in limit equilibrium analysis; therefore the failure surface is allowed to develop "naturally." Although the two failure surfaces are not arrived at by the same means, the results indicate that Cousins' charts agree well with finite element results.

4.2 *Effects of Dilation*

1. Overall factor of safety did not vary significantly (less than 5%), for the same slope geometry and soil strength parameters, between the two cases of the flow rule.

2. At failure, the yielding zones of the two cases appear are similar in size and shape (See Figures 1 and 2).

3. Although the failure zones were similar at failure, the spread of yield could vary significantly between the two cases.

4. Under normal gravity conditions for a given slope geometry and soil strength parameters (represented by a factor of safety equal to one in this analysis), regions within the slope may already be experiencing stress conditions that promote yielding of the slope material. These regions were exceptionally large in slopes with relatively small safety factors. Usually, these regions of yield appeared in the $\psi = 0$ case, and corresponded with a deep failure surface. For the associated flow rule case, this tendency was not as prominent; indicating an increased resistance to shear in the relatively more confined foundation material.

4.3 *Effects of applying gravity incrementally*

Selected results of this analysis are shown in figures 3 and 4. Here the trial safety factor was set to the value that caused failure in the analysis where gravity was applied all at once. The shear strength parameters were factored accordingly, and held constant. Gravity was then applied incrementally until failure occurred at 100%.

1. Overall factor of safety was virtually unchanged. For both dilation cases.

2. The general shape and location of the yielding zones appear to be equivalent at failure while some differences can be seen at other trial factor of safety values.

3. For both cases considered, a “deep” failure mechanism developed first, but apparently was resisted by frictional forces along the rigid foundation below. In general, when cohesion “dominates” the soils ability to resist shearing, a deep failure is expected. This is particularly evident in Taylor's chart for undrained clays ($\phi' = 0$), but is also apparent in Cousins' charts for frictional/cohesive materials.

4.4 *Effects of Embankment Procedure*

Selected results of this analysis are shown in Figures 5 and 6. Here the trial safety factor was set to the value that caused failure in the analysis when the height of the slope reached a certain value. The shear strength parameters were factored accordingly, and held constant.

1. The overall factor of safety was only slightly effected for both cases of the flow rule, but larger for the associated flow case as would be expected.

2. As noted previously, The general shape and location of the controlling yield zone appears to vary only slightly at failure while some differences can be seen during the initiation and spread of yield due to the effect of dilation.

3. The contours of F correspond well to those of the incremental gravity case with the associated flow analyses being the ‘smoothest’ in terms of spread of yield.

4. Overall, the spread of yield, and particularly the shape and location of the final toe mechanism, do not show much difference from the incremental gravity results. An exception to this would be the early appearance of yield zones at the top of the slope for the incremental case. For an embankment, at the gravity increment where they begin to appear, these regions would not exist yet (i.e., they would not have been placed yet).

References

[1] A.W. Bishop. Progressive failure-with special reference to the mechanisms causing it. In *Proc. Geotec. Conf., Oslo*, volume 2, pages 142–150. 1967.

[2] A.W. Bishop and N.R. Morgenstern. Stability coefficients for earth slopes. *Géotechnique*, 10:129–150, 1960.

[3] R.W. Clough and R.J. Woodward. Analysis of embankment stresses and deformations. *J Soil Mech Found Div, ASCE*, 93(SM4):529–549, 1967.

[4] B.F. Cousins. Stability charts for simple earth slopes. *J Geotech Eng, ASCE*, 104(2):267–279, 1978.

[5] D.V. Griffiths. *Finite element analyses of walls, footings and slopes.* PhD thesis, Department of Engineering, University of Manchester, 1980.

[6] D.V. Griffiths. Advantages of consistent over lumped methods for analysis of beams on elastic foundations. *Comm Appl Numer Methods*, 5(1):53–60, 1989.

[7] D.V. Griffiths and P.A. Lane. Slope stability analysis by finite elements. *Géotechnique*, 49(3):387–403, 1999.

[8] N. Janbu. Slope stability computations. In *Soil Mech. and Found. Engrg. Rep.* Technical University of Norway, Trondheim, Norway, 1968.

[9] T. Matsui and K-C. San. Finite element slope stability analysis by shear strength reduction technique. *Soils Found*, 32(1):59–70, 1992.

[10] I.M. Smith and D.V. Griffiths. *Programming the Finite Element Method.* John Wiley and Sons, Chichester, New York, 3rd edition, 1998.

[11] E. Spencer. A method of analysis of the stability of embankments assuming parallel interslice forces. *Géotechnique*, 17(1):11–26, 1967.

[12] D.W. Taylor. Stability of earth slopes. *J. Boston Soc. Civ. Eng.*, 24:197–246, 1937.

[13] R.V. Whitman and W.A. Bailey. Use of computers for slope stability analysis. *J Soil Mech Found Div, ASCE*, 93(SM4):475–498, 1967.

[14] S.G. Wright, F.H. Kulhawy, and J.M. Duncan. Accoracy of equilibrium slope stability analysis. *J Soil Mech Found Div, ASCE*, 99(SM10):783–791, 1973.

[15] O.C. Zienkiewicz, C. Humpheson, and R.W. Lewis. Associated and non-associated viscoplasticity and plasticity in soil mechanics. *Géotechnique*, 25:671–689, 1975.

Liquefaction And Dam Failures

B. Muhunthan[1] and A. N. Schofield[2]

Abstract

This paper is based on a critical state soil mechanics concept that liquefaction occurs when soil is on the dry side of critical states, near zero effective stress, and in the presence of high hydraulic gradients. In this view liquefaction is one of a group of phenomena; including piping, boiling, fluidization; with pipes and channels and hydraulic fractures, internal erosion and void migration. This paper will refer to some aspects of the failures of Fort Peck, Baldwin Hills, and Teton Dams in support of this view. Casagrande held an opposite view that liquefaction occurs by a chain reaction among sand grains on the wet side of critical states.

Cam-clay provides a model for ductile stable yielding and deformation of an aggregate of grains wetter than critical states. A layer of such sediment can form folds during deformation. If a soil aggregate is denser than the critical state, it can fail with fault planes on which gouge material dilates and softens, or it can fracture and crack into a clastic debris, or develop pipes and channels. The critical state explanation of rapid failure is rapid transmission of pore water pressure through such opening cracks or channels.

The Baldwin Hills and Teton dam failures were failures with cracks and pipes. In the case of the Fort Peck failure we suggest that high pore pressures from the core hydraulic fill were transmitted in the layer beneath the part of the dam that failed; Casagrande's view of the failure as evidence of a "chain reaction" is questioned. Selection and control of fills to ensure ductility and avoid over compaction and measures to ensure stability are discussed.

[1] Associate Professor, Civil Engineering Department, Washington State University, Pullman, WA 99164 (email: muhuntha@wsu.edu). Formerly, Visiting Professor, Engineering Department, Cambridge University.
[2] Emeritus Professor, Engineering Department, Cambridge University, Trumpington Street, Cambridge, CB 2 1PZ, UK. (email: ans@eng.cam.ac.uk)

Introduction

Castro (1969) referred to Roscoe et al. (1958) as being the first to prove the existence of the critical void ratio as hypothesised by Casagrande in 1936, but as not contributing to the understanding of the flow structure in liquefaction that Casagrande postulated. Some years before Castro wrote this, many critical state concepts including the Cam-clay model of yielding for soils had been set out in detail by Roscoe and Schofield (1963), Schofield and Togrol (1966), Schofield (1966), and the text book on critical state soil mechanics (Schofield and Wroth 1968). The way Cam-clay yields on the wet side of critical state fits Castro's data of slow load cycles in his undrained triaxial tests Nos. 1 to 6 with Ottawa sand. Each load increment Castro applied led to yielding and an increment of pore pressure, but tests were stable until critical state friction was almost fully mobilised. Rapid failure was expected in his load controlled tests near critical state. It was not evidence of chain reactions and special flow structures among grains.

Cambridge teaching and research after 1968 placed increasing emphasis on geotechnical centrifuge modelling, particularly after the ISSMFE conference in Moscow (see discussion in Schofield 1998) when it became clear how helpful static and dynamic centrifuge tests would be in solving liquefaction problems. Co-operation began in 1975 between the Cambridge group and the US Army Engineer WES with a view to the eventual creation of the Army Centrifuge. About that time at the Fifth Pan-American Conference in Soil Mechanics in 1975 Casagrande restated his position in a paper on liquefaction where he made no reference to the work that had been in progress in Cambridge for 20 years. In Section III of this paper Casagrande reiterated his belief that "the greater the effective confining pressure, e.g., the greater the depth of a sand stratum, the lower is the critical void ratio; or, in other words, the denser must the sand be to be safe against (actual) liquefaction. But when heavily loaded, even a medium dense sand may be susceptible to (actual) liquefaction."

After 15 years of static and dynamic centrifuge model testing in Cambridge, Schofield (1980, 1981) argued that liquefaction in models and in the field was not as Casagrande supposed. Sudden liquefaction events are not due to an effectively stressed soil aggregate structure changing to a flow structure, with a chain reaction among the grains analogous to the phase transition when a solid melts and becomes a liquid. Liquefaction is not an event that occurs at a point like melting. It involves the geometry of a failure mechanism and is more like the buckling of struts. In many cases the cause is cracks or pipes and channels opening up in very stiff soil. The presence of a high hydraulic gradient rapidly transforms crumbling ground into a clastic debris flow.

It is also not sufficient to state that soil which liquefies is near to zero effective stress (Seed 1979). Sand on the surface of the desert or on the sea bed is near to zero effective stress, but it is only when the wind blows in the desert, or current flows over the sea bed, that sand dunes or sea bed waves are formed. Liquefaction requires pore fluid gradients. The following section considers the critical state view of soil behavior to explain this in detail, and then the paper turns to the dams.

Critical States, Folds, Faults, And Fractures Of Soil Aggregates

Aggregates of soil grains form deposits that exhibit three distinct classes of behavior (Fig. 1). Pressures at large depths cause ductile yielding of the aggregates and layer of sediment *folds*. Above these depths and at lower pressures aggregates rupture and layers of sediment *fault* with the presence of gouge material along slip planes. Near the surface where the pressure is nearly zero, layers of sediment *fracture* or fissure and aggregates can disintegrate. Critical state soil mechanics (CSSM) captures these simple geological phenomena of folds, faults, and fractures of sedimentary deposits. It explicitly recognizes that soil is an aggregate of interlocking frictional particles and that the regimes of soil behavior depend in a major way on its density and effective pressure. Detailed accounts of the basic principles, the features, and finite element applications of the CSSM framework have been presented in a number of publications. We present here only the features of the framework relevant to folds, faults, and fractures in the context of soil failure.

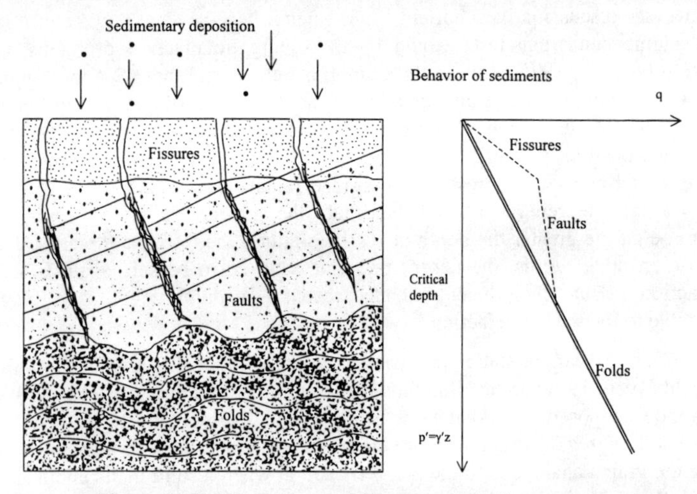

Figure 1. Folds, Faults, and Fissures of Sedimentary Deposits.

The two invariant stress parameters used in CSSM are the mean normal effective stress, $p' = 1/3(\sigma_1'+\sigma_2'+\sigma_3')$ and the deviator stress $q = 1/\sqrt{2}[(\sigma_2'-\sigma_3')^2 + (\sigma_3'-\sigma_1')^2 + (\sigma_1'-\sigma_2')^2)]^{1/2}$ where σ_1', σ_2', $\sigma_3'$are the principal effective compressive stresses. For triaxial test conditions, where $\sigma_2 = \sigma_3'$, they reduce to $p' = 1/3 (\sigma_1'+2\sigma_3')$, and $q = (\sigma_1'-\sigma_3')$, respectively. The two parameters p′ and q, and a third variable, the specific volume $v = (1+e)$, where e is the void ratio, define the state of a soil specimen.

Elastic compression and swelling of test specimens in general follow lines

$$v_\kappa = v + \kappa \ln p' = \text{const} \tag{1}$$

where v_κ is the value of the intercept of any specific line with the v axis. For example, in Fig. 2a the value of v_κ combines pressure p' and specific volume v to define the aggregate of grains which corresponds to the line through point A. The elastic compression and swelling characteristics of the aggregate defines the slope of this line. The packing density of the aggregate of grains defines the intercept v_κ. For the ideal soil defined as Cam-clay there is no slip among the grains while the aggregate experiences purely elastic changes. Any slippage results in small plastic deformation of the aggregate as a whole, with changes of many contacts between grains. Each time there is plastic deformation a new aggregation of particles is formed which has a swelling and compression line with the same slope but a different intercept. A shift between lines indicates a plastic volume change from one aggregation to the next. For illustrative purposes the plot of v_κ against $\ln p'$ gives a simple figure (Fig. 2b). Note that the line of critical states in the latter plot has slope $(\lambda - \kappa)$.

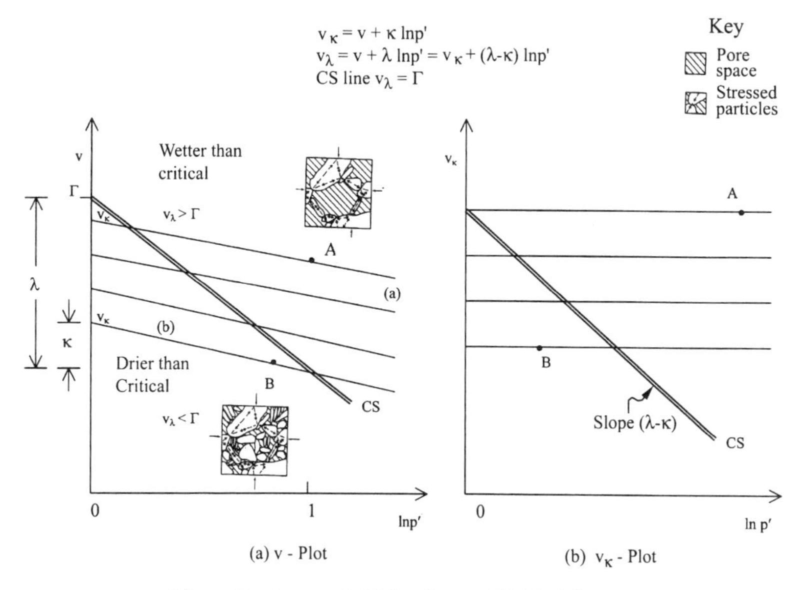

Figure 2. Aggregate Behavior and Critical States.

Consider two specimens with aggregates of grains at the same mean normal effective stress on lines (A) and (B) with identical lattices of highly loaded grains, but with different amount of lightly loaded grains (Fig. 2a). If line (A) has a higher value of v_κ than the line (B), then specimen (A) has fewer lightly loaded grains than specimen (B). If we now impose shear stresses on the aggregations represented by (A) and (B) and permit drainage of pore fluid, we may expect slippage of highly loaded particles and plastic volume change. This leads a transfer of load and other grains forming a highly loaded lattice.

The plastic volumetric response of the two specimens at the same mean effective stress will differ depending on the nature of packing of the lightly loaded grains. A specimen on line (A) with fewer lightly loaded grains loosely packed will compact with a reduction in v_κ and the dense one on line (B) will dilate with increase in v_κ during plastic shear distortion. Between these two limits there will be a density of packing at which, during shear distortion a succession of load carrying skeleton lattices of stressed grains will form and collapse with successive new structures being formed at about the same density of packing. In this shear strain increment a certain proportion of the grains which at one time formed the load carrying skeleton, now as individual grains become relatively lightly stressed or unstressed and play the role of "filler" particles filling voids. The notion of a critical state is that there exists one certain critical packing of grains or critical void ratio, at which continuous flow is possible at constant mean normal effective stress p', without damage to the grains, only with change of positions.

Roscoe, Schofield and Wroth (1958) quote-experimental evidence that the ultimate state of any soil specimen during a continuous remolding and shear flow will lie on a critical state line with equation:

$$\Gamma = v + \lambda \ln p' = v_\kappa + (\lambda - \kappa) \ln p' \qquad (2)$$

shown in Fig. 2. The critical state line with equation $(v + \lambda \ln p') = \Gamma$ can be seen as one of a family of parallel lines with equation $(v + \lambda \ln p') = v_\lambda$.

The critical state line can be used to distinguish the two different types of behavior of soils. There are states for which the combinations of specific volume v and mean normal effective stress p' lie further away from the origin than the line of critical states, so that,

$$v + \lambda \ln p' > \Gamma, \quad \text{or} \quad v_\kappa + (\lambda - \kappa) \ln p' > \Gamma, \quad \text{or} \quad v_\lambda > \Gamma \qquad (3)$$

and these states have been called "wetter than critical"; shearing there causes aggregates to compress to more dense packing and emit water with ductile stable yielding of a test specimen. There are also states of specific volume v and mean normal effective stress p' such that

$$v + \lambda \ln p' < \Gamma, \quad \text{or} \quad v_\kappa + (\lambda - \kappa) \ln p' < \Gamma, \quad \text{or} \quad v_\lambda < \Gamma \qquad (4)$$

and these states have been called "drier than critical"; where shearing causes aggregates to dilate and suck in water and ground slips at peak strength with unstable failures.

At the core of CSSM was the creation of the constitutive model called Cam-clay based on the theory of plasticity and the prediction of the successive ductile yielding states of specimens on the wet side of critical. The original Cam-clay model (Fig. 3) was synthesised from two basic equations.

The first says that if yielding obeys the stable associated plastic flow rule then the product of the plastic flow increment (dv, $d\varepsilon$) and any stress increment (dp', dq) outward directed from the yield locus is positive or zero - the zero applies to stress increments directed along the tangent to the yield locus.

The second equation says that when yielding occurs the work is purely frictional, as proposed by Taylor (1948). In his research thesis, Thurairajah (1961) reported the analysis of drained and undrained triaxial test data that confirmed Taylor's proposal. He did not begin the research with a prior intention of validating Taylor's equation, and his result came as a surprise. He took account of all work done by effective stresses on all moving boundaries and of all elastic energy released or taken up by a swelling or compressing aggregate under changes of p′. A lot of data were analysed and a simple result emerged. He found that the rate of dissipation during shear distortion was simply in the product of p′ times the friction coefficient M.

After eliminating the dilatancy rate dv/dε between these two equations (Fig. 3) a single differential equation is left which when integrated, predicts the form of the cam-clay yield curve (CD in Fig. 3). The specimens on this line CD are all at one v_κ on the same elastic compression and swelling line. Curve CD allows stress to extend a certain distance beyond the critical state line (Fig. 3) but there is a limit - when q = 0 the pressure cannot extend further than D, if the material is to remain stable. If there were soil in states beyond D they would be meta-stable. For example, when salt is leached out of quick clay it gets into this dangerous state and there is a risk of an avalanche.

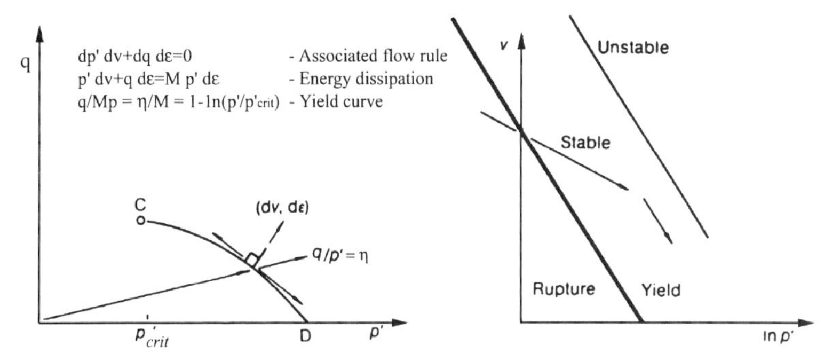

Figure 3. Cam-Clay Yielding

It was a strong outcome of the synthesis of the original Cam-clay model that it predicted an isotropic compression line $v_\lambda = \Gamma + (\lambda - \kappa)$ that bounded the region of wet clay behavior $\Gamma > v_\lambda > \Gamma + (\lambda - \kappa)$, exactly as was first observed by Casagrande and Albert (1930) and subsequently by Hvorslev (1937), Shibata (1963), and many others.

Soil in a state drier than critical such as point F in Fig. 4 has been observed to fail with well-defined rupture planes after reaching a peak strength fitting lines AB and GE. This behavior is very familiar to geotechnical engineers. In these states soil particles remain interlocked with each other and peak strength involves a contribution from dilatancy of the interlocked stressed grains. The dilating gouge material on the rupture planes will slowly soften to critical state plane strengths fitting lines OB and OE (Fig. 4), although suction can persist for many years provided the soil aggregate does not fissure or crumble.

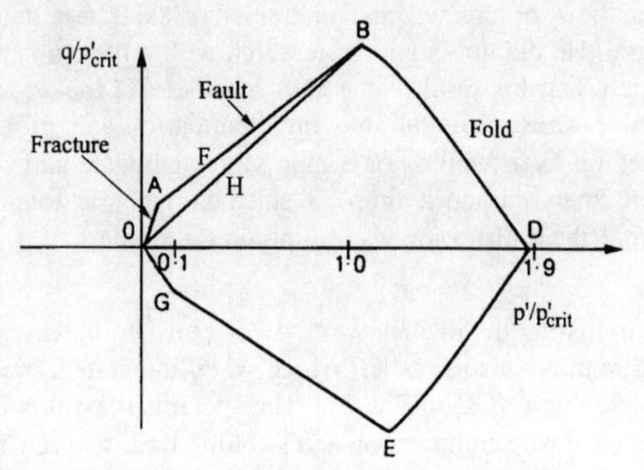

Figure 4. Limiting States of Soil Behavior

The critical state line also forms a bound to the region of faulting. There is a broad region of states where faults can occur and this region is bounded at low mean effective pressure by soil cracks in tension. A "no tension" or "limiting tensile strain" criterion is usually adopted to identify the initiation of tensile fracture (Schofield, 1980). For the triaxial specimen, the no tension criterion leads to $\sigma_a = 0$, which is the case of line OA, $p' = \sigma_a/3$, $q/p' = 3$, or to $\sigma_r = 0$ which is the case of line OG, $p' = 2/3\sigma_r$, $q = -\sigma_r$, $q/p' = -2/3$ (Fig. 4). Based on Weald clay data, Schofield (1980) has suggested that the change to tensile fracture from Coulomb rupture occurs in the vicinity of $p'/p_{crit} = 0.1$.

Critical state soil mechanics divides the soil behavior at limiting states into three distinct classes of failure. The limiting lines OA and OG (Fig. 4) indicate states limited by *fractures or fissures*; AB and GE indicate that Hvorslev's Coulomb *faults* on rupture planes will limit behavior; BD and ED indicate Cam-clay yield and sediment layer *folds*. The fractures, faults, and folds (FFF) diagram is useful to characterise all classes of observed mechanisms of large displacements in soils.

Critical States, the Harvard View of Liquefaction, and Seed's View

Liquefaction is one aspect of the undrained behavior of sands that has attracted attention for many decades. In a notable contribution to the Journal of the Boston Society of Civil Engineers, Casagrande (1936) described liquefaction of an aggregate in states more loose than the critical void ratio as if it were a phase transformation process such as the melting of a solid and the change to a fluid. On the other hand, based on undrained cyclic triaxial tests, Seed and Lee (1966) defined liquefaction as a phase transition condition when pore pressure approaches the confining stress and effective stress drops to zero. For Casagrande liquefaction had to be on the wet side of critical states while for Seed it had to be on the dry side.

Schofield and Togrol (1966) and Schofield (1981) have highlighted the difficulties with Casagrande's original notion of a constant critical void ratio.

Although Casagrande moved from this position to the steady state of sands put forward by Poulos (1981), many geotechnical engineers still use the word "critical" in the incorrect original sense.

Cam-clay is a model of uncemented soil aggregates on the wet side of the critical state line that can continue to yield in a ductile stable manner as a continuum. Quick clay is a lightly cemented or bonded soil aggregate, which can stand with vertical faces to small cliffs, and should be regarded as a soft rock. In an undisturbed state it contains a high water content and does not flow. When a quick clay avalanche occurs this soft rock disintegrates into a fissured debris, and as lumps of quick clay are remoulded their high water content become evident. When the debris is fully remoulded it forms a body of soil with even more water than the isotropic compression line, that is $v_\lambda >> \Gamma + (\lambda\text{-}\kappa)$. This is not a change of grain positions, but a loss of bonds. The notion of liquefaction as an event propagating with retrogressive slips in a meta-stable body of lightly cemented or bonded collapsing silt and causing quick clay flow slides is consistent with critical state theory. Centrifuge models performed at Cambridge on carefully sampled quick clay specimens did produce quick clay flow slides. A wide range of model tests at Cambridge also considered other so-called liquefaction phenomena where there were quite different mechanisms of failure. Schofield (1980) discussed such tests including those that modelled Mississippi river bank liquefaction.

When a soil aggregate is unloaded following a stress path its effective stress reduces leading to relaxation of stresses between grains. Such reduction in stress may be induced by imposing tensile strain, or by increasing the pore water pressure, or by cyclic loading. In each case, however, the soil particles remain geometrically interlocked with each other even though the effective stress falls. In this class of unloading paths if at any stage a large shear distortion were to be imposed on the interlocked but lightly stressed particles they would respond by dilation, with increase of effective stress and the stress path would head back towards the critical state line BH (Fig. 4).

But when an unloading effective stress path reaches the fracture regions OA or OG (Fig. 4) the continuum begins to disintegrate into a clastic body and unstressed grains become free to slide apart. In that case the average specific volume of the clastic mass and its permeability can increase greatly in a very short time. Whenever a soil is dug or is crumbled, for ease of handling or for mixing with water, an unloading stress path reduces a principal effective stress component to zero in a controlled manner. If, however, there is a hydraulic gradient across the soil body at the time it cracks or crumbles the event is less controlled and has the character of sudden hydraulic fracture or fluidisation.

Accordingly, Schofield (1981) in the St Louis conference defined liquefaction as a class of instability (channelling, piping, boiling, or fluidising) seen in soil far on the dry side of critical states near zero effective stress and in the presence of a high hydraulic gradient. This applies to the case of cyclic pore pressure in earthquake as well as to static hydraulic fracture.

The opening within the soil body may be an extensive crack or a local pipe or channel. In the case of a local pipe, water slowly following tortuous paths may be able to dislodge grains in a direction perpendicular to the axis in which the pipe is developing. If debris forms soft mud which blocks the channel the crack or pipe will heal itself. If hydraulic pressure are transmitted along a pipe or crack to regions where the pressure gradients cause cracking faster than cracks heal there is a sudden transmission of pressures, and a body of crumbling soil can disintegrate into a sort of soil avalanche. Or several pipes can break through a sand layer and vigorous sand boils can occur. This was the class of liquefaction with which Seed was concerned. It is important to note, however, that increase of excess pore pressure to the effective confining pressure is necessary but not sufficient. The formation of openings and the presence of high hydraulic gradient, which lead to disintegration of the continuum into clastic blocks of soil, is another important requirement.

Dam Failures

Casagrande began his Pan-American Conference lecture by saying that he found three common causes of disagreement with colleagues. Either he and they (i) looked at different aspects of the same problem, or (ii) generalised too much on the basis of different sorts of experimental data, or (iii) used the same terminology for different phenomena. Many disagreements about CSSM arise from these causes. This paper will ask if there is agreement about the word "liquefaction", and will consider only those aspects of three dam failures that relate to that word. The failure of a dam usually has several complex aspects, some of which are never fully understood. The concept of a primary cause or a triggering mechanism is in itself debatable when subsidiary causes of failure are needed. It is unusual to have reliable witnesses who report the events and during the event key elements involved in the process of failure may disappear.

The three dam failures that are discussed were interpreted in detail long ago. It is not feasible or necessary to present here all details of the site conditions, the design features, and the sequence of events preceding each failure, as excellent summaries of all these aspects are readily available. For example, Middlebrooks (1942) and subsequent discussions present a detailed account of the Fort Peck slide. The comprehensive volume of the international workshop on dam failures edited by Leonards (1986) gives detailed accounts of the Baldwin Hills reservoir and Teton Dam failures. Therefore, our review only asks if those conditions brought them into the class of instability discussed above. Were the Fort Peck Dam, Baldwin Hills Reservoir, and Teton Dam failures due to soil behavior on the wet side or the dry side of critical states?

Fig. 5 taken from Casagrande (1975) shows cross sections at Fort Peck before and after the slide. But an air photo (Fig. 6) shows an additional fact about the failure, that the upstream flank of the dam (or the shell) rotated in plan view. The appearance is that of a solid body rotation, consistent with their having been an uplift pressure below a rigid body on which a lateral force acted. The lateral pressure came from the hydraulic fill. The uplift pressure got below the shell because the sheet pile wall below the core (Fig. 5) allowed the full pore pressure at the base of the hydraulic fill to act as uplift. The existence of high uplift pressure below the shell was evident at the down stream toe where relief wells on the shell were observed to be flowing upwards. It is the nature of hydraulic fill that soil is in states on the dry side of critical. But the real danger to Fort Peck Dam was the high pore pressure at the base of the core which was a potential source of uplift, and the sheet pile wall delivered the pressure to all permeable layers below the dam. No doubt the designers thought of the sheet piles as preventing loss of water through permeable layers below the reservoir, but failed to realize that the very high pore pressure at the base of the core had this destructive potential. The upstream shell of the dam would have failed first because it was partially buoyant in the early reservoir filling. The hydraulic fill had practically no effective stress and so it would flow as slurry. There was no need to postulate a "flow structure", such as Casagrande supposed that allowed large cobbles to be carried along end pipes. This is not to say that his flow structure is not possible; simply that it is not essential to the explanation of the failure of the Fort Peck dam.

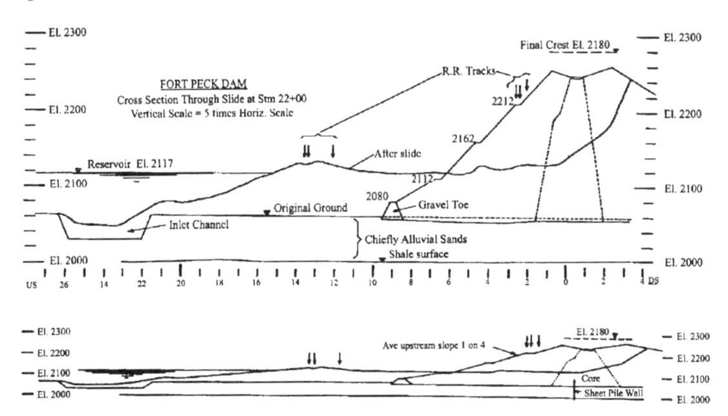

Figure 5. Schematic of the Cross-section of Fort Peck Dam Before and After Failure (Casagrande 1976)

Fort Peck was the work of the U.S. Army in a great river valley. Baldwin Hills and Teton were the works of the Los Angeles Division of Water and Power and the U.S. Bureau of Reclamation in the much more dry conditions out west. R.R. Proctor worked for Los Angeles Division of Water and Power and his ability to achieve very high compaction was evident in the steepness of the breach that was left after the Baldwin Hills Reservoir failed. The U.S.B.R had built a series of rather similar dams before Teton, and the control of compaction that was achieved became tighter in each successive dam. The great strength that was achieved in the final dam in Teton gorge is evident in the photograph of almost vertical strong faces on either side of the breach while the entire contents of the dam ran out. Both these embankment dams were built of soil in states which would be described in CSSM as very much on the dry side of critical. Both were made of low plasticity soil in a very brittle state.

Figure 6. Partial Failure of Fort Peck Dam As Seen From the Air

The weakness in both these two embankments was caused by quite small strains. Proctor and U.S.B.R. both economized in the linings to their reservoir and their cut off trench, not wishing to incur the cost of the use of graded filter materials. Proctor made a thin biscuit-like under drain of no-fines concrete with open pores blinded on the upper face and then bituminously sealed (Fig. 7). This underdrain was brittle and probably cracked as soon as the water was loaded into the reservoir. There were erratic seepage flows at once. The open cracks must have allowed transport of fines from the liner layer into pores in the no-fines concrete. Certainly after the failure it was clear that the liner had been ulcerated by voids rising from the most likely location of cracks in the cement blinding, where the load would cause differential settlement along the underlying fault line. The enquiry into the disaster was never carried to a conclusion because the oil companies settled, but if they had chosen to argue it they could well have claimed that there was no need to consider that continuing settlement on the fault due to extraction of oil caused the break. Once the no-fines concrete cracked it was simply a matter of time before the ulceration broke through.

Following the Teton Dam failure on June 5 1976 an Independent Panel reviewed the cause of failure and reported to the U. S. Department of the Interior and the State of Idaho in December 1976. The Independent Panel's report included an analysis of hydraulic fracturing and its possible role in the Teton Dam failure (Seed et al. 1976). The present review begins with a comment that the analysis did not consider the tensile strain conditions necessary for fracture.

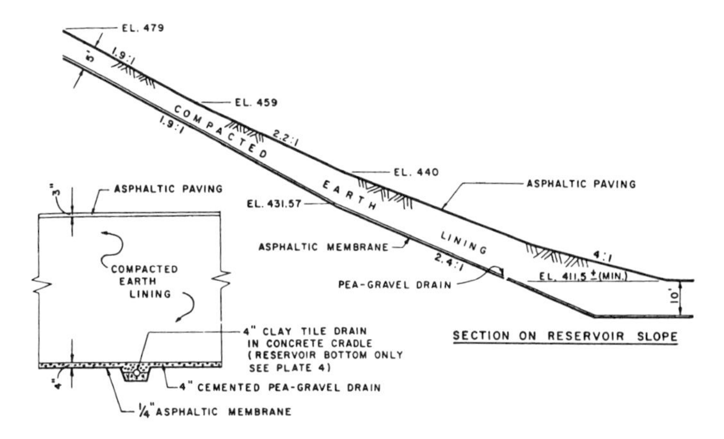

Figure 7. Main Features of Baldwin Hills Reservoir Lining

Seed et al. (1976) reported the FE analysis of the cause of cracking in the Teton Dam key trench. Their total stress computations implied certain lateral stress on the walls of the trench, and when they took away the pore pressure that they found for steady state seepage from the filled dam they obtained a tensile stress component. But this could not be a correct analysis. Tensile cracking results from lateral tensile strain, and the walls of the rock trench would have had to move apart to allow that to happen. An analogy would be if an odometer had soil compacted in it and there were lateral stresses on the walls of the cylinder. If a pore pressure exceeding the lateral stress were to be injected into the odometer it would not make a vertical crack in the soil inside unless it was able to cause internal pressure failure of the metal cylinder. The sidewalls would carry part of the stress. This comment was made by the second author of this paper to the late Professor Seed in April 1977.

Following discussion with U.S.B.R. in Denver various tests were undertaken both on the newly installed small centrifuge at UC Davis, and on the large beam centrifuge in Cambridge. In both test series tensile cracks were induced in tubs full of compacted Teton core material, and studies were made of erosion and void migration. The conclusion of those tests (Schofield's confidential report to USBR in 1980) was that the soil was highly susceptible to cracking. Even the strains caused by filling the dam probably were sufficient to cause extensive cracking. When such cracks were subjected to seepage with slowly increasing reservoir levels they tended to collapse, leaving a mud filled sealed crack and a rising void. The key to safety in such circumstances is to ensure that whenever cracks and rising voids occur there is material that can plug the voids. In this view it is important to have a downstream graded filter layer that can collapse and fill any void whenever it arises. It is also very significant that the actual rapid filling of Teton allowed no time for self-healing of cracks.

The 1980 Rankine lecture (Schofield 1980) mapped soil behavior described here (Fig. 4) on p'/p'_{crit} against q/p'_{crit} on axes of liquidity index against logarithm of pressure (Fig. 8). At any given liquidity, as effective mean-normal pressure increases, the mode of failure changes from fractures or fissures, to Coulomb faults or ruptures, and to yield or folds with plastic volume change. The stress ratio q/p' that can be carried by the soil increases as pressure and liquidity fall – the insert shows a section across the map at constant p'. This increased strength and stiffness tempts engineers to compact soil more and more, until they meet a new problem. When stiff soil becomes fissured its permeability increases very greatly. A fracture will involve open voids and channels, whereas Coulomb rupture preserves soil still in a relatively impermeable mass. In this sense it is safer to have softer and more ductile soil construction which remains water-tight even when ruptured.

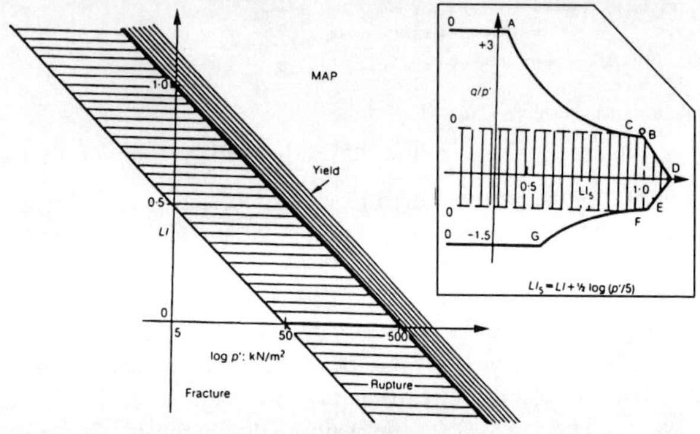

Figure 8. Liquidity and Limits of Soil Behavior (Schofield 1980)

Considering a body of soil initially at LI = 0.5 and subject to elastic compression the map suggests at shallow depths where $p < 5$ kPa there may be cracks, but for depths where $5 < p < 50$ kPa the soil will remain water-tight while deforming. In contrast a body of soil initially at LI ≤ 0 will be susceptible to fracture at depths for which $p < 50$ kPa; taking account of elastic compression it could require an overburden depth of say 50 m of drained soil or 100 m of buoyant soil to ensure that deformation caused water tight rupture planes rather than open permeable cracks. In this view the steep vertical face of the breach in Teton Dam can be seen as an open fracture in a very strong soil, standing to a height of 50 m to 100 m.

The emphasis in undergraduate teaching in Cambridge that arose as a result of many years of centrifuge model testing, was that over compaction should always be regarded as risky. Near critical states where equivalent liquidity (Schofield 1980) is $0.5 < LI_5$, compacted soil retains something of the toughness that is associated with ductile mild steel. At Teton the core was probably compacted to $LI_5 < 0$, which is as fragile as glass. The dam was bound to crack. It was disastrous to try to fill the dam more rapidly than ever been done before.

Summary and Conclusions

All classes of observed mechanisms of large displacements in soils could be characterised into three distinct classes; folds, faults, and fractures. An aggregate of grains wetter than critical states yield in a ductile stable manner. A layer of such sediment forms folds during deformation but it does not fail. Cam-clay describes the ductile stable yielding. If a soil aggregate is drier than the critical state, it can fail with fault planes on which gouge material dilates and softens, or it can fracture and crack into a clastic debris, or develop pipes and channels. Generalising too much on the basis of different sorts of experimental data without an understanding of the distinct classes of soil behavior has led to many disagreements on critical state soil mechanics.

Critical state soil mechanics associates rapid geotechnical disasters with soil on the dry side of critical states being brought near to zero effective stress while in the presence of a high hydraulic gradient. The three dam failures discussed here were all due to soil behavior on the dry side of critical. The Baldwin Hills and Teton dam were failures with cracks and pipes induced by over compaction. We see no evidence of Casagrande's "flow structure" of sand in the liquefaction of the Fort Peck dam failure. Rapid transmission of pore pressure gradients through soil near zero effective stress seems to us a better explanation of liquefaction failure on the "dry side" of critical states than Casagrande's transformation of the grain structure of sand by a "chain reaction" propagating through an aggregate of soil grains.

Acknowledgements

This study was performed while the first author was on leave at Cambridge University Engineering Department. An International Fellowship Award from the National Science Foundation (INT-9802887) and fellowship from Churchill College, Cambridge, sponsored the visit.

References

Casagrande, A. (1936a). "Characteristics of cohesionless soils affecting the stability of slopes and earth fills." *Journal of the Boston Society of Civil Engineers*, January; reprinted in *Contributions to Soil Mechanics 1925-1940*, BSCE, pp. 257-276.

Casagrande, A. (1975). "Liquefaction and cyclic deformation of sands, a critical review." *Proceedings of the Fifth Panamerican Conference on Soil Mechanics and Foundation Engineering*, Buenos Aires; reprinted as Harvard Soil Mechanics Series, No. 88, 27 pp. (1976).

Casagrande, A., and S. G. Albert (1930). "Research on the shearing resistance of soils." Massachusetts Institute of Technology,. Unpublished.

Castro, G. (1969). "Liquefaction of Sands." PhD Thesis, Harvard University; reprinted as Harvard Soil Mechanics Series, No. 81, 112 pp.

Hvorslev, M.J. (1937). *Über die Festigkeitseigenschaften Gestörter Bindiger Böden*, København.

Leonards, G.A. (Ed.) (1986). "Special issue on Dam failures." *Engineering Geology*, 24(1-4).

Middlebrooks, T.A. (1942). "Fort Peck slide." *Transactions*, ASCE, Vol. (107), pp. 723-764.

Poulos, S.J. (1981). "The steady state of deformation." *ASCE, Jour. Geot. Engrg.*,. 107(5), pp. 553-562.

Roscoe, K.H. and Schofield, A.N. (1963). "Mechanical behavior of an idealised 'wet clay'." *Proc. 2nd European Conf. on Soil Mechanics and Foundation Engineering, Wiesbaden*, Vol. 1, pp. 47-54.

Roscoe, K.H., Schofield, A.N., and Wroth, C.P. (1958). "On the yielding of soils." *Géotechnique*,. 8(1), pp. 22-53.

Schofield, A.N. (1966). "Original teaching on cam-clay." Lecture notes, Cambridge University Engineering Department.

Schofield, A.N. (1980). Cambridge geotechnical centrifuge operations, 20th Rankine Lecture, *Géotechnique*, 30(3), pp. 227-268.

Schofield, A.N. (1981). "Dynamic and earthquake geotechnical centrifuge modeling." State of Art lecture, Symposium on Recent Advances in Geotechnical Earthquake Engineering and Soil Dynamics, St. Louis, MO.

Schofield, A.N. (1998). "Geotechnical centrifuge development can correct soil mechanics errors." Keynote Lecture, Centrifuge 98, Tokyo.

Schofield, A.N. and Togrol, E. (1966). "Critical states of soil." *Bulletin of the Technical University of Istanbul*, No. 19, pp. 39-56.

Schofield, A.N. and Worth, P. (1968). *Critical State Soil Mechanics*, McGraw-Hill.

Seed, H.B. (1979). "Soil liquefaction and cyclic mobility evaluation for level ground during earthquakes." ASCE, *Jour. Geotech. Engr.*, 105(2), pp. 201-255.

Seed, H.B. and Lee, K.L. (1966). "Liquefaction of saturated sands during cyclic loading," ASCE *Jour. Soil Mech. Found.*, 92(6), pp. 105-134.

Seed, H.B., Leps, T.M., Duncan, J.M. and Bieber, R.E. (1976). "Hydraulic fracturing and its possible role in the teton dam failure," Appendix D of *Report to U.S. Dept. of the Interior and State of Idaho on Failure of Teton Dam by Independent Panel to Review Cause of Teton Dam Failure*, December, pp. D1-D39.

Shibata, T. (1963), "On the volume changes of normally-consolidated clays" (in Japanese), *Disaster Prevention Research Institute Annals, Kyoto University*, Vol. 6, 128-134.

Taylor, D.W. (1948) *Fundamentals of Soil Mechanics*, John Wiley & Sons.

Thurairajah, S. (1961) *Some Shear Properties of Kaolin and of Sand*, Ph.D. Thesis, Cambridge University.

ANALYSIS OF A FAILED EMBANKMENT ON PEATY GROUND

A. Porbaha[1], H. Hanzawa[2], and T. Kishida[2]

ABSTRACT

The difficulty in estimating engineering properties of peat has prompted the need to use combined field and laboratory tests coupled with an empirical procedure for stability analysis of embankments on peaty ground. Accordingly, this paper presents a case history of stability analysis for construction of an embankment on peaty ground for an expressway in multi-loading stages that failed at a height of 6.3 m. The method outlined here, which is the current state of the practice in Japan, employs cone penetration tests and direct shear tests using Mikasa's apparatus for the process of investigation, analysis, and quality control with a consistent approach.

INTRODUCTION

The project site was the Kameda Interchange for the Hokuriku Expressway in Niigata Prefecture. Fig. 1a shows the plan of the embankment and the locations of the borings. Soil conditions, natural water content, w_N, and unit weight, γ_t, obtained from the initial investigation are summarized in Fig. 1b. As can be seen, peaty soils found at the site vary in thickness from 2 m to 6 m, and can be classified into the two layers: (a) peat with w_N ranging from 100% to 400% and, (b) sand clay with peaty material with w_N from 45% to 75%.

[1] Office of Infrastructure Research, Department of Transportation, State of California, Sacramento, CA, USA(E-mail:ali@magical2.egg.or.jp)

[2] Technical Research Institute, TOA Corporation, Yokohama, Japan.

The procedure applied for this project is briefly outlined in the next section. Various relationships to be used for the analysis and quality control derived from the investigation are presented in detail. When the height of embankment rapidly increased from 4.5 m to 6.3 m large deformation took place together with tension cracks and heave. A detailed stability analysis using various shear strengths is presented.

METHODOLOGY

The procedure employed for this project is comprised of several steps, as summarized here:

1. Cone penetration tests (CPT) are carried out initially all over the project site to obtain point resistance values, q_c and to evaluate the subsoil conditions.
2. Undisturbed sampling is undertaken at representative locations using a hydraulic type stationary piston sampler because of its known reasonable quality and higher working rate compared with the standard stationary piston sampler.
3. Direct shear tests (DST) using Mikasa's apparatus (Takada, 1993) are carried out to obtain the in-situ direct shear strength, $S_{uo(d)}$, with the recompression method (Hanzawa and Kishida, 1982), and the direct shear strength in the normally consolidated (NC) state, $S_{un(d)}$. When settlement with time is measured in the DST to determine $S_{un(d)}$, it is also possible to obtain the coefficients of consolidation, c_v and volume compressibility, m_v in the NC state.
4. Correlation between effective point resistance (q_c-σ_{vo}) and $S_{uo(d)}$ is estimated (note that σ_{vo} is the total overburden stress), and then the mobilized strength $S_{un(mob)}$ are determined based on the approach by Hanzawa (1991) by taking strain rate into consideration.
5. Analysis is carried out using $S_{un(mob)}$, $S_{un(d)}/\sigma'_{vc}$, c_v, m_v and quality control is carried out with CPT and the use of the correlations obtained at stage (4).
6. Delineation of hazard zone during construction.

The details of these steps are described in the following sections along with the results of stability analysis.

ESTIMATION OF PARAMETERS

Just before the start of the construction, CPT, sheathed field vane test (FVT) and undisturbed sampling with the hydraulic piston sampler were carried out at three locations, No. 2, No. 3 and No. 4 shown in Fig. 1a. The values of (q_c-σ_{vo}) obtained from CPT are indicated in Fig. 1c. The values of (q_c - σ_{vo}) of peaty soils ranged from 100 to 300 kPa and a trend was indicated that the q_c value of the sandy clay is somewhat greater than those of the peat. A series of DST was carried out to determine $S_{uo(d)}$. For this test, the undisturbed samples were reconsolidated at an effective

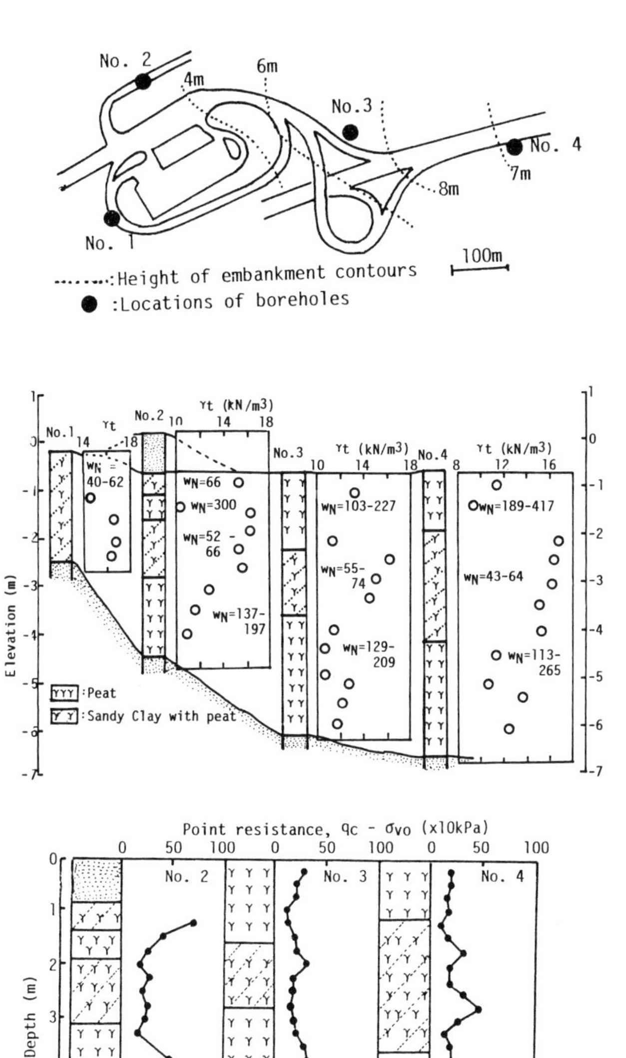

Fig.1: Description of the expressway project (a) borehole location, (b) Soil conditions at the site, (c) Variation of CPT with depth

overburden stress, σ'_{vc} for 5 minutes and then sheared under the constant volume condition at a deformation rate of 0.25 mm/min.

The values of $S_{uo(d)}$, $S_{uo(v)}$ from the sheathed FVT and unconfined compression test (UCT) $q_u/2$ already measured in the 1st soil investigation are plotted versus depth in Fig. 2. The values of $S_{uo(d)}$ ranged from 10 to 25 kPa and showed higher values at some locations in the sandy clay, which is consistent with the characteristics observed in $(q_c\text{-}\sigma_{vo})$ values. The values of $(q_c\text{-}\sigma_{vo})$, $q_u/2$ and $S_{uo(v)}$ are plotted versus $S_{uo(d)}$ in Figs. 3 and 4, and the following correlations were obtained:

$$\begin{aligned} (q_c - \sigma_{vo}) &= 10\ S_{uo(d)} \\ q_u/2 &= 0.6\ S_{uo(d)} \\ S_{uo(v)} &= 1.5\ S_{uo(d)} \end{aligned} \qquad (1)$$

Appropriate evaluation of the strength increment ratio in the NC state, S_{un}/σ'_{vo} is an important relationship for peaty soils because embankments are generally constructed in multi-loading stages where each new loading is implemented after an increase in the shear strength has occurred by the consolidation under the previous loads. In order to obtain the $S_{un(d)}/\sigma'_{vo}$ ratio, a series of DST was carried out on representative samples of both the peat and the sandy clay. In this test, representative samples are first consolidated at σ'_{vo} much greater than σ'_{vo} for 10 minutes, which is enough to achieve primary consolidation, and then sheared under constant volume condition at a deformation rate of 0.25 mm/min. The values of $S_{un(d)}$ obtained from the tests are plotted versus σ'_{vo} in Fig. 5, and a value of 0.45 was obtained as the $S_{un(d)}/\sigma'_{vo}$ ratio irrespective of the peat and the sandy clay. This ratio is much higher than that of marine clay but lower than that of typical peat found in Hokkaido which was investigated by FVT (Miyakawa, 1962) although the $S_{un(d)}/\sigma'_{vo}$, ratio can not be directly compared with that from FVT as indicated by Eq. (1).

$$S_{un(d)}/\sigma'_{vo} = 0.45 \qquad (2)$$

The values of $S_{u(mob)}$ during the initial stage with various methods can also be determined from $S_{un(d)}$ as shown in Fig. 5 and correlations given by Eq. (1). Since it is considered dangerous to adopt the average $S_{un(d)}$ value in determining $S_{u(mob)}$ because of the higher inhomogeneity of peat, a relatively lower $S_{un(d)}$ value should be adopted and a value of 15 kPa was selected as the $S_{un(d)}$ value for the initial stage through its depth. In addition, the strain rate effect of 0.85 was considered for $S_{uo(d)}$ as as discussed in detail by Hanzawa (1983).

$$\begin{aligned} S_{u(mob)} &= 0.85\ S_{uo(d)} = 13\ \text{kPa} && \text{(from DST)} \\ &= q_u/2 = 0.6\ S_{uo(d)} = 9\text{kPa} && \text{(from UCT)} \\ &= S_{uo(v)} = 1.5\ S_{uo(d)} = 23\ \text{kPa} && \text{(from FVT)} \end{aligned} \qquad (3)$$

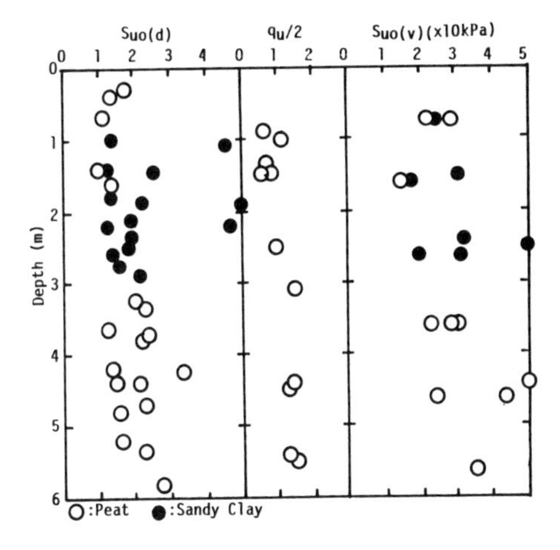

Fig.2: Variation of strength with depth

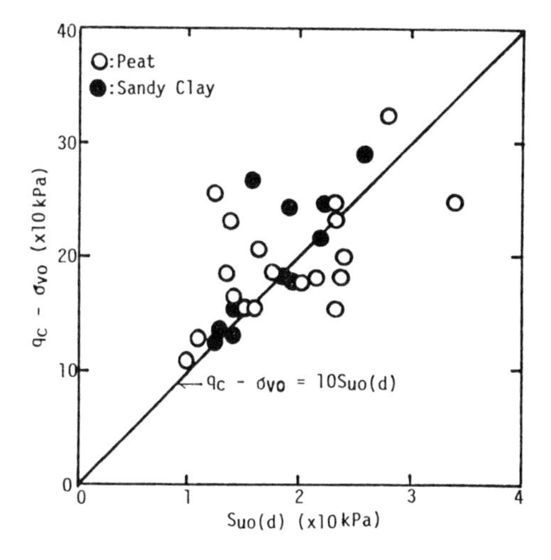

Fig.3: Correlation of CPT and DST

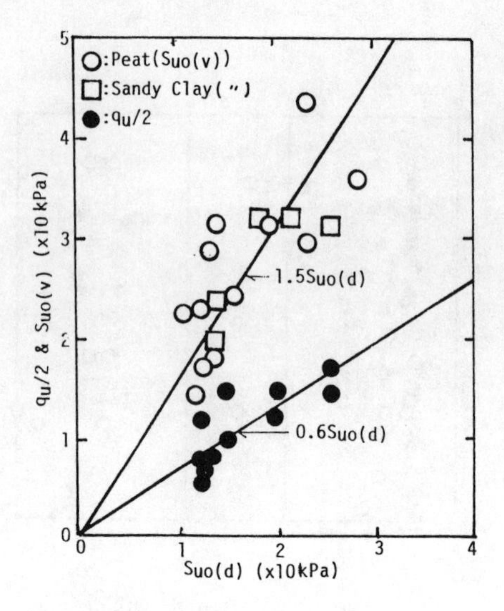

Fig.4: Correlation of q_u with CPT

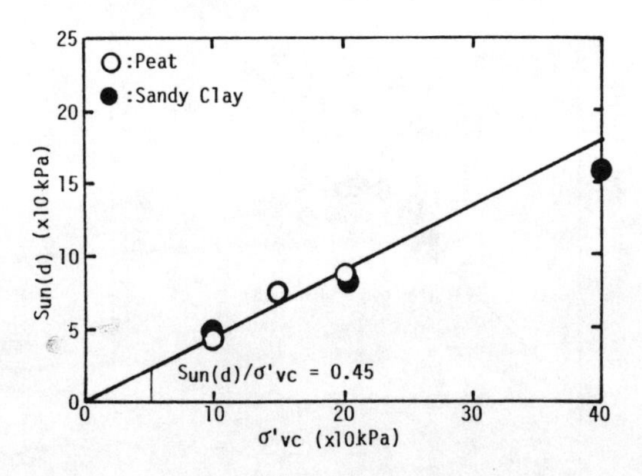

Fig.5: Stress increment ratio

When the strength increase induced by consolidation under multi-loading construction is considered, the $S_{u(mob)}$ from different methods derived from the correlations given by Eq. (1) are as follows:

$$
\begin{aligned}
S_{u(mob)} &= 0.85 \times 0.45(\sigma'_{vo} + \Delta\sigma \times U) && \text{(from DST)} \\
&= 0.45(\sigma'_{vo} + \Delta\sigma \times U) \times 0.6 && \text{(from UCT)} \qquad (4) \\
&= 0.45(\sigma'_{vo} + \Delta\sigma \times U) \times 1.5 && \text{(from FVT)}
\end{aligned}
$$

where $\Delta\sigma$ is stress increment induced by embankment and U is degree of consolidation.

It is also possible to obtain c_v and m_v values in the NC state from the DST to determine the $S_{un(d)}/\sigma'_{vo}$ ratio when settlement with time is measured. The values of c_v and m_v obtained are as follows:

$$
\begin{aligned}
c_v &= 2{,}000\ \text{cm}^2/\text{day (peat)} \\
&= 6{,}000\ \text{cm}^2/\text{day (sandy clay)} \\
m_v &= 0.0013\text{-}0.0027/\text{kPa (peat)} \qquad (5) \\
&= 0.0002\text{-}0.0005/\text{kPa (sandy clay)}
\end{aligned}
$$

STABILITY ANALYSIS

In the first stage the embankment was constructed to a height of 4.5 m, and maintained for 6 months to allow completion of the primary consolidation under the embankment. The second embankment (1.8 m high) making a total height of 6.3 m was quickly constructed in 3 days. On the final day, a large deformation took place together with settlement, causing tension cracks and heave as shown in Fig. 6. During this stage, an excavation 2.5 m deep for an irrigation canal was under construction along one side of the embankment as also shown in Fig. 6. In this section, stability analysis for this failure is presented in detail using various shear strengths.

The direct shear strength, $S_{u(d)}$ is obtained as follows:

$$S_{u(d)} = 0.45(\sigma'_{vo} + \Delta\sigma_1) \qquad (6)$$

where $\Delta\sigma_1$= stress increment induced by the 1st embankment. In order to evaluate the shear strength decrease in the excavated zone, a series of DST was carried out. In this test, a sample was first consolidated in the NC state, then allowed to swell at a given overconsolidation ratio, OCR, and finally sheared under the constant volume condition at a deformation rate of 0.25 mm/min. The correction factor for strength decrease due to swelling obtained from these tests, which is given by α in this paper, is plotted versus OCR in Fig. 7. The value of $S_{u(d)}$ in the excavated zone is as follows:

$$S_{u(d)} = S_{u(d)} \times \alpha = 15 \times \alpha\ \text{(kPa)} \qquad (7)$$

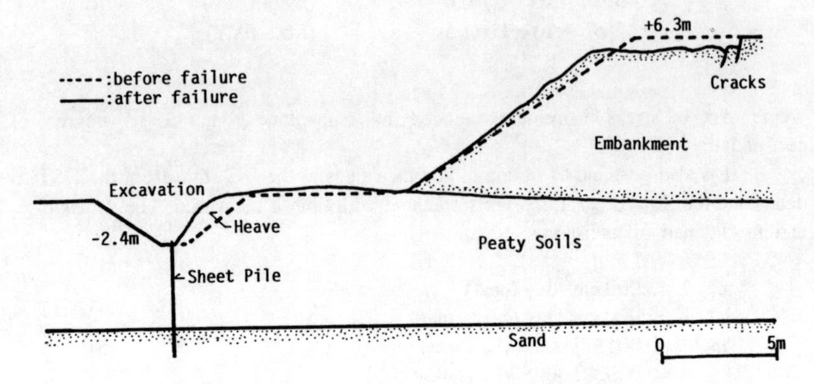

Fig.6: Cross section of the embankment before and after failure

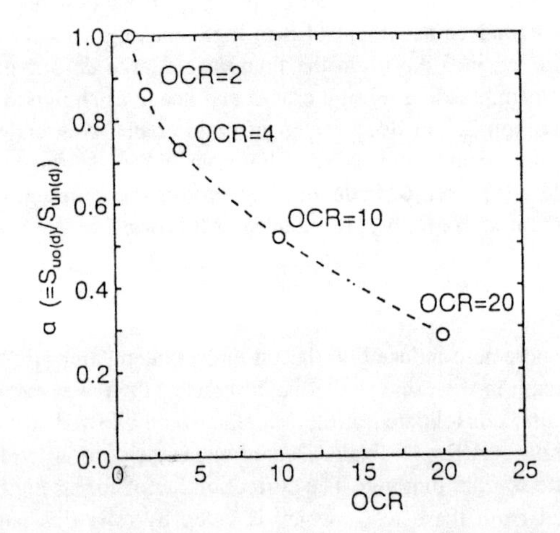

Fig.7: Variation of correction factor with OCR

In determining $S_{u(mob)}$, the peaty ground is divided into three zones (see Fig. 8), as follows:

Zone I:

$$S_{u(mob)}=0.85x0.45(\sigma'_{vo}+\Delta\sigma_1) \text{ (DST)}$$
$$=0.45(\sigma'_{vo}+\Delta\sigma_1) x0.6 \text{ (UCT)} \quad (8a)$$
$$=0.45(\sigma'_{vo}+\Delta\sigma_1) x 1.5 \text{ (FVT)}$$

Zone II:

$$S_{u(mob)}=0.85x S_{uo(d)}=13 \text{ (kPa) (DST)}$$
$$=0.6x S_{uo(d)} =9 \text{ (kPa) (UCT)} \quad (8b)$$
$$=1.5x S_{uo(d)} =23 \text{ (kPa) (FVT)}$$

Zone III:

$$S_{u(mob)}=0.85x S_{uo(d)}x\alpha=13\alpha\text{(kPa) (DST)}$$
$$=0.6x S_{uo(d)}x\alpha =9\alpha \text{ (kPa) (UCT)} \quad (8c)$$
$$=1.5x S_{uo(d)}x\alpha =23\alpha \text{ (kPa) (FVT)}$$

Although there are various problems to be clarified in evaluating the shear strength of the embankment, two shear strength were assumed for the stability analysis: 1) $\varphi = 30°$ and 2) $\varphi= 35°$ with $\gamma_t = 19$ kN/m^3 (measured at the site).

Stability analyses were carried out for 3 cases: 1) 1st embankment 4.5 m high, 2) 2nd embankment 6.3 m high without excavation and 3) 2nd embankment with excavation. The results of the stability analysis for each case are summarized in Table 1 . Fig. 9 indicates the critical slip surface giving $(F.S.)_{min}$ and Fig. 10 shows the effect of the thickness of peaty soils on the $(F.S.)_{min}$ for the failed embankment. Both of them were evaluated with $0.85S_{no(d)}$ from DST. The following conclusions can be drawn:

1) $(F.S.)_{min}$, values from 0.85 $S_{uo(d)}$ for failed embankment with excavation are 0.98 to 1.02, while those from $q_u/2$ and $S_{uo(v)}$ are too low and too high compared to the field behavior of the embankment.
2) The circular slip surface giving $(F.S.)_{min}$ agrees quite well with the actual sliding surface estimated from tension cracks and heave as also indicated in Fig. 9.
3) $(F.S.)_{min}$ value varies with the thickness of peaty soils, and this will be a useful tool in determining the hazard zone for stability of the embankment related to the thickness of peat.

Table 1: Results of stability analysis

Method	$(F.S.)_{min}$		
	Case I	Case II	Case III
DST(0.85 $S_{uo(d)}$)	1.13-1.17	1.03-1.07	0.98-1.02
UCT($q_u/2$)	0.78-0.81	0.71-0.74	0.69-0.73
FVT($S_{uo(v)}$)	1.68-1.72	1.53-1.57	1.44-1.48

Case I: 1st embankment of 4.5m in height (stable condition); Case II: 2st embankment of 6.3 m in height (stable condition); Case II: 2st embankment of 6.3 m in height (failed condition)

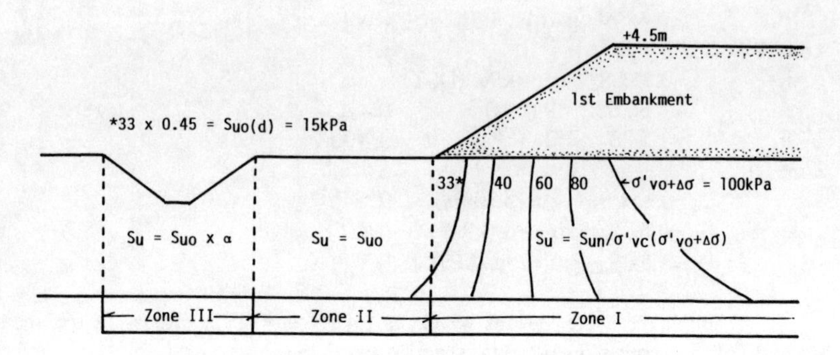

Fig.8: $S_{u(mob)}$ at different zones for stability analysis

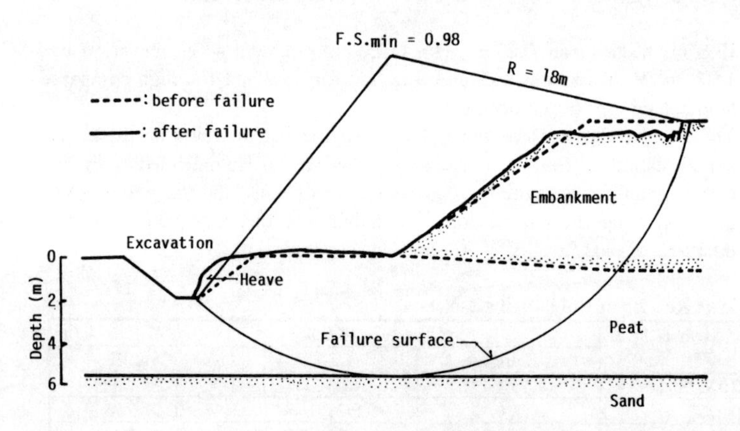

Fig.9: Slip surface for $(F.S.)_{min}$

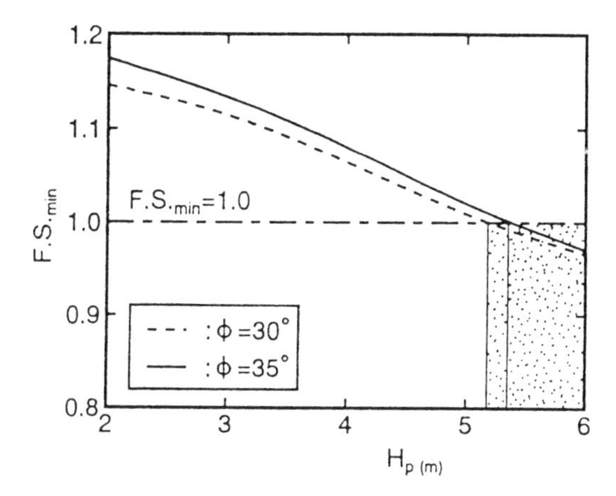

Fig.10: Variation of $(F.S.)_{min}$ with thickness of peat

DELINEATION OF HAZARD ZONES

Immediately after the failure of the embankment, CPT were carried out at 25 points within the area, where the height of the embankment exceeded 6 m. This was done to determine the hazard zone, which is closely related to the thickness of peaty soils, when the construction rate for the 2nd embankment is maintained at the same rate as for the failed embankment. Contour lines showing the thickness of peaty soil and the height of embankment are presented in Fig. 11 together with the failed zone. The hazard zone shown in this Figure was determined as follows: (a) $(F.S.)_{min}$ value for the 2nd embankment (6.3 m high) should be equal to or more than 1.05, which could still be dangerous but was so determined based on the fact that catastrophic, flow failure such as that with a marine clay has not been observed for peaty soil (Noto, 1991); and (b) the critical F. S. value is satisfied when the thickness of the peat is equal to or less than 4 m as shown in Fig. 10. For the hazard zone, it was determined that the rate of embankment construction is decreased by CPT to check the shear strength using the correlation of $(q_c\text{-}\sigma_{vo}) = 10S_{uo(d)}$ in addition to checking with the stability charts proposed by Matsuo and Kawamura (1977).

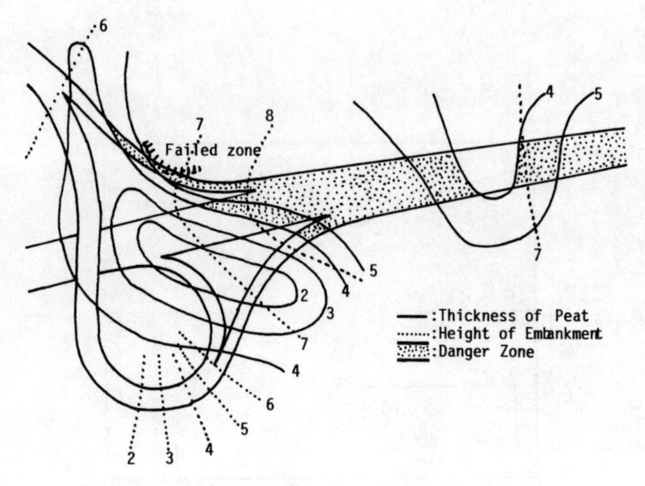

Fig.11: Delineation of hazard zone

CONCLUDING REMARKS

The implementation of a methodology using combined Mikasa's direct shear apparatus and cone penetration tests in soil investigation, design and quality control for an embankment on peaty subsoils are presented. Relationships are developed to correlate field and laboratory tests. Results of stability analyses for the failed embankment clearly indicate that $(F.S.)_{min}$ values from 0.85 $S_{uo(d)}$ show good agreement with the field behavior of embankment, while $(F.S.)_{min}$ values from $S_{uo(v)}$ and $q_u/2$ are not safe for the former and conservative for the latter. Consequently, CPT is a very useful tool for quality control as well as for determining the hazard zone of instability when the correlation between $(q_c-\sigma_{vo})$ and $S_{un(d)}$ and the $S_{un(d)}/\sigma'_{vc}$ ratio are available from the site investigation.

REFERENCES

Hanzawa, H. (1983) Three case studies for short term stability of soft clay deposits, Soils and Foundations, Vol. 23, No. 2, pp. 140-154.

Hanzawa, H. (1991) A new approach to determine the shear strength of soft clay deposits, Proc., GEO-COAST '91, Vol. 1, pp. 23-28, Yokohama.

Hanzawa, H. and Kishida, T. (1982) Determination of in-situ undrained shear strength of soft clay deposits, Soils and Foundations, Vol. 22, No. 2, pp. 1-14.

Matsuo, M, and Kawamura. K. (1977) Diagram for construction control of embankment on soft ground, Soils and Foundations Vol. 17, No. 3, pp. 37-52.

Miyakawa, I. (1962) Some geotechnical problems as related to peat, Hokkaido Branch. JSCE (in Japanese).

Noto, S. (1991) Geotechnical engineering for peat, Gihodo, Tokyo (in Japanese).

Takada, N. (1993) Mikasa's direct shear apparatus, testing procedures and results, Geotechnical Testing Journal, GTJODJ, Vol. 16, No.3, 314-322.

Yoshida, N., Kanaya, Y. and Hanzawa, H. (1987) Large scale direct shear test on mud stone fill, Report of the Technical Research Institute, Toa Corporation. GR-13-87 (in Japanese).

Sea-Cliff Erosion at Pacifica, California Caused by 1997/98 El Niño Storms

Charles B. Snell, Affiliate Member[1]
Kenneth R. Lajoie[2]
Edmund W. Medley, Member[3]

Abstract

Twelve homes were constructed in 1949 at the top of a sea cliff along Esplanade Drive in the City of Pacifica, located on the northern coast of San Mateo County, California. The rear yards of those properties were bounded by an approximately 20-meter (70-foot) high cliff that has retreated episodically at an average rate of 0.5 to 0.6 meter (1.5 to 2 feet) per year over the past 146 years. During the heavy storms of the 1997/1998 El Niño winter, a severe episode of cliff retreat undermined seven homes and threatened three others. All ten homes were condemned and demolished by the City of Pacifica.

In this study we analyze geologic, tide, wave, rainfall and wind data in an attempt to determine the causes of this most recent erosion event. We identify the following possible contributory causes of the cliff retreat:

- wave-induced undercutting of the cliff landward of an old revetment,
- reduction in beach width over time,
- reduction in cliff-face stability owing to infiltration from heavy rains,
- erosion of the cliff face by groundwater piping, and
- wind-induced erosion of loose dune sand at the top of the cliff.

While these factors may explain the retreat of the cliff below the twelve homes along Esplanade Drive, the question remains as to why other geologically

----

[1] Gilpin Geosciences, 25 Evergreen Avenue, Suite 8, Mill Valley, CA 94941 (Formerly of Exponent Failure Analysis Associates, Menlo Park, CA)
[2] U.S. Geological Survey, Mail Stop 975, Menlo Park, CA 94025
[3] Exponent Failure Analysis Associates, 149 Commonwealth Drive, Menlo Park, CA 94025

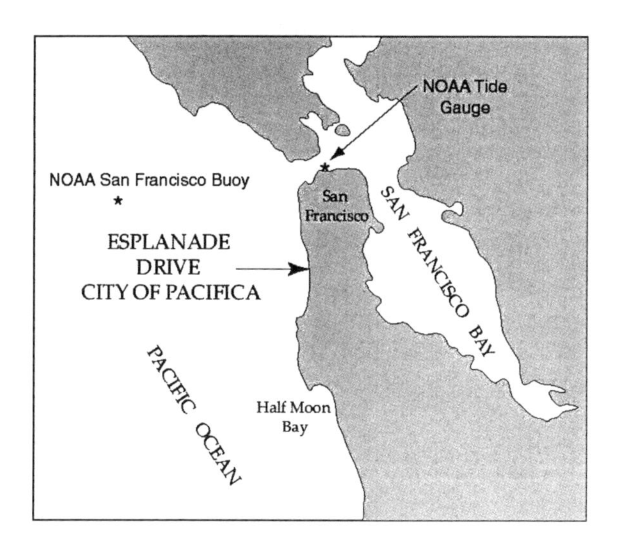

Figure 1. Index map showing location of Esplanade Drive and NOAA San Francisco tide gauge and buoy. The NOAA Point Reyes and Point Arena buoys are located along the California coast northwest of the area shown in this figure.

similar sites in the region were not severely eroded during the 1997/1998 El Niño winter.

Introduction

A low, unstable sea cliff backs the 4-kilometer-long, north-trending beach in the northern part of the City of Pacifica, in northwestern San Mateo County, California (Figure 1). In 1949, twelve houses were built near the cliff top along the west side of Esplanade Drive (Figures 2 and 3). The houses were numbered 528 to 572 from north to south. Initially, the separation between the houses and the cliff top was 30 meters (100 feet) at 528 and 65 feet at 568. Between 1949 and January 1998, the cliff top retreated by incremental erosion to within a few meters of the houses at 564 and 568 Esplanade Drive. During the same period, erosion moved the cliff top to within 24 meters (80 feet) of the house at 528. In February 1998, severe El Niño storms triggered the largest increment of cliff erosion observed since the homes were built. This erosion event partially undercut seven houses (548 to 572) and threatened three others (536 to 544), resulting in their condemnation and demolition by the City of Pacifica.

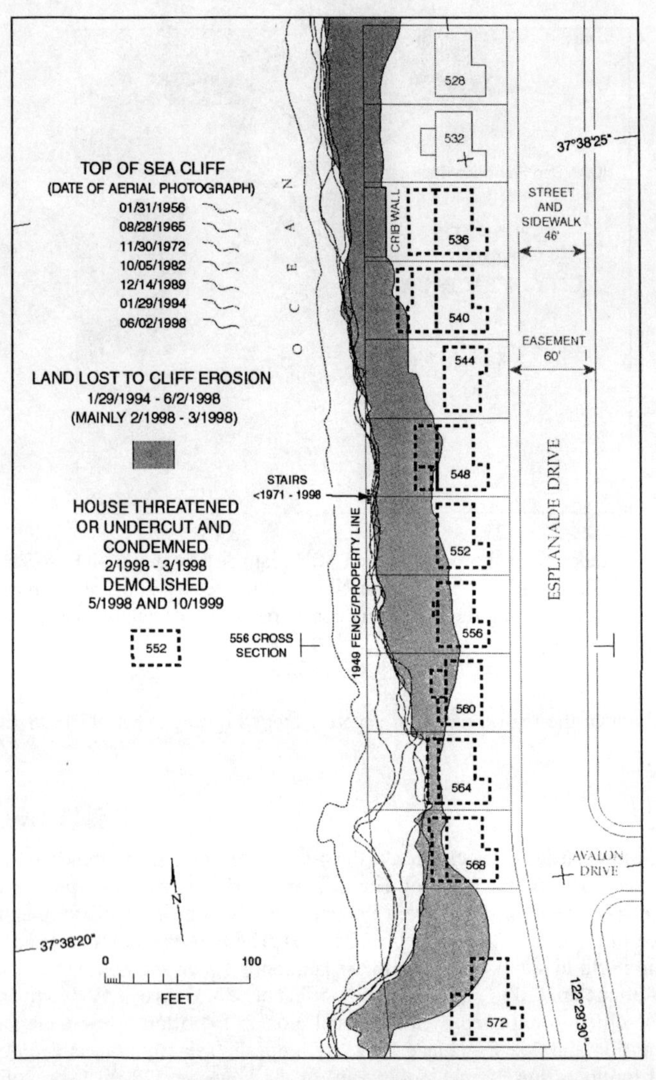

Figure 2. Map showing the history of sea-cliff retreat along the 500 block of Esplanade Drive. The lines representing the seven successively younger cliff-top locations were derived from large-scale, unrectified, vertical aerial photographs (California Department of Transportation, 1:24,000 and 1:4,800 scale, 1/31/56 to 1/29/94) and a USGS field survey (6/2/98). The minor crossings of these lines mainly reflect uncertainties in defining the top of the sea cliff on the photographs, not distortions in the photographs or inaccuracies in collating the data.

Figure 3. A 1954 aerial view to the southeast of the low sea cliff behind the 500 block of Esplanade Drive. The light-colored geologic unit exposed along the cliff top is loose Holocene dune sand, which is underlain by Pleistocene alluvium containing two thin beds of beach sand. The dark band along the lower portion of the cliff face is due to groundwater seepage wetting the Pleistocene sediments. The cliff face is close to its location in 1949, when the houses were built. The lower dashed line marks the location of the cliff top in January 1998, just before the 1998 erosion event. The upper dashed line marks the location of the cliff top after the erosion event. (Pacifica Tribune photograph; Floyd Easterby, exact date unknown, but most likely November 1954.)

History of Coastal Erosion

The approximate locations of the cliff top along the 500 block of Esplanade Drive between 1956 and 1998 is documented by historical aerial photographs (Figure 2). The retreat of the cliff is also illustrated by Figure 3, which is a photograph taken only six years after the homes were built.

Historical Cliff Retreat

Historical erosion rates for sea cliffs in northern San Mateo County generally range from about 0.2 to 0.6 meters per year (0.5 to 2 feet per year), although rates on the order of 1.5 to 3 m/yr (5 to 10 ft/yr) have been observed in a few localized areas (Lajoie and Mathieson, 1985). Along the 500 block of Esplanade Drive, the erosion rate averaged about 0.2 m/yr (0.7 ft/yr) from the mid-1800's to the mid-1900's, based on historical topographic maps and photography (LaJoie and Mathieson, 1985).

Figure 4. A March 1998 ground view of 556 Esplanade Drive. In February 1998, up to 15 meters (49 feet) of sea-cliff retreat undercut the foundation. This and six other houses (544 to 572) undercut by cliff retreat in February 1998 were condemned and demolished by the City of Pacifica on May 6, 1998. Note the 3-meter (10-foot) thick bed of loose dune sand exposed in the uppermost cliff face. (USGS photograph; K. Lajoie, 4/4/98).

When the twelve houses were built along Esplanade Drive in 1949, the cliff top was at least 15 and 3 meters (50 and 10 feet) west of the rear fence/property line behind 528 and 564 Esplanade Drive, respectively. By 1954, the cliff top had retreated to the fence line behind 564 (Coastside Tribune, 8/11/54). However, vegetation on the upper part of the beach and the lower cliff face in 1954 (Figure 3) suggests that waves had not significantly eroded the cliff north of 564 for at least a year or two. Consequently, the location of the cliff in 1954 was probably close to its location in 1949.

During the 26-year period between 1956 and 1982, the cliff top retreated 8.5 to 11.5 meters (28 to 38 feet) along the northern half of the block, and up to 22 meters (72 feet) behind 564 and 568. The corresponding average retreat rates are 0.33 to 0.45 m/yr (1.1 to 1.5 ft/yr) and up to 0.85 m/yr (2.8 ft/yr), respectively. However, most of this retreat occurred during the winter storms of 1962 and 1963 (Pacifica Tribune, 2/15/62 and 2/15/78; San Francisco Chronicle, 10/14/63).

Wooden steps from the cliff top to the beach behind 548 and 552 were built sometime before April 20, 1971 (Figure 5). The steps were not destroyed by cliff retreat until February 1998, indicating that most of the post-development cliff retreat had occurred prior 1971. Between 1982 and January 1998, the cliff was relatively stable, except for 5.8 meters (19 feet) of retreat behind 568 Esplanade Drive that occurred during the heavy El Niño storms of 1982 and 1983 (Pacifica Tribune, 2/23/83).

In summary, from the 1950s until the winter of 1997/1998, the cliff is estimated to have locally retreated about 10 to 20 meters (30 to 60 feet). Erosion tended to be episodic rather than continuous, with relatively broad sections of ground being eroded over a short period of time followed by periods with little or no erosion. The effect this retreat had on diminishing the rear yards of the Esplanade Drive houses was compounded by loss of yard space resulting from construction of substantial additions at the rear of some residences.

Recent Cliff Retreat

Since January 1994 up to about 15 meters (50 feet) of additional cliff retreat occurred, mostly during February 1998. Erosion caused an estimated 10 to 12 meters (30 to 40 feet) of cliff retreat during a five-day period near the end of February 1998. Consequently, many homes were declared as unsafe for occupancy by the City of Pacifica (Pacifica Tribune, 2/25/98). In marked contrast to the previous erosional patterns, the cliff retreated only 3 to 5 meters (12 to 17 feet) behind 564 and 568 at that time. Elsewhere along the Esplanade Drive shoreline, cliff retreat during the 1998 erosion event exceeded the total retreat over the previous 49 years since the houses were built.

Figure 5. A March 1983 ground view to the north of the sea cliff along the 500 block of Esplanade Drive, taken from a bedrock-supported headland just south of 572 Esplanade Drive. The house on the right with the prominent horizontal siding is 564. The sagging concrete slab behind 564 is the foundation of a room addition that was undercut by cliff erosion on February 7, 1978, and subsequently demolished (Pacifica Tribune, 2/15/78). The rip-rap at the base of the cliff below 548, 552 and 556 was placed prior to April 20, 1971. The wooden stairs descending the cliff face between 548 and 552 (see Figure 2 for location) were built prior to April 20, 1971, and were destroyed by wave erosion in early February 1998. (USGS photograph; K. Lajoie, 4/10/83).

The seven houses from 548 to 572 Esplanade were undercut by cliff retreat in February 1998, and were condemned at that time and demolished in May 1998. The three houses from 536 to 544 were threatened by cliff retreat and condemned in February 1998, and were demolished in October 1999.

Beach Erosion

Based on historical topographic maps and aerial photographs, the beach appears to have narrowed in the last 40 years, thus providing the cliff with less protection from breaking waves. Erosion of the beach below the 500 block of Esplanade Drive was significantly greater at the south end of the beach than at the north end. In late March 1998, the beach consisted mainly of sand but became gravelly near the south end of Esplanade Drive. The transition from sand to gravel indicates a transition from lower wave-related energy at the north end of the beach to higher energy at the south end.

A bedrock headland at the south end of the beach (below 572 Esplanade Drive) became more pronounced in the last 50 years. This headland focuses wave energy, resulting in greater erosion at the south end of the beach. Based on the apparent northward displacement of debris that had fallen onto the beach from the residences in February 1998, sediment transport (littoral drift) along the beach is northward during winter months. Many years ago, this northward sediment transport probably replenished sand along the beach below Esplanade Drive. As the headland grew more pronounced, it may have disrupted the replenishment of sand to the beach north of the headland, below Esplanade Drive.

Mitigation Efforts

Efforts have been made in the past to mitigate erosion of the cliff. In the late 1960's to early 1970's, rip-rap was placed at the base of the cliff below 556 through 548 Esplanade Drive. Following severe erosion during the winter of 1982/1983, a rip-rap revetment was constructed in 1984 to protect about 500 meters (1,650 feet) of shoreline, including the base of the cliff below the 500 block of Esplanade Drive (Earth Investigation Consultants, 1991). The portion of the revetment below Esplanade Drive was installed at a cost of $400,000.

The revetment, 9 meters (30 feet) wide and between 1.5 and 2 meters (5 and 7 feet) thick, was composed of nominally 1-meter (3-foot) diameter boulders and was designed to be keyed 1 meter (3 feet) into a hard-pan layer below the beach. However, the revetment was apparently not keyed nor was it maintained following its construction. Consequently, it began to fail within a few years. By the winter of 1997/1998, gaps were reported between the revetment and the cliff base against which it had originally been built. The revetment was essentially destroyed during the 1997/1998 winter, with boulders being re-distributed on the beach.

Geological Setting

The sea cliff below Esplanade Drive is 20 meters (70 feet) high and its slopes are inclined between 60 and 85 degrees. The cliff is cut into about 18 meters (60 feet) of weakly indurated Pleistocene alluvium containing a few thin beds of beach sand (Figure 6). At the top, about 3 meters (10 feet) of Holocene dune sand overlies a well developed soil formed in the the alluvial sediments. Highly fractured greenstone of the Franciscan complex crops out in the lower cliff face below 572 at the south end of Esplanade Drive. The bedrock outcrop, which is the formerly-buried northwestern end of a prominent ridge (Milagra Ridge) that climbs inland to the southeast, now forms a small headland.

Along the northern Pacifica coastline groundwater flows southwestward toward the open cliff face where it is seasonally expressed by moist sediments and numerous seeps along permeable sand beds. However, the buried northwest end of Milagra Ridge is a partial hydrologic barrier that concentrates and redirects groundwater flow westward, toward the cliff face at the south end of Esplanade Drive. During winter months the water table rises, and numerous seeps flow from permeable beds, weakening the already unstable sediments exposed in the cliff face.

When the area was developed in 1949, surface runoff was concentrated at the south end of Esplanade Drive and flowed over the cliff top through an open ditch, and later through a small pipe. That drainage system often failed, flooding the three houses at 564, 568 and 572 Esplanade Drive. More significantly, the flood waters flowed between the houses, eroding the loose dune sand at the top of the cliff. Both the natural concentration of ground water and the artificial concentration of surface runoff at the south end of Esplanade Drive significantly destabilized the cliff face at that location.

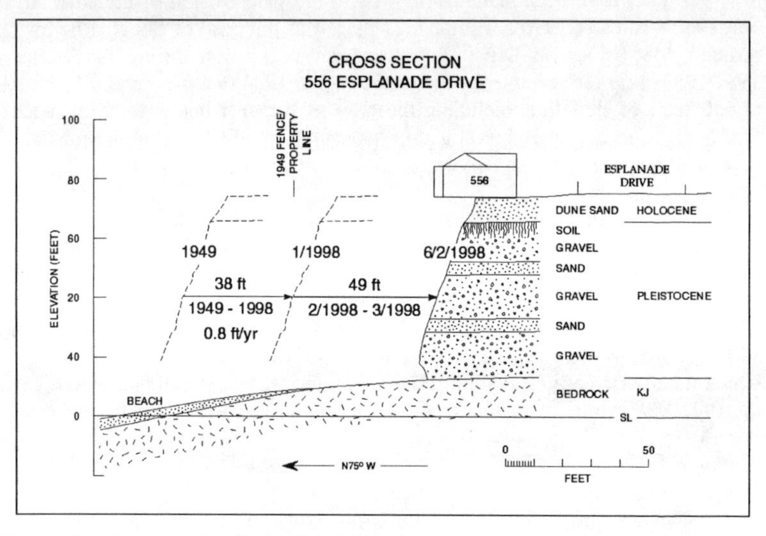

Figure 6. Cross section through 556 Esplanade Drive showing soil stratigraphy and history of sea-cliff retreat. See Figures 3 and 4 for aerial and ground views of this property in 1954 and 1998, respectively.

Atmospheric and Oceanic Conditions

California's coastal Mediterranean climate has a distinct rainy season that peaks during the winter months. While the onset of rain usually begins in late October or November, the heaviest rainfall typically occurs during the months of January and February. Afterwards, storm intensities and frequencies diminish until late spring when the rains virtually stop. The winter of 1997/1998 was no exception, with the heaviest rains falling in January and February 1998 following a relatively wet November and December.

Precipitation

Daily rainfall data for the 1997/1998 rainy season were obtained from the National Weather Service (NWS) Cooperative Observer site in Pacifica. These data were generally consistent with rainfall data from other NWS Cooperative Observer sites in the region, including Half Moon Bay, the Richmond District of San Francisco, and east-central San Francisco (Mission Dolores).

During the 1997/1998 rainy season, there was high cumulative rainfall prior to and during February 1998 (Figure 7). Approximately 46 centimeters (18 inches) of rain fell at Pacifica in that month alone, compared to a mean annual precipitation of approximately 53 centimeters (21 inches) for nearby San Francisco (Null, 1992). Most of the rain fell during the first and third weeks of February, increasing the moisture content and weight of the cliff-face sediments. More importantly, the cumulative rainfall prior to February was significantly greater than the long-term average (Figure 7). This near-record precipitation produced higher groundwater levels, greater flow out of the cliff face, and the resultant increased piping of the cliff face.

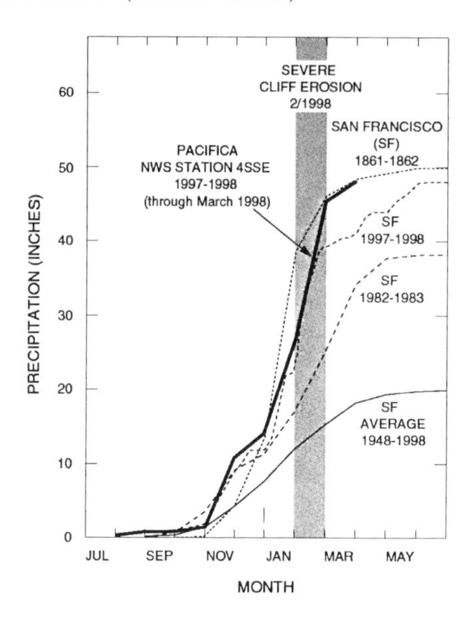

Figure 7. 1997-1998 cumulative precipitation for Pacifica (NWS Cooperative Observer Site 4SSE, located about 6 kilometers south of Esplanade Drive at elevation 144 meters) and San Francisco (Mission Dolores). Data from NOAA and NWS web sites. Data for Pacifica site are unavailable for April and June 1998.

Wind and Barometric Pressure

A series of large winter storms passed over the northern California coast in early February 1998, then again during the middle and later parts of the month. The near-record negative barometric pressure anomalies (shown in black on Figure 8) are associated with the passage of these storms. In addition, the early part of February was characterized by southerly winds that, while oblique to the coast, had a landward component that piled up water along the coast. The net effect of the low barometric pressure and southerly winds was to raise the overall sea level along the California coast. Such elevated conditions are indicative of storm-related surge, which tends to raise overall water depths close to shore, thereby allowing waves to break closer to the sea cliff.

Waves

Data indicating the characteristics of waves breaking along the Pacifica coastline during the 1997/1998 El Niño storms were not available. Instead, deep-water wave height data were obtained to estimate relative overall wave conditions during the period of most severe sea-cliff retreat (February 1998) with historical trends. Sources for wave data for February 1998 were limited to the nearest three operating buoys at Point Arena, Point Reyes and San Francisco (Figure 1).

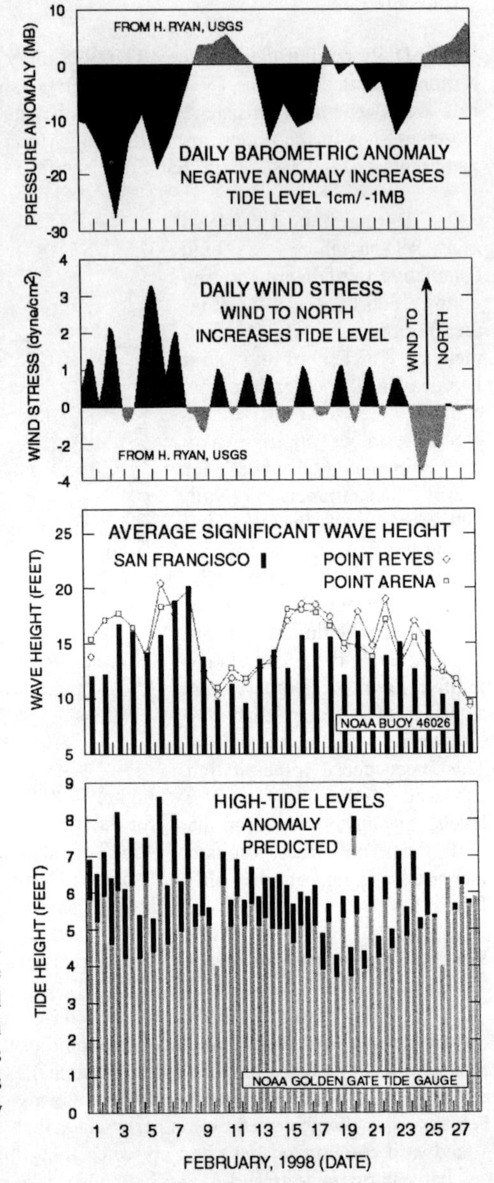

Figure 8. February 1998 climatic and marine conditions for the central California coast. Barometric and wind-stress data from Ryan (1999). Wave and tide data for Point Arena and San Francisco buoys from NOAA web site. Pt. Reyes buoy data from the Scripps Institution of Oceanography web site.

Wave data from 1982 to 1998 yield average significant wave heights during the months of November through March, when the largest waves are most common. The significant wave height is the average of the highest one-third of a measured group of waves. It should be noted that these data are statistical indicators, and that maximum and minimum wave heights for individual waves were often considerable departures from the indicators. Long-term wave data were obtained from records for the Point Arena and San Francisco buoys; no long-term wave data were available for the Point Reyes buoy. The mean monthly wave height was 2.8 meters (9.2 feet) for Point Arena and 2.1 meters (6.9 feet) for San Francisco.

The maximum winter wave heights at the Point Arena buoy generally ranged from 3.0 to 3.6 meters (10 to 12 feet), as shown in Figure 9. In contrast, the maximum winter wave height during the winter of 1997/1998 was almost 4.5 meters (15 feet) at the Point Arena buoy. Maximum winter wave heights between 1982 and 1998 recorded at the San Francisco buoy were generally about 0.7 to 0.9 meters (2.5 to 3 feet) lower than those recorded by the Point Arena buoy. However, the overall annual pattern of these data is generally consistent with the Point Arena data for those winters when there were above-normal wave heights. The available data show that deep-water wave conditions during the winter of 1997/1998 were

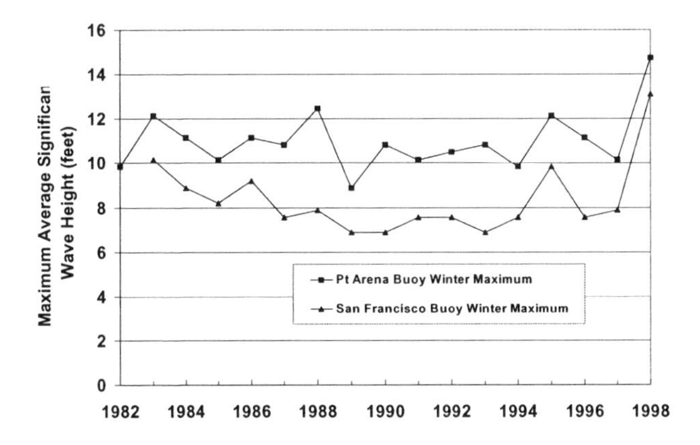

Figure 9. Maximum average significant wave heights during 1981 through 1998. The data point for each year represents the time period from November of the previous year through March of the indicated year. The corresponding wave height value is the highest of the monthly average heights recorded during the given November-through-March time period. Wave and tide data for Point Arena and San Francisco buoys from NOAA web site.

more severe than during the previous 15 years. This was primarily due to severe wave conditions in February 1998.

Data from the San Francisco buoy correlate reasonably well with those from the other buoys (Figure 8). The mean value calculated for the November through March storm season for 1982 through 1998 was 2.1 meters (6.9 feet) for the San Francisco buoy. Although the actual sizes of waves reaching the Pacifica coast were not recorded, the similarity in wave height trends at the three buoys indicate that wave-height data from the San Francisco buoy are indicative of the wave height trends along the Pacifica coast.

Tides

The San Francisco Bay area experiences two tidal cycles per day. The higher high tides were as much as 0.7 meters (2.2 feet) higher than predicted levels in early February 1998 (Figure 8). Such elevated conditions are indicative of storm-related surge caused by low barometric pressure and the landward component of winds. The higher sea-water temperature associated with the 1997/1998 El Niño event contributed slightly to the elevated tides by causing a corresponding increase in water volume.

Summary

The net effect of the elevated water temperature, low barometric pressure, and southerly winds was to raise the overall sea level along the coast by as much as 0.7 meters (2.2 feet). The relatively large height of the deep water waves that occurred persistently during February 1998 indicates that the energy of the waves breaking along the Pacifica coastline was considerably higher than had been experienced there for many years. The deepening of near-shore water depths allowed waves to break closer to shore. The resulting zone of highly turbulent surf adjacent to the cliff increased the rate and extent cliff erosion. The erosion occurred due to persistent sapping of the cliff base by turbulent surf rather than waves breaking at the cliffs. Thus, the combination of the large waves and elevated tide levels allowed severe erosion to take place along the base of the cliffs below Esplanade Drive.

All of the oceanic and atmospheric conditions discussed above had the potential to influence the cliff erosion process. The periods of time during which the various oceanic and atmospheric conditions may have combined to accelerate erosion of the cliff are indicated by the concurrence of anomalous conditions (Figures 7 and 8). The most severe conditions occurred during the first and third weeks of February, when elevated water levels, high waves, rainfall and wind all combined to erode the cliffs. During those periods, the erosive power of the combination of high tides and high waves was greatest, and was the most significant contributor to the erosion that occurred during February 1998. Heavy rainfall during February 1998, combined with above-average precipitation during the preceding months, increased seepage and subsequent piping of soil layers within the cliff and made the soils at the cliff face heavier and weaker. In addition to the wind's effect on tide levels, it also eroded loose dune sands at the top of the cliff.

While the severity of oceanographic and atmospheric conditions during February 1998 is readily apparent, it does not explain why other geologically similar areas along the northern San Mateo County coast survived the 1997/1998 winter relatively unscathed. Nor does it explain the minor sea-cliff erosion along Esplanade Drive during the 1982/83 El Niño winter, when heavy winter storms induced flooding and landslides in the greater San Francisco Bay area on a scale similar to the 1997/1998 El Niño winter storms. Obviously, other factors unique to the Esplanade Drive area influence erosion. These factors probably include:

- Periodic fluctuations in beach width and near-shore bathymetry due to movement of sand at variable superimposed frequencies (seasonal and storm-related).
- Gradual exposure of the small bedrock headland as the weaker Pleistocene sediments were eroded landward. During the winter months, the headland might block northward littoral drift.
- The possible reduction in beach width due to the focusing of wave energy on a poorly maintained rip-rap revetment placed along the base of the sea cliff in 1984.

The reduction in the beach width, possibly in conjunction with certain near-shore seafloor geometries, reduces the beach's ability to dissipate wave energy, thereby rendering the sea cliff more vulnerable to wave attack. It is possible that a different configuration of beach and near-shore bathymetry in February 1998 might have prevented the severe 1997/1998 erosion event along Esplanade Drive. However, history shows that it would only be a matter of time before the right conditions combined to again erode the Pacifica coastline.

Mitigation and Future Land Use

In early 1998 the Federal Emergency Management Agency (FEMA) awarded the City of Pacifica a $1.4 million hazard-relief grant to build a rip-rap revetment at the base of the cliff along the entire 500 block of Esplanade Drive. This wall was designed to provide protection for the street for at least 20 years. With the cliff-erosion problem mitigated for the immediate future, the city originally planned to narrow the street, deed part of the public easement to the ten property owners who had lost their homes, and allow them to rebuild on their shifted, though still diminished, lots.

Ultimately, however, both the city and the ten affected property owners realized that rebuilding along the west side of Esplanade Drive was neither prudent nor economically feasible. The city received a $1.0 million grant from the FEMA hazard-mitigation program to buy out the ten property owners, eight of whom had accepted the offer as of August 1999. Interestingly, this mitigation grant was the first awarded by FEMA for property loss not caused by flood inundation. The buy-out rate of $105 per square foot of heated floor space was about half the market value of the houses before they were undercut by erosion. As required by FEMA, the property owners who accepted the buy-out deeded the remainder of their property to the city as perpetual open space.

References

Earth Investigation Consultants, 1991, Geotechnical Investigation: Evaluation of Instability Mitigation, 536 Esplanade Drive, Pacifica, February 20.

Lajoie, Kenneth R., and Mathieson, S. A., 1985, San Francisco to Año Nuevo, in Griggs, Gary, and Savoy, Lauret, editors, Living With the California Coast, National Audubon Society, Duke University Press, Durham, North Carolina, pg. 140-177.

Null, J., 1992, A Climatology of San Francisco Rainfall, 1849-1991, San Jose State University, M.A. Thesis, UMI, Ann Arbor, Michigan.

Ryan, H., U.S. Geological Survey, 1999, written communication.

MONITORING THE LONG TERM STABILITY OF THE FRESH KILLS LANDFILL IN NEW YORK CITY

Thomas G. Thomann,[1] Majed A. Khoury,[1] Aaron D. Goldberg,[1] and Richard A. Napolitano[2]

ABSTRACT: The stability of many embankments on soft soils is typically assessed during construction by monitoring the behavior of key parameters through the installation of geotechnical instruments. The measurements from the instruments are typically presented in graphical form and then compared to the results from engineering analyses. This procedure generally works well if there are not a lot of instruments to be monitored, the stratigraphy and embankment configuration are relatively uniform, and construction occurs over a relatively short period of time thereby limiting the amount of analyses that need to be performed. However, for many large landfills over soft soils this procedure can be rather cumbersome because the subsurface conditions and side slopes are likely to change over time, many instruments are needed over a large area, and landfill construction may occur over many years. This paper presents a unique methodology that has been implemented at the Fresh Kills Landfill in New York City to rapidly assess the stability conditions of the landfill during construction based on the large number of installed instruments and computer engineering analyses.

Introduction

The Fresh Kills Landfill, located in Staten Island, a borough of the City of New York (see Figure 1), serves as the repository of the majority of the municipal solid waste from the five boroughs of New York City. The landfill is operated by the New York City Department of Sanitation (NYCDOS) and has received as much as approximately 115,650 kN (13,000 tons) of municipal refuse per day. Construction of the landfill began in 1948 and now encompasses more than 9.7 sq. km. (2,400 acres) of former marshland. The landfill is comprised of four sections (see Figure 2) identified as the Fresh Kills Bargefill (Sections 1/9 and 6/7), the Victory Boulevard Truckfill (Section 3/4) and the Muldoon Avenue Truckfill (Section 2/8). Sections 2/8 and 3/4 are no

[1] Senior Project Manager, Senior Principal, and Project Engineer, respectively, URS Greiner Woodward Clyde, 363 Seventh Ave., New York, New York.
[2] Senior Project Manager, New York City Department of Sanitation, New York, New York

longer receiving refuse fill and have undergone final closure. Section 6/7 has stopped receiving refuse fill; however, Section 1/9 currently receives refuse fill. Originally, the landfill operations were to continue into the year 2015 to reach a maximum height of about 152 m (500 ft). However, present plans are to close the landfill by the year 2002.

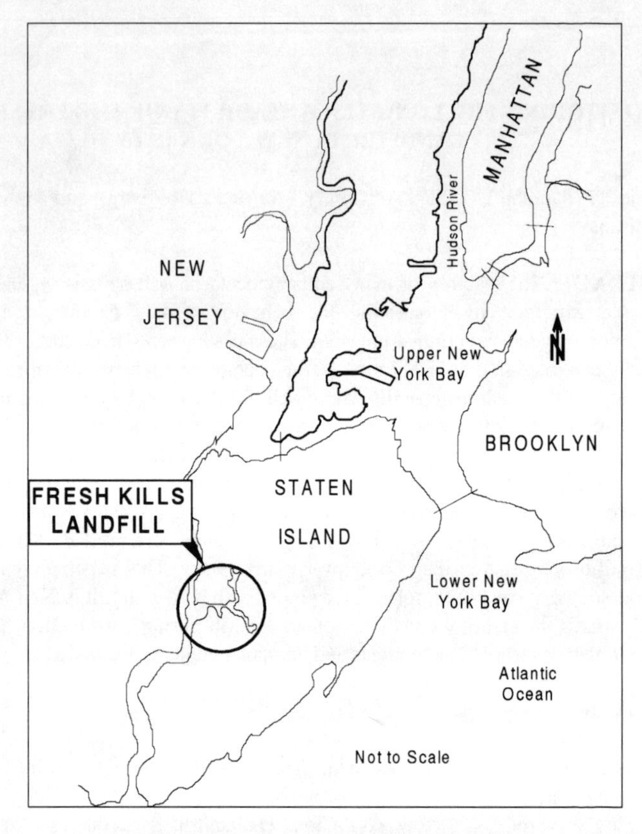

Figure 1 - ***General Site Location Plan***

Considerable importance is being placed on the relationship between filling operations and the safety and stability of the landfill because of 1) the planned size of the landfill (Section 1/9 when completed will be about 76 m (250 ft) high based on the most recent plan); 2) the existence of compressible soils under most filling areas, and 3) the presence of an urban environment in the vicinity of the landfill. As a result of these concerns, the NYCDOS has undertaken a program of geotechnical site characterizations, stability and deformation analyses, design and installation of a semi-automated monitoring instrumentation system, including its operations and maintenance. This program was also designed to fulfill a portion of an agreement between the NYCDOS and the New York State Department of Environmental Conservation (NYSDEC).

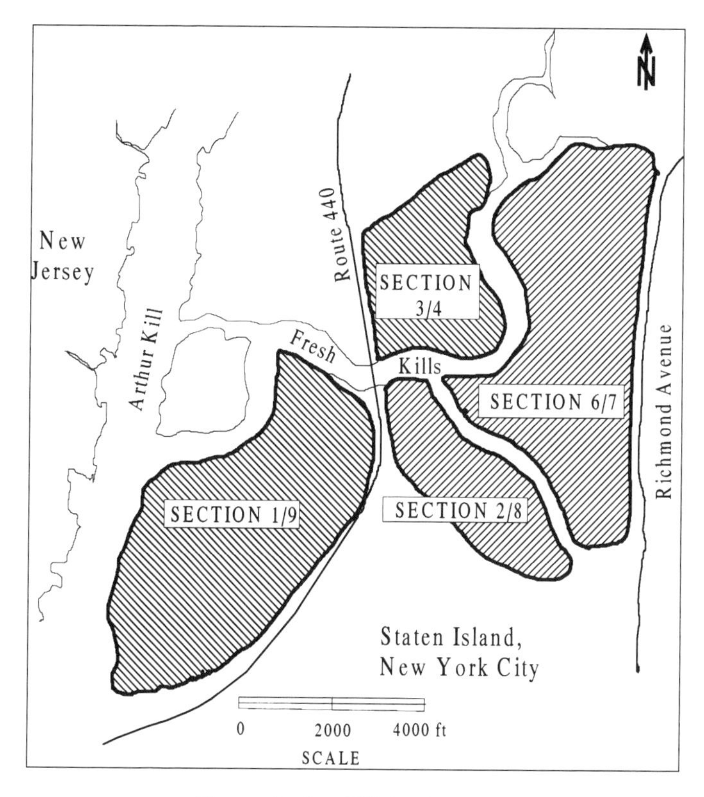

Figure 2 - ***Landfill Section Plan***

Because of the complex stratigraphy at the site and the uncertainties regarding the short-term and long-term behavior of the refuse and foundation soils (in terms of their field behavior versus laboratory behavior), an observational methodology has been adopted (Peck, 1969). This entails installing a monitoring system to obtain measurements of key parameters (e.g, pore pressure and deformation) during landfilling to ascertain the performance of the landfill, and then to modify, if necessary, landfilling operations (e.g., location of filling, rate of filling, schedule of filling) to maintain the satisfactory performance of the landfill. Similar approaches have been adopted for other projects involving earthen embankments and landfills (Handfelt, et. al. 1987, Withiam, et. al. 1995, Oweis, et. al. 1985, Duplanic 1990).

This paper presents an overview of the subsurface conditions, the overall slope stability monitoring system approach for the landfill, a description of the various instruments of the monitoring system, and how the measurements from these instruments are used to determine action levels associated with stability concerns.

Site Characterization

Fresh Kills Landfill is located in a very complex lithology due to the past existence of two and, perhaps, three known glacial end-moraines found south of the site. As a result, the soil stratigraphy under the landfill is extremely variable in layer thickness, extent, and material properties. The soil stratigraphy generally consists of Recent Silt and Clay (Qrc), Recent Sand (Qrs), Glacial Sand (Qgs), Glacio-lacustrine Clay (Qgl), Glacial Till (Qgt), Cretaceous sequence of Clays and Sands (Kc and Ks). However, not all strata are present throughout the landfill complex.

In terms of landfill stability, the Qrc and the Qgl are likely to have the greatest influence. The Qrc consists primarily of marine tidal marsh deposits predominantly organic in nature. It is composed of brown to black peat with traces of silt and clay, grading downward through gray organic silt, to dark gray silty clay with traces of vegetation. The maximum thickness of this layer is approximately 15 ft and is typically absent from historic upland areas. The soils are typically normally consolidated or slightly overconsolidated. The Qgl consists primarily of reddish-brown silt and clay. The thickness of this layer is generally about 10 ft to 20 ft with some areas reaching a thickness of approximately 35 ft. The soils are highly overconsolidated due to past glacier activity.

Instrument Types

To date, over 260 geotechnical instruments have been installed within the refuse fill and foundation soils to monitor both the magnitude and rate of change of pore pressure, lateral and vertical movements, and temperature. The instruments installed include vibrating wire and open standpipe piezometers, inclinometers, magnetic extensometers, and temperature probes. These instruments and their associated components (e.g., electrical cables, inclinometer casing, readout locations, etc.) had to be designed to survive on-going landfilling and closure activities, and environmental hazards at and below the landfill surface. These hazards include temperatures within the refuse fill in excess of 65°C (150°F), temperatures within the foundation soils near the refuse fill in excess of 32°C (90°F), lateral and vertical movements within the refuse fill on the order of tens of centimeters and meters, respectively, and the harsh chemistry of the landfill leachate. An Automated Data Acquisition System (ADAS) was also installed to obtain readings from the piezometers because of the large number of instruments to be installed, difficult access conditions, large areal coverage, and the necessity to make timely changes in landfill operations in case of instability concerns.

Additional information regarding details of the instrument designs, the performance of the instruments, and a summary of the measurements from the instruments are included elsewhere (Thomann, et.al., 1999).

Monitoring System Approach

The results from preliminary stability analyses and geotechnical site characterization studies were used to select representative landfill cross-sections, or profiles, and locations for installation of instrumentation. The criteria used for selecting representative profiles and instrument locations included the presence and characteristics of the Qrc and Qgl layers, the planned height and side slopes of the landfill, and the provision for adequate and somewhat redundant instrumentation coverage for the complex and diverse stratigraphy at a reasonable cost.

Along a profile, the instruments are typically located in groups referred to as clusters. In general, the cluster locations were selected so that the instruments would be positioned within a potential failure surface zone. Consideration was also given to locating the clusters near an existing or future landfill side slope bench so as to minimize the amount of lateral trenching across the landfill (to protect instrument signal cables) and to facilitate installation, manual readings, and future maintenance. Typically, three to four instrument clusters are installed within a profile (see Figure 3). One cluster is generally located near the toe, one near the mid-point of the

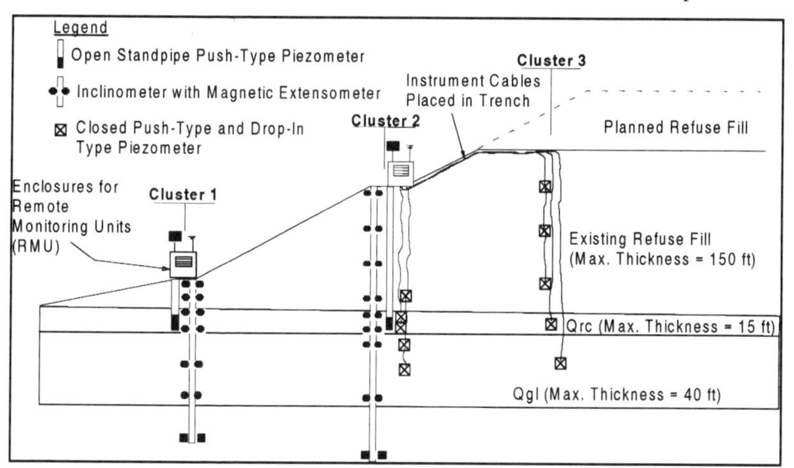

Figure 3 – *Typical Profile Instrumentation*

existing slope, and one or two clusters are located inward of the present landfill crest, so that they will be below the edge of the crest of the future landfill stages.

In general, the distribution of instrument types within the clusters at each profile is designed to measure the following:

- Pore pressures in the cohesive foundations soils (Qrc and Qg1);
- Pore pressures in the sand layer(s) (Qrs and Qgs);
- Pore pressures in the refuse fill;
- Vertical deformation (settlement or heave) and horizontal deformation of the refuse fill and the foundation soils; and,
- Temperature within the refuse fill.

The instruments were designed and installed in a phased approach to follow the progress and location of refuse filling. This phased approach also spread the cost over several years and allowed for the refinement of instrument designs and installations based on experience gained during previous phases. A total of three (3) instrument installation phases were performed between 1992 and 1997 resulting in a total of sixteen (16) instrument profiles. The majority (13) of these profiles have been located in Sections 1/9 and 6/7 since these sections either continue to receive refuse fill or have in the recent past.

Use of Instrumentation Measurements for Monitoring Stability

Measurements from the instruments are obtained automatically or manually on a regular basis and stored in several data management system computer programs. These computer programs take the raw readings and convert them into meaningful values (e.g., pore pressure, deformation). A report is then prepared that provides graphical information (primarily as a function of time) of each slope stability monitoring instrument installed at the landfill. Due to changes in the configuration of the landfill side slopes and landfill height because of active refuse filling and the large volume of data collected, it is difficult to rapidly assess the stability condition of the instrumented profiles. As such, a technique was developed whereby key measurements are used to determine an action level associated with stability conditions at each instrumented profile. The action levels provide a relatively straightforward and frequent method of informing the landfill engineer/operator and the client of the stability conditions so that grading plans can be developed that take into consideration these conditions, or on-going landfilling operations can be adjusted.

Since the action levels only use part of the measured geotechnical instrument data, a review of other measured data that may have an impact on the landfill stability is performed on a regular basis to determine if any unfavorable trends are developing. In addition, a comprehensive assessment of the landfill stability is performed on an annual basis.

Determination of Action Levels

The determination of the action levels is based on the magnitude and trends of the factor of safety against instability and the shear strain rate in the foundation soils. Four action levels are used. Action Level A corresponds to a situation where the factor of safety (FOS) against slope instability is above the established minimum and the maximum shear strain rate (MSSR) from the inclinometers is below the established maximum and the FOS and MSSR have improved since the previous readings. Action Level B1 corresponds to a situation where the FOS and the MSSR are at acceptable levels but have worsened since the previous reading. Action Level B2 corresponds to a situation where the FOS is close to the established minimum FOS or the MSSR from the previous two inclinometer measurements has worsened but is less than the established maximum. Action Level C corresponds to a situation where the FOS is below the established minimum or the MSSR is above the established maximum. Associated with each action level are contingency measures that range from maintaining normal operating procedures to stopping refuse fill placement and moving operations to a contingency area.

Factor of Safety Determination

The factor of safety used in determining the action level is based on limit equilibrium slope stability analyses using the computer program UTEXAS3 (Wright, 1991). In order to perform these analyses, the strength of the refuse fill and the foundation soils is needed along with other parameters (e.g. subsurface conditions and landfill profile). The determination of the strength of the cohesive foundation soils (i.e., Qrc and Qgl) is performed using the Undrained Strength Analysis (USA) methodology (Ladd, 1991), which is appropriate for conditions during construction of the landfill. In this method, the in-situ undrained shear strength is estimated based on the effective stresses in the foundation soils and laboratory determined strength relationships. The effective stresses in the foundation soils are based on the pore pressures in the foundation soils, as measured by the piezometers located along the instrumented profiles. The refuse fill and foundation soil strength properties used in the slope stability analyses are based on information contained in the literature and the results of site specific field and laboratory investigations.

Based primarily on a review of commonly accepted minimum factors of safety, the established minimum factor of safety, for this project, during construction is taken as 1.35.

Maximum Shear Strain Rate Determination

The shear strain rate (expressed as a percentage per year) obtained from the inclinometers installed in the foundation soils is computed as the difference between shear strains measured at two different times divided by the time between the readings.

The shear strain is computed as the change in horizontal deformation over a given measuring interval along the inclinometer casing and divided by the interval distance. The location of the maximum shear strain rate within the foundation soils gives an indication of where the greatest rate of horizontal deformations are occurring and where the critical shear surface may be developing. The measurements to date indicate that the maximum shear strain rate in the foundation soils typically occurs in the Qrc or the Qgl layer. In addition, the results of slope stability analyses indicate that the critical shear surface will most likely develop in the Qrc or the Qgl. Therefore, the action level is based on the maximum shear strain rate in the Qrc and/or the Qgl layer.

In general, the MSSR may initially increase when refuse fill is placed along a profile but will gradually decrease with time after refuse fill placement is stopped or moves away from a profile. It is anticipated that if the maximum shear strain rate is maintained at or below the historical maximum shear strain rate, the concern regarding the deformations of the slope are reduced. This is because the slope has already experienced these rates of deformations and remained stable. Therefore, selecting the acceptable maximum shear strain rate can be conservatively based on the maximum shear strain rates measured in the past. The following values, based on the shear strain rate data collected to date, are used as the maximum shear strain rates within the Qrc and the Qgl :

Location of Inclinometer	Soil Layer	Established Maximum MSSR (%/yr.)
Toe of Landfill	Qrc	1
	Qgl	0.5
Within the Landfill	Qrc	3
	Qgl	1.5

These values are reviewed and modified, if necessary, on an annual basis, as additional field and laboratory data becomes available.

Automated Slope Stability Assessment Program (ASSAP)

The Automated Slope Stability Assessment Program (ASSAP) allows for a rapid assessment of the action level on a pre-set schedule or an on-demand basis. The ASSAP program primarily consists of two modules, as shown in Figure 4, that perform the following operations for each instrumented landfill profile: 1) computes a factor of safety against slope instability and shear strain rates in the foundation soils, and 2) identifies an action level based on the current and previous factor of safety and maximum shear strain rates.

The factor of safety of the instrumented profiles is determined by selecting the slope stability module button, as shown in Figure 4. After selecting this option, the user sees the screen shown in Figure 5. The user selects the desired instrumented profile by using the pull down menu at the top of the screen. The data needed to assess the pore pressure conditions within the foundation soils comes from the piezometers and is obtained automatically from the instrumentation database by specifying the reading date. Since the landfill configuration changes because of active filling, it is also necessary to input the maximum landfill elevation on the date that the readings were obtained. The user runs the slope stability program and then has the option to save the result to a database or cancel the results. The program also has the ability to produce plots of the factor of safety, piezometric elevation within the foundation soils, and maximum landfill surface elevation as a function of time.

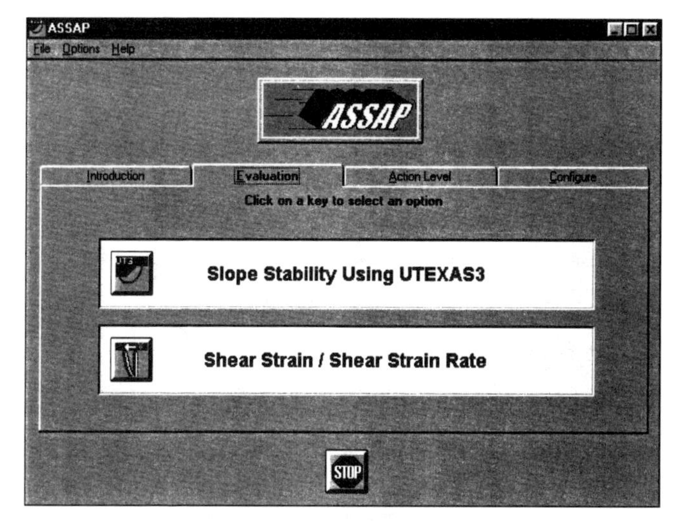

Figure 4 - Evaluation Screen

The shear strain rate from the inclinometers installed at a profile is determined by selecting the shear strain / shear strain rate module button, as shown in Figure 4. After selecting this option, the user sees the screen shown in Figure 6. The user initially selects the desired inclinometer from the pull down menu. Because the shear strain rate is a comparison between inclinometers readings performed on different days, two inclinometer files (referred to as RPP files) are needed. Once inputting the proper information, this module is run and the user has the option to save the results or cancel the results. The program also has the ability to produce plots of the maximum shear strain rate in the Qrc and/or Qgl, and the location of the shear strain rate within the foundation soils as a function of time.

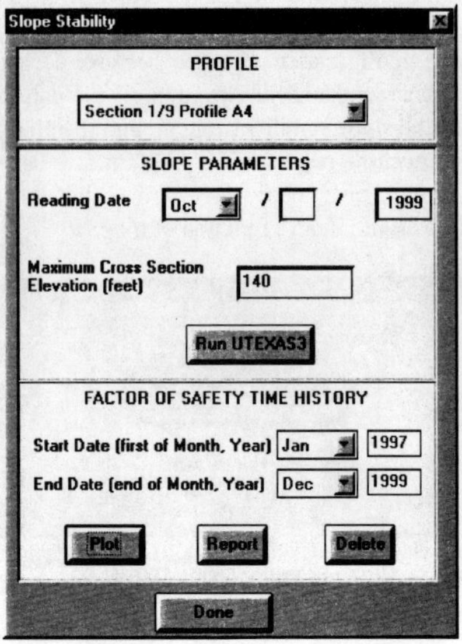

Slope Stability
PROFILE
Section 1/9 Profile A4
SLOPE PARAMETERS
Reading Date Oct / / 1999
Maximum Cross Section Elevation (feet) 140
Run UTEXAS3
FACTOR OF SAFETY TIME HISTORY
Start Date (first of Month, Year) Jan 1997
End Date (end of Month, Year) Dec 1999
Plot Report Delete
Done

Figure 5 - Slope Stability Main Screen

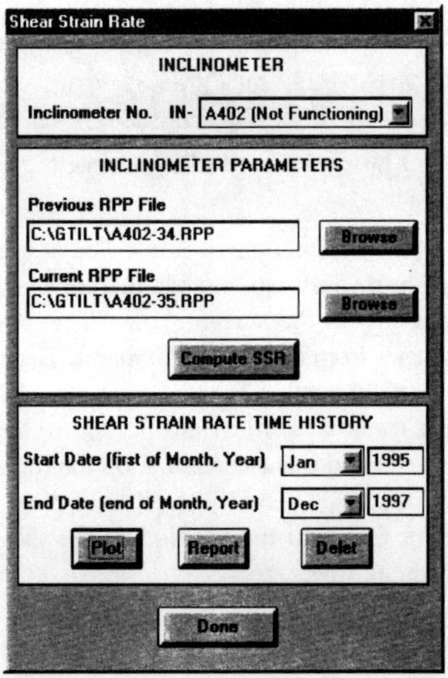

Shear Strain Rate
INCLINOMETER
Inclinometer No. IN- A402 (Not Functioning)
INCLINOMETER PARAMETERS
Previous RPP File
C:\GTILT\A402-34.RPP Browse
Current RPP File
C:\GTILT\A402-35.RPP Browse
Compute SSR
SHEAR STRAIN RATE TIME HISTORY
Start Date (first of Month, Year) Jan 1995
End Date (end of Month, Year) Dec 1997
Plot Report Delet
Done

Figure 6 - Shear Strain Rate Screen

The action level screens are used to identify the action level for the selected instrumented profile based on the factor of safety and the maximum shear strain rates in the foundation soils. The user selects the desired instrumented profile from the screen shown in Figure 7. A date can either be specified or the default date can be used. If the default date is specified, the program will obtain the necessary data from the most recent date that measurements were made. After selecting the evaluate button, the action level is presented along with a summary of the values used in determining this action level, as shown in Figure 8. The results can be saved to the project database thereby maintaining a record of the action levels for the selected profile.

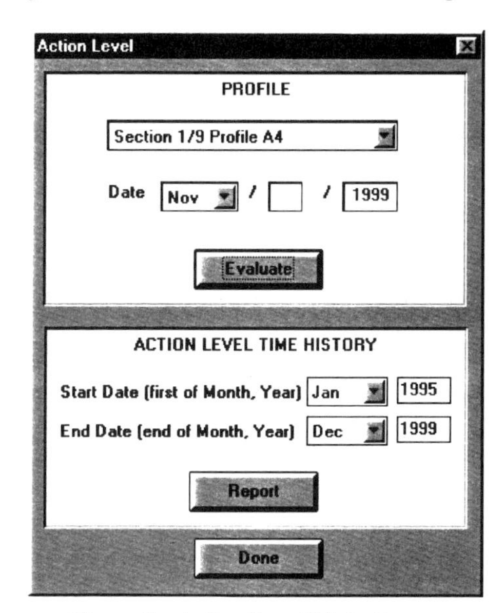

Figure 7 - Action Level Main Screen

A schematic of the slope stability monitoring system in operation at Fresh Kills Landfill is shown in Figure 9. The raw readings from the instruments are downloaded, manually or automatically, and stored in several databases that also have the capability to produce plots of deformations and piezometric elevation versus time. In order to avoid the need for double entry of information thereby reducing the possibility of human error, the ASSAP program obtains the necessary data to perform its evaluation directly from the database of stored information.

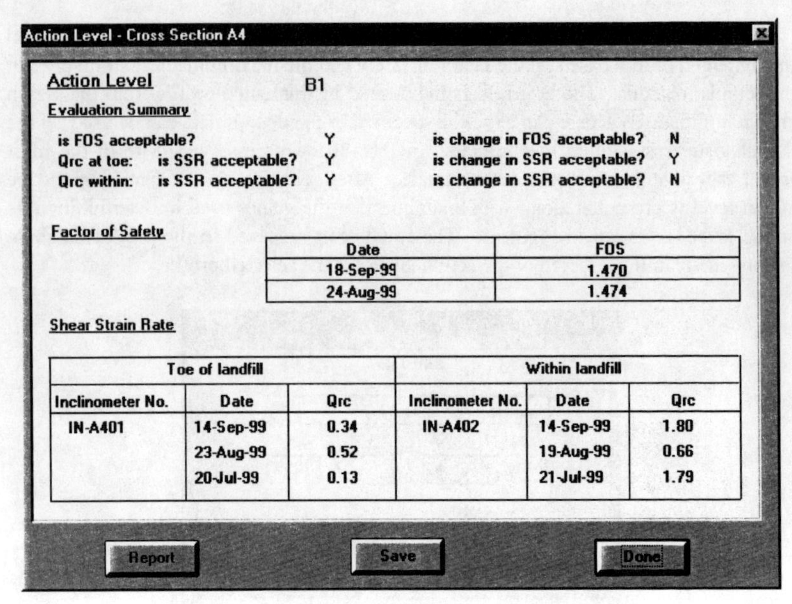

Action Level - Cross Section A4

Action Level B1

Evaluation Summary

is FOS acceptable?		Y	is change in FOS acceptable?	N
Qrc at toe:	is SSR acceptable?	Y	is change in SSR acceptable?	Y
Qrc within:	is SSR acceptable?	Y	is change in SSR acceptable?	N

Factor of Safety

Date	FOS
18-Sep-99	1.470
24-Aug-99	1.474

Shear Strain Rate

Toe of landfill			Within landfill		
Inclinometer No.	Date	Qrc	Inclinometer No.	Date	Qrc
IN-A401	14-Sep-99	0.34	IN-A402	14-Sep-99	1.80
	23-Aug-99	0.52		19-Aug-99	0.66
	20-Jul-99	0.13		21-Jul-99	1.79

Report Save Done

Figure 8 - Action Level Summary Screen

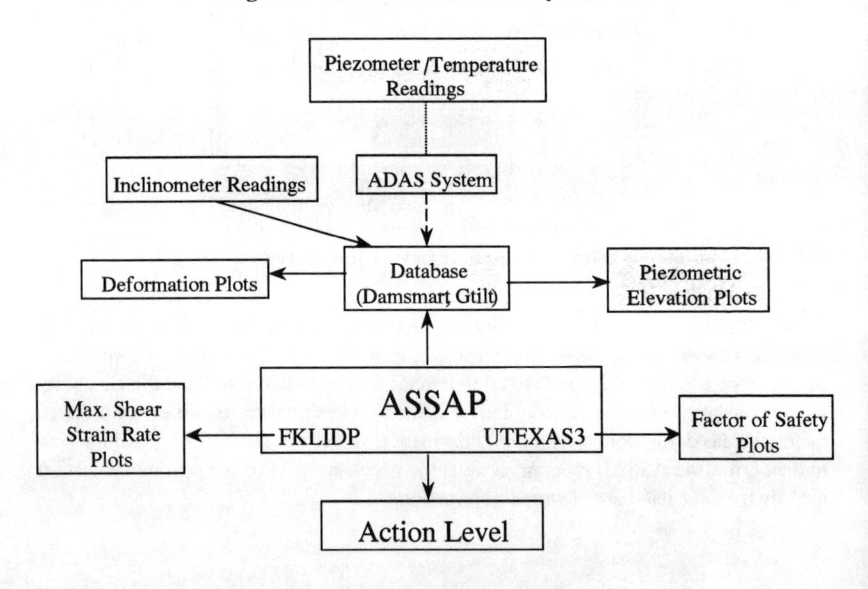

Figure 9 - Schematic of Monitoring System

Conclusions

A slope stability monitoring system has been designed and implemented for the Fresh Kills Landfill in New York City. To date, over 260 geotechnical instruments have been installed within the refuse fill and foundation soils for the primary purpose of monitoring both the magnitude and rate of change of pore pressure, and the lateral and vertical deformations. The instruments, which include closed vibrating wire piezometers, open standpipe piezometers, inclinometers, magnetic extensometers, and temperature probes, have been successfully designed and installed to accommodate on-going landfilling and closure activities, high temperatures, large movements, and the landfill leachate.

A unique system and computer program has been developed that uses the measurements from some of the instruments to determine the factor of safety against slope instability and shear strain rate in the foundation soils. The magnitude and trends with time of these values are then used to determine an action level associated with stability conditions for an instrumented profile. These action levels and associated contingency measures have been successfully used to monitor the stability of the landfill and to provide information to the landfill engineer/operator and the client for use in the development of grading plans, or adjustment to on-going landfilling activities.

The action levels provide an initial indication of the stability of a given profile and can assist the landfill engineer/operator in determining stability trends. However, it is important to realize that since the action levels only use part of the measured geotechnical instrument data, other measured data must be reviewed on a regular basis to determine if any unfavorable trends are developing. It is also important to perform a comprehensive stability assessment on a regular but less frequent basis to assess possible changes to the parameters and threshold levels being used based on new information.

Acknowledgments

The design, installation, and monitoring of the instrumentation system for the Fresh Kills Landfill started in 1989 and is still on-going. URS Greiner Woodward Clyde has performed all aspects of this program under contract with the New York City Department of Sanitation. The authors wish to acknowledge the contribution to this work by their colleagues: Drina Ferreira, Niels Jensen, Nicky Ng, Jack Rosenfarb, and Phalkun Tan. Also, Charles Ladd of the Massachusetts Institute of Technology, John Dunnicliff - Consultant, Harry Horn - Consultant, and Melvin Esrig - Consultant, have provided valuable discussions as members of the Technical Review Board. Finally, the authors want to express their appreciation to Mr. Phillip Gleason, NYCDOS Director of Landfill Engineering, for his support and guidance on this program. His continued interest has been instrumental to the overall success of the project.

References

Duplanic, N., 1990, "Landfill Deformation Monitoring and Stability Analysis," *Geotechnics of Waste Fill, ASTM STP 1070*, Landva and Knowles, Eds., pp. 303-239.

Handfelt, L.D., Koutsoftas, D.C., Foott, R., 1987, "Instrumentation for Test Fill in Hong Kong," *Journal of Geotechnical Engineering*, Vol. 103, No. 2, pp. 127-146.

Ladd, C.C., 1991, "Stability Evaluation During Staged Construction," *Journal of Geotechnical Engineering,* Vol. 117, No. 4, pp. 540-615.

Oweis, I.S., Mills, W. T., Leung, A., Scarino, J., 1985, "Stability of Sanitary Landfills", *Geotechnical Aspects of Waste Management*, American Society of Civil Engineers Metropolitan New York Section, pp. 1-30.

Peck, R. B., 1969, "Advantages and Limitations of the Observational Method in Applied Soil Mechanics : 9th Rankine Lecture," *Geotechnique*, Vol. 19, No. 2, pp.171-187.

Thomann, T.G., Khoury, M.A., Rosenfarb, J.L., Napolitano, R.A., 1999, "Stability Monitoring System for the Fresh Kills Landfill", *Field Instrumentation for Soil and Rock, ASTM STP 1358*, G.N. Durham and W.A. Marr, Eds., (to be published)

Withiam, J.L., Tarvin, P.A., Bushell, T.D., Snow, R.E., Germann, H.W., 1995, "Prediction and Performance of Municipal Landfill Slope", *Geoenvironment 2000,* ASCE Special Publication No. 46, pp. 1005-1019.

Wright, S.G., 1991, "UTEXAS3 - A Computer Program for Slope Stability Calculations", Shinoak Software.

RESIDUAL SHEAR STRENGTHS OF BENTONITES ON PALOS VERDES PENINSULA, CALIFORNIA

Stephen M. Watry[1], M. ASCE & Poul V. Lade[2], M. ASCE

ABSTRACT

Several large active and ancient landslides have failed along seams of bentonite located on the Palos Verdes peninsula, California. This is a desirable location for residential development due to its commanding ocean views. Numerous geotechnical reports regarding the residual shear strength of bentonites have been prepared over the years. The shear test procedure used has primarily been the direct shear test, but triaxial and ring shear testing have been performed. In-situ testing of block samples has also been performed on at least two occasions. Data indicates that Palos Verdes bentonite samples that have been tested for shear strength have similar mineralogical properties and similar Atterberg Limits. It would be expected that the residual shear strength values obtained would be similar. However, the actual residual shear strength values measured by various researchers vary widely. Direct shear and ring shear tests on samples of remolded bentonite performed for a study by the first author indicate that testing methods can be a significant source of variation in the measured residual shear strength. The residual shear strength determined by various methods was used in slope stability analyses of the Abalone Cove landslide, for which the slide geometry and groundwater conditions at the time of movement are known. The analyses indicate that the residual shear strength of the bentonite determined by ring shear testing resulted in the best fit as determined by a factor of safety of 1.0.

INTRODUCTION

The premise for comparing residual shear strength values from different locations is that soils with the same mineralogical composition and grain size distribution under the same effective stress conditions will have similar residual shear strengths. Relationships have been developed correlating residual friction angle with liquid limit (Mesri & Cepeda-Diaz, 1986; Jamiolkowski & Pasqualini, 1976), plasticity index (Voight, 1973; Kanji & Wolle, 1977; Lupini et al., 1981; Chandler, 1984) and clay fraction (Lupini et al., 1981; Skempton, 1985). Highly plastic clay has a curvilinear

[1]Eng., Grover-Hollingsworth & Assoc. 31129 Via Colinas, Westlake Village, CA
[2]Prof., Aalborg University, Sohngaardsholmsvej 57, 9000 Aalborg, Denmark

failure envelope with a steeper slope at lower effective normal stresses and a flatter, essentially linear slope at higher effective normal stresses. The curved portion of the failure envelopes for many clays is noted to occur below an effective normal stress of 200 kPa (4000 psf) (Hawkins & Privett, 1985). The residual friction angle as measured at a point on the curved section of the failure envelope at a low effective stress and at a point on the linear section of the failure envelope can vary by several degrees. Stark & Eid (1994) developed a relationship that correlates liquid limit, clay fraction, and the effective normal stress with a secant residual friction angle.

The development of a correlation requires that high quality experimental testing be available or be performed. High quality shear testing of highly plastic clays can be difficult. In this paper it is attempted to determine the accuracy of various testing methods used to measure the residual shear strength of bentonite from the Palos Verdes peninsula. The measured residual shear strengths are used in slope stability analyses of landslides on the Palos Verdes peninsula to ascertain their validity.

GEOLOGIC SETTING

Large inactive and active landslides, as outlined on Figure 1, mantle the southern portion of the Palos Verdes peninsula. These landslides failed along seams of bentonite that are interlayered with marine shale. The thickness of the bentonite seams ranges from a few centimeters to about 12 meters (40 feet). The bentonite seams and shale bedding are subparallel to the ground surface that slopes gently to the south toward steep wave-cut seacliffs. Bentonite seams that are inclined out of the 30 to 45 meter (100 to 150 feet) high seacliffs or toe out in the surf zone below the seacliffs form the basal slip surfaces of several large translational landslides, the most well-known of which are the Portuguese Bend and Abalone Cove landslides.

INDEX PROPERTIES

Watry & Ehlig (1995) made a compilation of the physical, mineralogical, and chemical properties of bentonite samples from the Palos Verdes peninsula. The majority of the data was derived from geotechnical reports prepared by consulting firms addressing the stability of properties underlain by bentonite seams. The bentonite samples were obtained from different geographic and stratigraphic locations on the peninsula. The results of Atterberg limit tests are plotted on a plasticity chart as shown in Figure 2. The liquid limit of the samples generally ranges from 80% to 110% with the plasticity index ranging from 40% to 70%. Clay fractions measured range from 2% to 80%. Some of the variations in the liquid limit, plastic limit, and clay fraction, are undoubtedly due to physical differences in the samples, but the variation would likely be much less if uniform sample preparation and testing procedures had been used. The presence of clusters and peds (silt and fine sand size aggregates of clay particles) in the bentonite if not disaggregated by washing with distilled water and/or milling will result in lower liquid limits, plastic limits and clay fractions.

Mineralogical and chemical analyses of bentonites from the Palos Verdes peninsula as

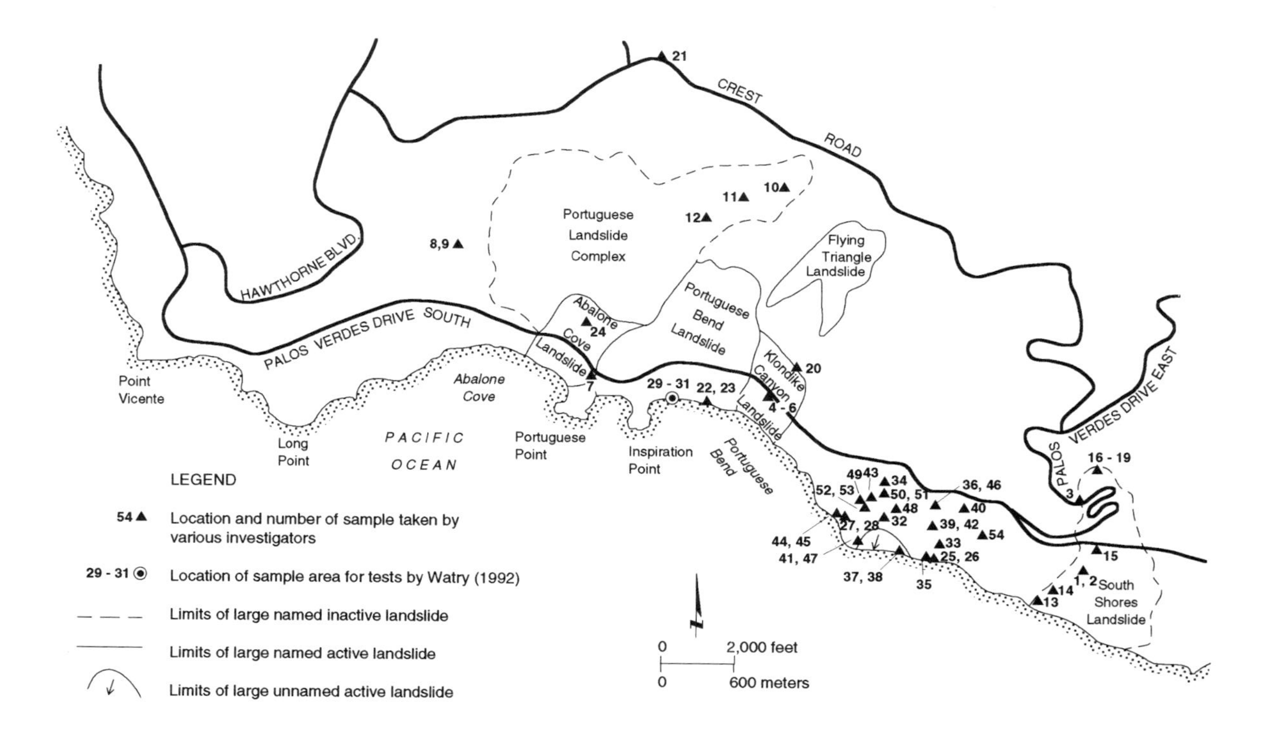

Figure 1. Location Map of Landslides and Residual Shear Strength Samples on the Palos Verdes Peninsula, Southern, California (modified from Watry & Ehlig, 1995).

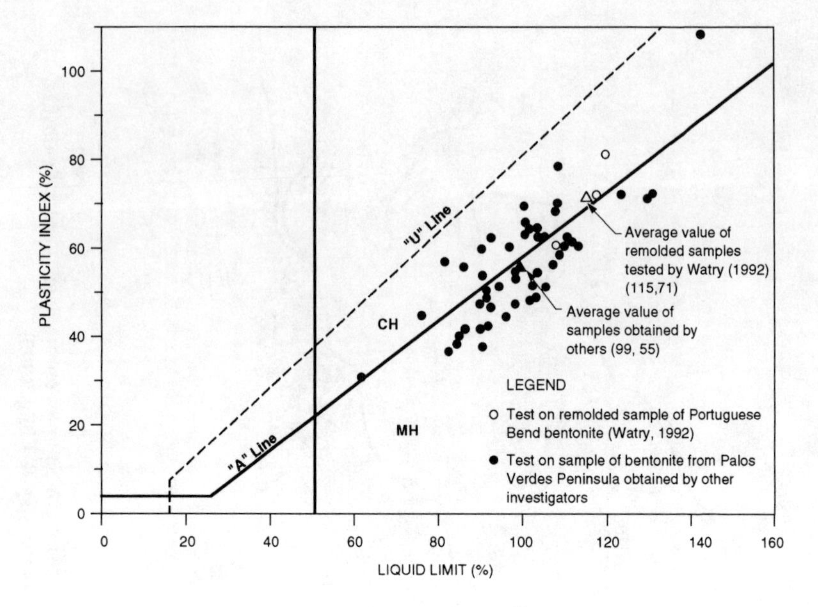

Figure 2. Plasticity Chart with Bentonite Samples Obtained by Various Investigators (modified from Watry & Ehlig, 1995)

performed by Kerr & Drew (1969), Novak (1982), Leighton & Associates (1989 & 1990) and Converse Consultants West (1991a) show the bentonites to be very similar and suggests that the bentonites are predominantly a calcium montmorillonite.

PREVIOUS EXPERIMENTS

The results of residual shear strength tests performed by various consultants on bentonites from various geographic and stratigraphic locations on the Palos Verdes peninsula are provided in Figure 3. Residual friction angles for the bentonite range from 3.5 degrees to 27 degrees with residual cohesion values ranging from 0 to 191.5 kPa (0 to 4000 psf).

As previously noted, the failure envelopes for highly plastic clay, such as the bentonite from the Palos Verdes peninsula, are curved at lower effective normal stresses, and becomes linear at effective normal stresses in excess of about 200 kPa (4000 psf). Most of the shear testing by the various consultants was performed over an effective normal stress range that included points on both the curved and linear portions of the failure envelope. However, in all but a few instances a straight-line approximation was made using all the points in the series by performing a linear regression analysis or by visual approximation. The use of a single straight-line approximation of a curvilinear failure envelope can lead to serious errors when performing slope stability analyses.

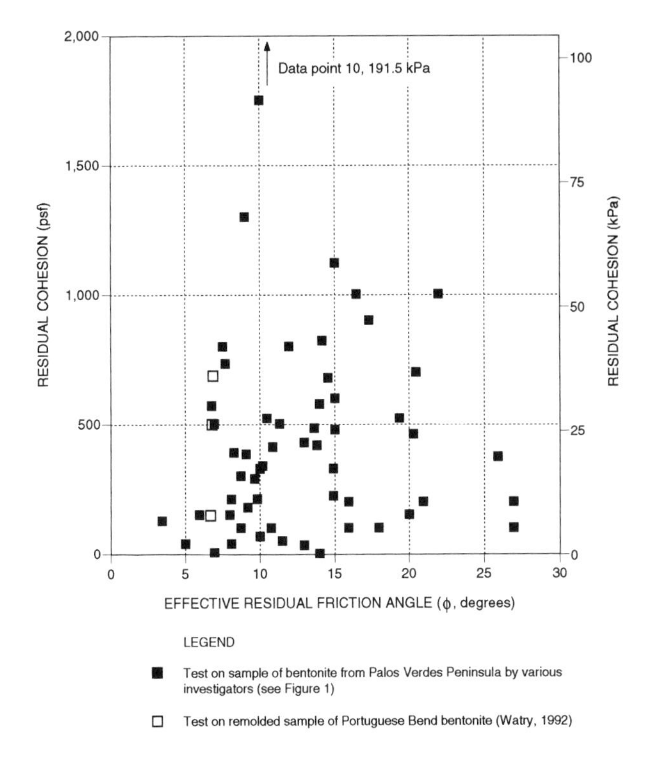

Figure 3. Residual Shear Strength of Bentonite Samples from the Palos Verdes Peninsula as Determined by Various Investigators

TEST METHODS FOR RESIDUAL SHEAR STRENGTH

The residual shear strength of earth materials can be determined by laboratory shear test methods that include triaxial compression, direct shear, and ring shear. The residual strength can also be determined in the field by testing an in-situ block. A brief discussion of each of the methods is given below.

Triaxial Compression Test

The triaxial compression test may provide satisfactory residual strengths for cohesionless soils and low plasticity clays, where minor nonuniformity in the stress condition may allow for a shear band to develop at small strains. However, the triaxial test is rarely used to determine the residual shear strengths of highly plastic clays, because the strains to reach the residual strength are very large. Attempts to assess the residual shear strength of highly plastic clay have been made by using a sample with a natural, well-defined shear surface or by creating a shear plane by cutting the sample with a wire saw. The natural or preformed shear plane was to be oriented at 45 + ϕ/2 (ϕ = the anticipated residual friction angle of the soil) to the horizontal which is the plane of the major principal stress. Chandler (1966) reports that small inaccuracies in the estimation of the residual friction angle used to determine the orientation of the shear plane in a sample being tested does not significantly affect the measured residual friction angle.

F. Beach Leighton and Associates (1974) references an earlier investigator (Altmeyer) from 1957 who "by means of triaxial tests on the slide plane material and by back-calculation of the active slide" defined residual shear strength parameters for the bentonite forming the Portuguese Bend landslide as residual friction angle $\phi_r = 6.0$ degrees with cohesion C_r=13.4 kPa (280 psf).

Direct Shear Test

The direct shear test is commonly used for determination of the residual shear strength of soils. The advantages of the direct shear system are its simplicity of operation and its economy.

The amount of displacement available in a direct shear device is typically 0.65 cm (0.25 inch). This amount of displacement is sufficient to obtain the peak shear strength of most soils and the residual strength of many soils. However, moderate to high plasticity clay requires more displacement than can be obtained in one full pass to reach residual strength. Direct shear tests of highly plastic clay typically require several passes before the measured strength value approximates the residual strength. Some researchers have created a preferred shear surface by slicing the sample horizontally and parallel to the application of the shear force to eliminate the necessity of multiple passes (Kenney, 1967). Another method used to decrease the amount of displacement required to attain a residual strength is to place a polished rock or metal surface on top of the soil sample and to shear the soil sample along the interface (Kanji, 1974; Kanji & Wolle, 1977). The reduction in the amount of displacement is attributed to a more rapid orientation of clay particles into a parallel alignment adjacent to the hard, polished surface.

The direct shear test is by far the most common test used in assessing the shear strengths of bentonites from the Palos Verdes peninsula.

Ring Shear Test

The ring shear test for determining the residual shear strength of a sample is performed by applying rotational shearing to an annular soil specimen that is laterally confined. Various ring shear type devices have been developed since the 1930's.

Bromhead (1979) describes a ring shear device that has been simplified both in design and operation when compared to earlier ring shear devices. The Bromhead ring shear device utilizes annular specimens inside a solid confining ring. Porous bronze stones are placed both above and below the sample. The normal load is applied through the upper stone. The magnitude of the friction between the sample and the confining ring increases with depth and so the minimum resistance to shear is developed at or near the soil/top porous stone interface (Stark & Vettel, 1992).

The primary advantage of the ring shear test method in determining residual shear strength values are that it allows the sample to be continuously sheared in one direction, allowing for optimal orientation of the soil particles, while maintaining a constant cross sectional area of the shear plane (Bishop et al., 1971; Stark & Vettel, 1992). The main disadvantage of the ring shear is the potential for nonuniform stress distribution across the sample. The potential for nonuniform stress distribution is minimized by accurate centering of the normal load upon the sample and the use of a narrow sample width (the inside and outside diameters of the sample are not grossly different). The stress distribution is uniform by the time the specimen has been sheared to residual strength.

The ring shear device has not yet been extensively used to determine the shear strengths of bentonites from the Palos Verdes peninsula.

In-Situ Block Test

The in-situ block test consists of isolating a block of soil in a pit and then applying normal and shear forces by dead weight and/or hydraulic jacks.

The minimum size of in-situ samples to be utilized for an accurate determination of shear strength, as recommended by the International Society of Rock Mechanics, is 700 mm x 700 mm (28 inches x 28 inches) (Petley, 1984). In some cases rectangular specimens are used where the potential failure horizon, such as a bedding plane or joint, is well defined and is isolated from other potential failure horizons. The long dimension of the rectangular sample is oriented parallel to the proposed direction of shear allowing for greater displacement along the shear failure surface. Channels are cut around the block to allow for movement of the sample block when the shear load is applied. The channels are also necessary to allow encapsulation of the sample block with concrete, construction of reaction pads, and installation of testing equipment.

The orientation of the encapsulated sample block, the reaction pads and ties must be carefully chosen in those cases where the shear strength of the material along a natural discontinuity is to be tested. Shear strength tests perpendicular to the dip of a bed, joint or slip surface may yield a different result than that obtained from a test parallel to the dip. Fault planes and landslide slip surfaces frequently have striations or mullion structure, which could produce different shear strengths when, sheared at different orientations.

Tests requiring drained shear strengths must allow for saturation of the specimen. This is achieved by filling the pit surrounding the sample with water. The rate of shear to allow for drained conditions to develop can sometimes be determined by noting displacement of the block under the first normal load applied. The

International Society of Rock Mechanics suggests that a displacement rate be determined on the basis of a minimum of 6 x t_{100} for clay bearing and soil like rocks (Franklin et. al, 1974). An empirical upper displacement rate of 0.2 mm/min (0.079 inches/min) and an empirical lower displacement rate of 0.02 mm/min (0.0079 inches/min) were established.

The maximum amount of displacement in an in-situ test is typically on the order of 10 to 20 percent of the length of the block. Displacements beyond this range may cause significant changes in the normal load distribution, instability of the equipment and rupture of the encapsulated sample.

It is desirable to have at least three test points to define a shear strength envelope. Multi-stage tests with different normal stresses applied to the same sample block after a residual shear strength value is obtained at each normal stress can yield several test points. However, the potential for variations in pore pressures due to the change in normal and shear stresses must be evaluated at each point to make sure that drained conditions still exist.

The main advantage of the in-situ test is that it allows determination of the shear strength of materials, which possess irregular, natural discontinuities that are of such a scale that they will not be represented when a sample is trimmed to the required size for a laboratory shear device.

There are several disadvantages of the in-situ test for residual shear strength determinations. Foremost are the cost and time required to setup the test. It is also difficult to assure that in-situ clay seams are fully saturated during testing. The determination of residual shear strength in intact, unsheared clay seams may require more shear displacement than can be provided. The reversal of the shear direction as sometimes performed in laboratory direct shear tests, has been performed in in-situ tests, but it is likely that this causes some particle reorientation along the shear surface as has been noted in laboratory direct shear tests.

The authors are aware of only two in-situ block shear tests of bentonites on the Palos Verdes peninsula.

SYSTEMATIC STUDIES OF RESIDUAL SHEAR STRENGTH

Two studies have been performed that attempted to systematically evaluate factors influencing the measured residual shear strengths of bentonites from the Palos Verdes peninsula. Converse Consultants West (CCW) (1991a, 1991b) performed the first study. This study utilized triaxial compression, direct shear, and in-situ block testing to evaluate the properties of bentonite seams. Both undisturbed and remolded samples were tested.

The second systematic study was performed at essentially the same time as the CCW study as part of a masters thesis by the first author (Watry, 1992). The second study involved the use of remolded bentonite samples that were derived from slurry. The intent of using samples remolded from a slurry was to create a uniform sample for testing so that any differences in measured residual shear strength using different shear testing devices would be from differences/errors in the testing devices, methods, and procedures and not differences in the physical properties of the bentonite.

Study by Converse Consultants West

The study by CCW evaluating the residual shear strength of bentonite seams is contained in reports for a residential tract and golf course above the seacliffs about 1500 meters (one mile) southeast of the Portuguese Bend landslide (CCW 1991a, 1991b). Sample numbers from 32 to 54 indicates the location in Figure 1.

Two bentonite seams, described as the upper and lower bentonite were the primary focus of the CCW study. These bentonite seams are stratigraphically above the bentonite that forms the slip surface for the Portuguese Bend and Abalone Cove landslides that were the foci of the second study. Undisturbed and remolded bentonite samples were used in the shear tests performed for the CCW study.

The liquid limit of the upper bentonite was measured at 85-98% with plasticity indices of 53-59%. The liquid limit of the lower bentonite was measured at 91-100% with plasticity indices of 49-69%. The clay fractions of two samples of the upper bentonite were 46% and 50%, while the clay fraction was 48% for a sample of the lower bentonite.

Direct shear testing to determine the residual shear strength was performed by CCW and two outside consultants in order to "eliminate equipment or procedural bias". Direct shear tests and ring shear tests were later performed on samples of the upper and lower bentonite by another consultant (Leighton, 1991).

The direct shear tests show a wide range of residual shear strength values, even when the results are sorted into upper and lower bentonite results, as shown in Figures 4(a) and (b). It is believed that the variation is mostly due to differences in test procedures rather than variation in the physical properties of the bentonite tested. The residual friction angle for the bentonite samples determined by the direct shear tests cover a much greater range than would be anticipated when compared against the correlation of the residual friction angle with the liquid limit, clay fraction, and effective normal stress prepared by Stark & Eid (1994) as shown on Figure 5.

Straight-line approximations of the failure envelope were used by CCW to determine the residual shear strength parameters. The effective normal stress range used in the direct shear testing of the bentonite extends over the range from the curved failure envelope and into the range above 200 kPa (4000 psf) where the failure envelope may be better characterized by a straight line. Of the 18 direct shear test series in the CCW study with three or more points to determine the residual shear strength, 12

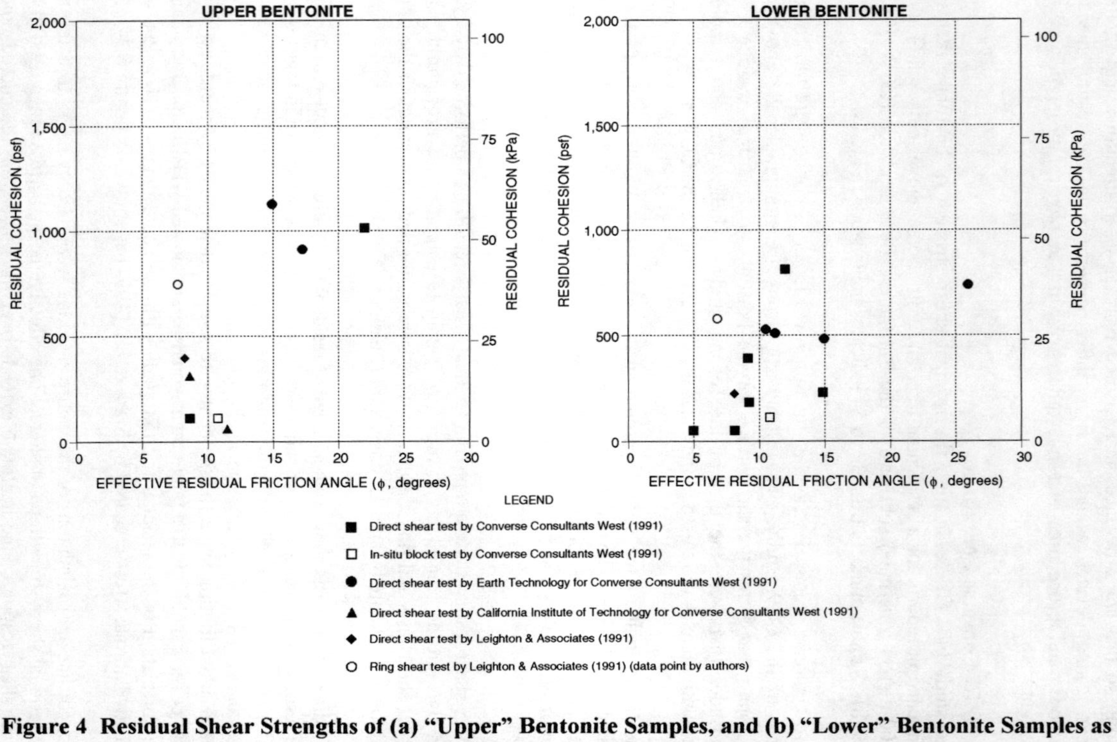

Figure 4 Residual Shear Strengths of (a) "Upper" Bentonite Samples, and (b) "Lower" Bentonite Samples as Determined by Various Investigators

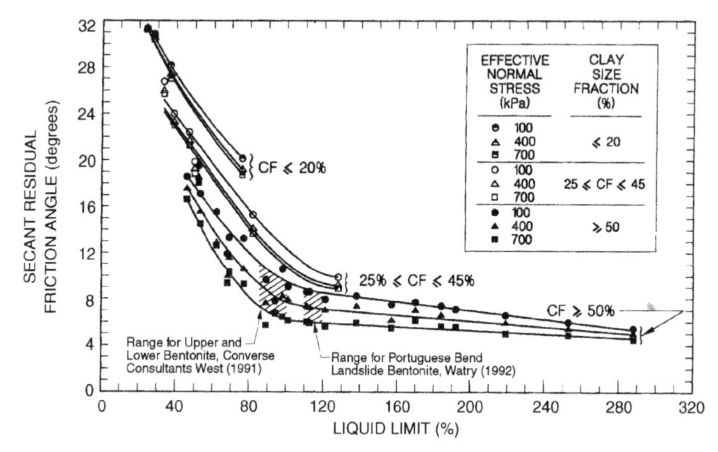

Figure 5. Relationship of liquid limit, clay fraction, effective normal stress and residual friction angle (Stark & Eid, 1994)

had shear strength points that plotted above the straight-line failure envelope selected. The effect of including these points more correctly in the strength characterization would be to increase the residual friction angle and reduce the cohesion. It appears that the use of a straight-line approximation may be inappropriate in this case, because the failure envelope is curved.

It is possible that the cohesion may have been overestimated in some tests due to the measurement of resistance of bentonite that squeezed between the halves of the shear box. CCW notes that "laboratory test results which fell below the in-situ test results were all on samples which were run to the end of the shear box, the samples taken apart, smoothed on the slip surfaces, put back together and reconsolidated before reshearing. This technique appears to give lower results than shearing, pushing the samples back to the origin, and reconsolidating before reshearing." The lower values may therefore be the result of creation of a preformed slip surface and the removal of bentonite that squeezed between the shear box halves during the initial shear pass.

In-situ block shear tests of the upper and lower bentonite were performed by CCW. The in-situ tests were performed by isolating a block of rock containing a seam of the lower or upper bentonite. The dimensions of the block were about 0.6 meter (2 feet) square by about 0.6 meter (2 feet) high. The test block was encapsulated in concrete with concrete reaction pads constructed on the sidewalls of the pit for the jacks to apply shear loads. A steel reaction beam anchored into the ground was constructed to allow for application of a normal load. The normal load was applied to steel rollers that rested in an oil bath on top of the test block. Gauges attached to the reaction frame measured deformation.

CCW describes three different test procedures for the in-situ block shear tests: 1) Parallel to the strike, applied shear load; 2) Up-dip, passive shear load; and 3) Up-dip, applied shear load. Tests were performed with the shear surface submerged. The effective normal stresses with which the in-situ shears tests were performed ranged from about 47.9 to 135.3 kPa (1000 to 2800 psf).

The first procedure involved shearing the test block parallel to the strike of the bentonite seam. The advantage noted was that the shear and normal forces would act on a plane that was horizontal in the direction of movement. The shear load was increased incrementally and the strain measured until the applied shear load moved the block at a constant rate while the pressure on the shear ram gauge simultaneously dropped. This point in the test was considered to be a point on the shear strength envelope. The normal load was then increased and the process repeated. The process was repeated for additional points until the block began to move down-dip rather than perpendicular to the dip.

The second procedure used, the "up-dip, passive shear load procedure", was intended to limit the possibility of two components of horizontal strain as could have occurred during the first procedure where strain was applied parallel to the strike, but at the end of the procedure the block was moving down dip. Down-dip testing was not considered feasible as there was concern that the test block could fail simply under the applied vertical load, which imposes a down dip shear load on the inclined failure surface. A constant normal load was applied while the shear load was allowed to self adjust. Strain readings were recorded until the strain and load readings became roughly constant. This was considered to correspond to a point on the strength envelope. However, inspection of the test results by CCW revealed the data to be faulty as negative cohesion and friction angles were being recorded, and so this testing procedure was stopped.

The third procedure, the "up-dip, applied shear load procedure" involved applying a constant normal load while an incrementally increasing shear load was applied up-dip. A point on the shear strength envelope was considered reached when the applied shear load moved the block at a constant rate.

The results of the parallel to strike test procedure on the lower bentonite and the up-dip applied shear load procedure on the upper bentonite were combined by CCW to establish the residual shear strength parameters for the bentonite shown in Figure 6. The residual shear strength parameters as determined by CCW are friction angle of 10.8 degrees and cohesion of 4.79 kPa (100 psf). It appears that the testing did not yield residual shear strength values and the results as plotted by CCW are too scattered to reasonably determine residual shear strength parameters.

Present Study

Watry (1992) performed shear strength tests on remolded samples of bentonite. Bulk samples of bentonite were obtained from the toe of the Portuguese Bend landslide.

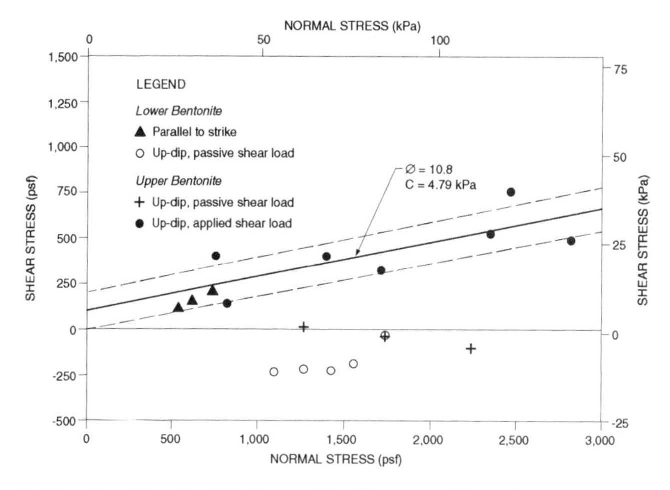

Figure 6. Results of in-situ block tests by Converse Consultants West (1991a)

The bulk samples were, mixed, air-dried, and made into slurry by adding distilled water. The slurry was consolidated in a 25.4 cm (10 inch) diameter consolidation tank to create firm clay that was trimmed to create samples for the various shear test devices. The liquid limits of the remolded bentonite samples ranged from 108 to 119% and the plasticity indices varied from 61 to 81%. The clay fraction determined for the remolded bentonite was 80%.

Consolidated-undrained (CU) triaxial compression tests were performed to determine stress-strain characteristics, pore-pressure development during shear, peak shear strength, the coefficient of consolidation, and permeability of the remolded bentonite samples, but not the residual shear strength of the bentonite. The peak shear strength shown in Figure 7 was attained at axial strains of 10.5 to 11.5%. The tests were continued to about 20% strain by which time the samples had failed by buckling or bulging. The shear strength values at 20% axial strain had not decreased significantly below the peak strength as may occur for intact normally consolidated clay.

The peak and residual shear strengths of the remolded bentonite were determined using a conventional direct shear device, a modified direct shear device that could accommodate several centimeters of travel, and a ring shear device. The peak and residual shear strength envelopes determined by these devices are shown on Figures 7 and 8, respectively.

Conventional direct shear tests were performed as done by other investigators, except a thinner sample of 1.27 cm (0.5 inch) was used to allow for quicker drainage. The shear tests were performed using very slow rates of deformation (0.000254

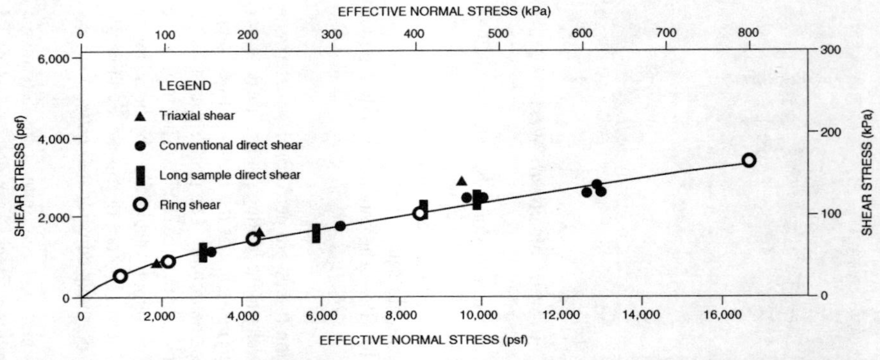

Figure 7 Drained Peak Failure Envelope for Combined Tests on Remolded Bentonite from Portuguese Bend Landslide for Present Study (Results of Ring Shear Tests by Stark)

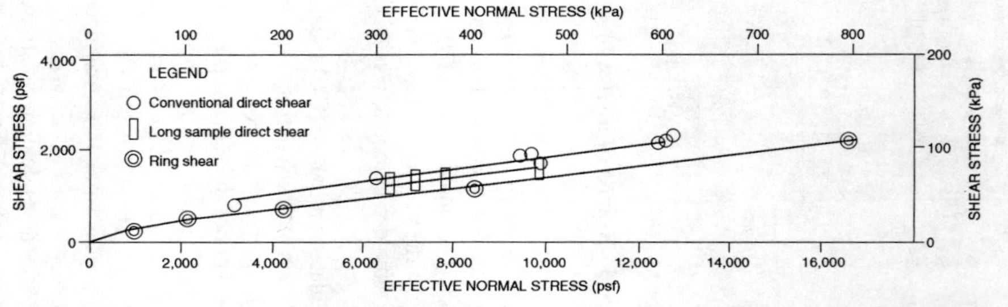

Figure 8 Drained Residual Failure Envelope for Combined Tests on Remolded Bentonite from Portuguese Bend Landslide for Present Study (Results of Ring Shear Tests by Stark)

mm/min or 0.00001 inches/min). The samples were deformed in one direction about 0.64 cm (0.25 inch), which is almost the full amount of deformation available, before reversing direction. The samples were repeatedly sheared by reversing the direction of deformation until a residual shear strength value was approached. Total deformations were between 28 and 56 mm (1.1 and 2.2 inches).

Shear surfaces were observed in the samples after testing. The shear surfaces in the samples, with one exception, were observed entirely within the clay just below the roughened surface of the upper filter stone. The shear surface in one test was observed near the center of the sample. The location of shear surface just below the filter stone is considered analogous to the formation of the slip surface at the soil/polished metal or polished stone interface described by Kanji & Wolle (1977).

A second series of residual shear tests was performed utilizing a modified direct shear device that can shear rectangular samples 5.08 cm by 25.40 cm (2.00 inches by 10.00 inches). The prototype device, informally called the Long Sample Direct Shear Device, was developed at the UCLA Soil Mechanics Laboratory. The device was designed to be able to continuously shear a sample in one direction over 20 cm (8 inches) while maintaining a constant normal stress centered over the portion of the sample being sheared. In a test on the remolded bentonite, the residual shear strength value was reached after a deformation of about 10.7 cm (4.2 inches). The normal stress was then adjusted to increase with continued deformation so that a failure envelope could be plotted over a range of normal stresses.

A third series of residual shear strength tests was performed using a Bromhead ring shear apparatus on samples of bentonite from the Portuguese Bend landslide prepared at UCLA as explained above. The ring shear testing was performed under the direction of Professor Tim Stark at the University of Illinois. The inner diameter of the ring shear sample is 7.00 cm (2.756 inches) and the outer diameter 10.00 cm (3.937 inches). The nominal thickness of the sample is 0.5 cm (0.196 in). The shear surface was induced at the top of the sample just below the porous bronze stone. Techniques were used to limit sample loss due to squeezing during the testing.

A plot of the effective normal stress versus the peak and residual shear stress of the remolded bentonite as determined by the various shear test methods produced curved failure envelopes, as seen in Figures 7 and 8. The curved portion of the failure envelope is below about 200 kPa (4,000 psf).

The residual shear strengths of the remolded bentonite as determined by the three different testing devices resulted in three separate failure envelopes with similar friction angles, but different cohesions over the effective normal stress range that was evaluated. The residual friction angle for the bentonite samples determined by all three procedures correlates well with the secant residual friction angle determined by the relationship with the liquid limit, clay fraction, and effective normal stress as shown on Figure 5. The lower cohesion value determined by the ring shear test is likely the result of the ability of the ring shear device to shear the sample in one direction over large shear displacements while maintaining a constant effective normal stress, the inducing of the failure surface at the top of the sample just below the porous stone, and the limiting of sample loss due to squeezing. The long sample shear device, while capable of large displacements in one direction, was not as capable of maintaining a constant effective normal stress as the ring shear device. Also the shear surface in the long sample shear device formed within the center of the sample and so the relatively rapid development of a preferred orientation of clay particles by shearing the sample along a hard smooth surface, as suggested by Kanji & Wolle (1977), may not have been realized. The conventional direct shear device was limited to a small amount of displacement and therefore was not able to develop the degree of preferred orientation of the clay particles obtained by the ring shear and long sample direct shear devices. The reversal of the shear direction has been shown to disturb the orientation of clay particles developed by the preceding shear displacement (Kenney, 1967). The area of the shear surface decreases as the sample is displaced resulting in a change in normal stress. The variation in sample size and normal stress produce nonuniform strains that hinder the development of a shear surface bounded by highly orientated clay particles. The conventional direct shear tests, and to a lesser extent the modified direct shear tests, also are hampered by the loss of sample by squeezing of the material through the gap between the upper and lower halves of the shear box. The squeezing of material between the shear box halves may have led to the measurement of shear resistance developed from cohesion of the bentonite squeezed between the shear box halves. Preforming a shear surface by cutting the sample, as performed by others, would likely have resulted in a decrease in the cohesion value measured.

BACK CALCULATION OF BENTONITE SHEAR STRENGTHS

The credibility of the residual shear strength parameters determined by the three shear test methods performed for the present study was checked by performing three-dimensional slope stability analyses on the active Portuguese Bend and Abalone Cove landslides. In addition, two assumed sets of strength parameters were employed to assess the sensitivity of the stability analyses of the landslides to variation in the residual friction angle. The computer program TSLOPE 3 (TAGA, 1991) was used to perform the analyses.

The Portuguese Bend landslide covers 108 ha (270 acres) and is generally between 30 and 45 m (100 to 150 feet) thick. The failure surface of the Portuguese Bend landslide is smooth with large-scale undulations approximately perpendicular to the direction of movement. The Abalone Cove Landslide covers 32 ha (80 acres) and ranges from 30 to 60 m (100 to 200 feet) in thickness. The failure surface is smooth and dips gently to the south before turning sharply upward at the toe. The basal failure surface of the Abalone Cove landslide is shown in Figure 9. Both of the failure surfaces developed along seams of bentonite that dip shallowly seaward. These landslides can be reasonably evaluated with a single set of shear strength parameters as the effective normal stress on the slip surfaces is above the curvilinear portion of the failure envelope, with the exception of very small areas near the toes of the slides. A uniform total density of 17.43 kN/m^3 (111 pcf) was assumed for earth materials above the slide plane.

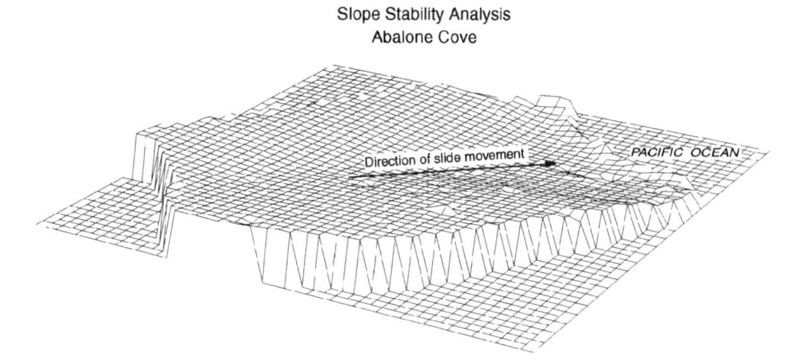

Figure 9. Plot of Basal Surface of Abalone Cove Landslide

Data points describing the geometry of the failure surface, phreatic surface, and the ground surface, in the three-dimensional analyses were obtained by establishing a grid of points at 30 m (100 feet) on center across the landslide mass. Additional points were assigned where abrupt changes occur in the failure, phreatic, and ground surfaces. The phreatic surfaces used are considered representative of the condition present during periods of slide movement so that the factor of safety calculated for the slide masses should be very near 1.0. Table 1 shows the results of the analyses. Typically the factor of safety determined by slope stability analyses is reported only to tenths place. The factor of safety is reported to hundredths place to emphasize the change in factor of safety with the variation of shear strength parameters.

Residual Shear Strength Parameters	$\phi_r = 6.7$ $C_r = 7.18$ kPa Ring Shear	$\phi_r = 6.8$ $C_r = 23.80$ kPa Long Sample Direct Shear	$\phi_r = 6.9$ $C_r = 33.04$ kPa Conventional Direct Shear	$\phi_r = 7.0$ $C_r = 0$ Assumed	$\phi_r = 10.0$ $C_r = 0$ Assumed
Portuguese Bend Landslide	FS = 1.11	FS = 1.42	FS = 1.60	FS = 1.03	FS = 1.48
Abalone Cove Landslide	FS = 1.00	FS = 1.23	FS = 1.36	FS = 0.94	FS = 1.35

Table 1. Results of three-dimensional slope stability analyses of Portuguese Bend and Abalone Cove landslides

The high factor of safety determined for the Portuguese Bend landslide is considered to be the result of assumptions and simplifications of the failure surface, phreatic surface, and boundary conditions in areas where subsurface information was sparse. The failure and phreatic surfaces, and boundary conditions for the Abalone Cove landslide are much better defined and so the factors of safety yielded by the three-dimensional analyses of this landslide are considered more appropriate. The assumed values of residual friction angle of 7 and 10 degrees and no cohesion demonstrate the large variation in the factor of safety with small changes in the friction angle for these deep slides.

CONCLUSIONS

Shear testing of the Palos Verdes bentonites by different investigators has resulted in vastly differing residual shear strengths. Differences in shear test procedure results in some of the disparity. Failures to deform the sample sufficiently to attain residual strength, and sample squeeze and loss during testing are considered the primary causes of the measured disparity. There does not appear to be any advantage of performing in-situ block tests, as the residual shear strength of the bentonite is not likely effected by any scale factor or contact discontinuities that can be accounted for by the testing.

Past studies of the residual shear strength of the Palos Verdes bentonites largely did not consider the curvilinear nature of the failure envelope and used straight-line approximations of shear test data points to determine the residual shear strength parameters, which can lead to erroneous data. It is important when presenting residual shear strength parameters for highly plastic clays that the effective normal stress under which the sample was tested be noted.

Slope stability analyses involving the bentonite seams should consider the effective normal stress on the potential (or actual) failure surface for determination of appropriate residual shear strength parameters. For those analyses involving a failure surface with a range of overburden that will impart both low and high effective normal stresses on the bentonite, it may be appropriate to assign two or more sets of residual shear strength parameters depending on the shape of the failure envelope. For shallow landslides with low effective normal stresses it may be easier to assign several effective shear strengths from the curvilinear portion of the failure envelope at effective normal stresses that correspond to the overburden pressure, or to describe the curved failure envelope by an appropriate mathematical expression.

Relationships involving Atterberg limits, clay fraction, and effective normal stress can be used to approximate a residual friction angle for a particular soil and/or evaluate the credibility of a measured laboratory value. Techniques used to prepare the sample for testing should be noted in the report.

Acknowledgment
The experiments for the present study were performed while both authors were associated with the Department of Civil Engineering at the University of California, Los Angeles. The authors thank an anonymous reviewer.

REFERENCES

Bishop A.W., Green, G.E., Garga, V.K., Andresen, A., & Brown, J.D., 1971. A New Ring Shear Apparatus and its Application to the Measurement of Residual Shear Strength, Géotechnique 21, pp. 273-328.

Bromhead, E.N., 1979, A Simple Ring Shear Apparatus, Ground Engineering, Volume 12, No. 5, pp. 40-44.

Chandler, R.J. 1966, Measurement of Residual Strength in Triaxial Compression, Géotechnique 16, No. 3, pp. 181-186.

Chandler, R.J., 1984, Recent European Experience of Landslides in Over-Consolidated Clays and Soft Rocks, in Proceedings International Symposium on Landslides, pp. 61-81.

Converse Consultants West, 1991a, Geotechnical Feasibility Investigation South Shores Parcels 1 and 1A Tentative Tract 49470, Rancho Palos Verdes, California: Unpublished Consulting Report, Project No. 88-31-131-02(03).

Converse Consultants West, 1991b, Response to RPV Geotechnical Review South Shores parcels 1 and 1A Tentative Tract 49470, Rancho Palos Verdes, California: Unpublished Consulting Report, Project No. 88-31-131-04.

F. Beach Leighton and Associates, 1974, Report of Geotechnical Investigation, Abalone Cove, Palos Verdes Peninsula, Los Angeles County, California: unpublished consulting report, Project No. G3395-12-1.

Franklin, J.A., Manailoglou, J., and Sherwood, D., 1974, Field Determination of Direct Shear Strength, Proceedings of the 3d International Congress Rock Mechanics, Denver, Colorado, Volume IIA, pp. 233-240.

Hawkins, A.B., and Privett, K.D., 1985, Measurement and Use of Residual Shear Strength of Cohesive Soils, Ground Engineering, pp. 22-29.

Jamiolkowski, M., and Pasqualini, E., 1976, Sulla scelta dei parametri geotecnici che intervengono nelle verifiche di stabilita dei pendii naturali e artificiali: Atti Istitutodi Scienza delle Costruzioni, Politecnico di Torino, v. 319, 53 p.

Kanji, M.A., 1974, The Relationship Between Drained Friction Angles and Atterberg. Limits of Natural Soils, Géotechnique 24, No. 4, pp. 671-674.

Kanji, M.A. & Wolle, C.M., 1977, Residual Strength - New Testing and Microstructure. Proc. 9th Int. Conf. Soil Mech., Tokyo 1, pp. 153-154.

Kenney, T.C., 1967, The Influence of Mineral Composition on the Residual Strength of Natural Soils. Proceedings of Geotechnical Conference, Oslo 1, pp. 123-129. Oslo: Norwegian Geotechnical Institute.

Kenney, T.C., 1977, Residual Strengths of Mineral Mixtures, Proc. 9th Int. Conf. Soil Mech. 1, pp. 155-161.

Kerr, P.F., and Drew, I.M., 1969, Clay Mobility, Portuguese Bend, California: California Division of Mines and Geology, Special Report 100, pp. 3-16.

Leighton and Associates, Inc., 1989, Geotechnical Feasibility Investigation of Proposed Golf Course, Golf Course Clubhouse, and 25th Street Widening, South Shores Landslide, City of Rancho Palos Verdes, California: unpublished consulting report, Project No. 1870265-10.

Leighton and Associates, 1990, Supplementary Feasibility Investigation of Peacock Hill, a Portion of Parcel 10, City of Rancho Palos Verdes, California: unpublished consulting report, Project No. 1881922-07, Volume I of II.

Leighton and Associates, 1991, Geotechnical Analysis of Gross Stability of Tentative Tract No. 50667, Between Paseo Del Mar Avenue, Palos Verdes Drive South and La Rotonda Drive, West of the Former Hotel Site, Subregions 7 and 8, City of Rancho Palos Verdes, California: unpublished consulting report, Project No.1870265-24, Volume I of II.

Lupini, J.F., Skinner, A.E. & Vaughan, P.R., 1981, The Drained Residual Strength of Cohesive Soils. Géotechnique 31, pp. 181-213.

Mesri, G., and Cepeda-Diaz, A.F., 1986, Residual Shear Strength of Clays and Shales, Géotechnique, Volume 36, Number 2, pp. 269-274.

Novak, G. A., 1982, Mineralogy and Chemistry of Bentonite from the Abalone Cove and Portuguese Bend Landslides Southern California, in Cooper, J. D., (comp.), Landslides and Landslide Mitigation in Southern California: Assoc. of Engineering Geologists, Southern California Section, p. 27.

Petley, D.J., 1984, Ground Investigation, Sampling and Testing for Studies of Slope Instability: in Brunsden, D., and Prior, D.B., eds. Slope Instability: New York, New York, John Wiley and Sons, pp. 67-101.

Skempton, A.W., 1985, Residual Strength of Clays in Landslides, Folded Strata and the Laboratory, Géotechnique, Vol. 35, No. 1, pp. 3-18.

Stark, T.D., and Vettel, J.J., 1992, Bromhead Ring Shear Test Procedure, ASTM Geotechnical Testing Journal, Volume 15, No. 1, pp. 24-32.

Stark, T.D. and Eid, H.T., 1994, Drained Residual Strength of Cohesive Soils, Journal of Geotechnical Engineering Division, ASCE, Volume 120, No.5, pp. 856-871.

TAGA, Inc., 1991, TSLOPE3 Computer Program for Three-Dimensional Slope Stability Analyses, TAGA Engineering Systems & Software, Lafayette, California.

Voight, B., 1973, Correlation Between Atterberg Plasticity Limits and Residual Shear Strength of Natural Soils. Géotechnique 23, No. 2, pp. 265-267.

Watry, S. M., 1992, Shear Strength of Bentonite from the Toe of the Portuguese Bend Landslide: California State Univ. Los Angeles, unpublished. M.S. thesis, 263 p.

Watry, S.M. and Ehlig, P.L., 1995, Effect of Test method and Procedure on Measurements of Residual Shear Strength of Bentonite from the Portuguese Bend Landslide, in Haneberg, W.C., and Anderson, S.A. eds. Clay and Shale Slope Instability: Boulder Colorado, Geological Society of America Reviews in Engineering Geology, v. X, pp. 13-38.

A LIMIT EQUILIBRIUM STABILITY ANALYSIS OF SLOPES WITH STABILIZING PILES

Takuo Yamagami[1], Member, ASCE, Jing-Cai Jiang[2] and Katsutoshi Ueno[3]

Abstract

This paper presents a design method for slopes with a row of piles to enhance slope stability or to prevent slope failure. The basic idea of the method is to allow that two different slip surfaces can be assumed in upslope and downslope earth masses of the row of piles. Contrary to the conventional limit equilibrium procedure where a factor of safety is calculated for a given slip surface, in the present method forces acting on the stabilizing piles are estimated based on a given slip surface whose factor of safety is prescribed to ensure the slope stability. By assuming a slip surface in upper and lower earth masses respectively, the forces acting on the piles can be calculated from an existing limit equilibrium method with a prescribed value of the factor of safety. The Bishop method is employed, and a repeated trial procedure is used to find the most dangerous situation for the piles and the corresponding two critical slip surfaces in the upside and downside of the pile row. It may be understood that if the stabilizing piles are still sound under the most dangerous situation, then the slope will have, at least, the prescribed (target) factor of safety.

In the proposed method, we clearly distinguish between forces which piles in a row can bear, i.e. the horizontal bearing capacity as passive piles, and forces which piles in a row should bear under the target factor of safety. The consideration in which the latter does not exceed the former makes possible a more rational design for stabilizing piles. The results of some examples illustrate the effectiveness of the proposed method.

Introduction

A limit equilibrium design method is proposed for a slope with a row of piles,

[1] Prof., Dept. of Civ. Engrg., The Univ. of Tokushima, Tokushima 770-8506, Japan
[2] Assist. Prof., Dept. of Civ. Engrg., The Univ. of Tokushima, Japan
[3] Lecturer, Dept. of Civ. Engrg., The Univ. of Tokushima, Japan

what is called passive piles, to enhance the stability of the slope. In the past, many studies have been conducted on slopes having stabilizing piles. They are classified into two groups: (1) studies based on the limit equilibrium approach (e.g. Ito & Matsui, 1975; Ito, Matsui & Hong, 1981, 1982; Reese, Wang & Fouse, 1992; Hassiotis, Chameau & Gunaratne, 1997), and (2) those using numerical approach such as FEM and/or BEM (e.g. Poulos, 1995; Lee Hull & Poulos, 1995; Chow, 1996). The present method falls within the first group as well; however, it is based on a novel manner markedly different from existing methods.

It is common knowledge that any design procedure for piles subjected to vertical loads consists of three steps: (i) evaluation of the load (V_D) that will act on each pile, (ii) determination of the allowable vertical bearing capacity (V_R), and (iii) design of each pile itself on condition that $V_D \leqq V_R$. We should follow the same procedure even for the passive pile design as suggested by some researchers (Viggiani, 1981; Popescu, 1991; Poulos, 1995):

i) Evaluation of the load (P_D) that the piles should sustain;
ii) Determination of horizontal bearing capacity (P_R) of passive piles;
iii) Design for the piles themselves, i.e. determination of the necessary pile geometries, material strengths, etc. on condition that $P_D \leqq P_R$.

In this paper we will focus only on the first two steps, i) and ii). Specifically a rapid and effective approach is proposed to evaluate load (P_D), based on a limit equilibrium method. A unique feature of this approach is that two different slip surfaces are allowed to exist in the upslope and downslope soil masses of the pile row. This is because a single slip surface that passes through the installed pile row cannot take place due to the high rigidity of the piles. The other feature of this approach is, unlike the conventional limit equilibrium procedure, that interactive forces acting on the pile row is estimated by prescribing a required value of the factor of safety for each of the assumed slip surfaces. It may be understood that, if the stabilizing piles can sustain the interactive forces between the pile row and soil masses, the pile-installed slope will have, at least the prescribed value of the factor of safety.

Based on the above concepts, the Bishop method is employed for formulation and a repeated trial procedure is used to find the most dangerous situations for the piles and the corresponding two critical slip surfaces in the upslope and downslope soil masses of the piles.

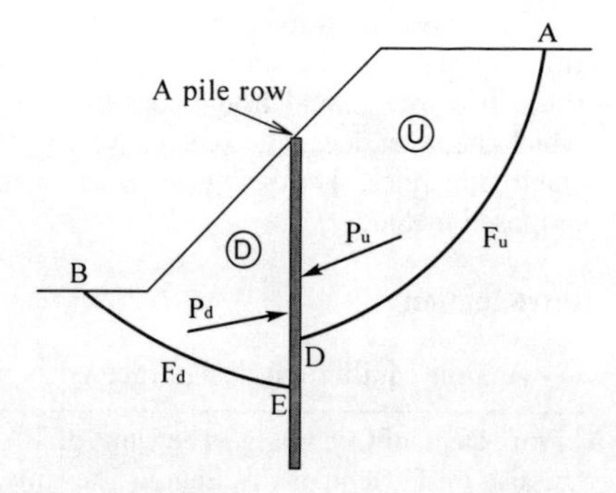

Fig. 1 Basic concepts of the method

Basic idea for evaluation of P_D

A typical cross section of a slope with a row of installed piles is illustrated in Fig.1, where two slip surfaces, *AD* and *BE,* are assumed respectively in the upslope and downslope soil masses of the pile row. The proposed method consists of the following two characteristic concepts:

1) Upslope and downslope sliding masses of the pile row may have different values of the factors of safety.
2) Interactive forces between the pile row and soil masses are estimated by prescribing target values of the factors of safety.

The first concept means that the upslope and downslope sliding masses bounded by the pile row are allowed to possess different factors of safety and hence different slip surfaces. It is quite natural to make such a consideration as different slip surfaces can appear respectively in the two soil masses due to the presence of the piles in a row. In order to justify this concept, let's consider an example in which sufficiently strong piles are to be installed deeply enough in an active landslide slope. The upslope sliding mass will be stopped due to sufficient resistance provided by the piles, and therefore become stable. However, the downslope sliding mass of the piles still may be in an unstable state, and thus it is possible that the sliding movement will continue. From this it is obvious that the upslope and downslope sliding masses may have different factors of safety. The magnitude of these two factors of safety, F_u and F_d (Fig.1), are generally unknown.

The second concept implies that desired, target values of the factors of safety are first specified by the designer, and then a design procedure is adopted such that the specified factors of safety are actually achieved. In this design procedure, contrary to the conventional limit equilibrium method where a factor of safety is calculated for a given slip surface, interactive forces P_u and P_d acting on the upslope and downslope faces of the piles in a row are calculated using the prescribed values of the factors of safety.

Now suppose that, in Fig. 1, the interactive forces P_u and P_d correspond respectively to the target factors of safety F_u and F_d of the assumed slip surfaces *AD* and *BE*. These two forces will be derived from an existing limit equilibrium method in the following section. At this stage, however, it is sufficient to suppose that P_u and P_d act at appropriate points and in appropriate directions on the piles in a row. As shown in Fig.1, P_u has an overturning effect on the piles in a row while P_d has a resisting effect. Therefore, if the pile row can sustain the net difference of the two effects, then the prescribed values of safety factors F_u and F_d will be ensured. In other words, in order to achieve the desired safety factors (F_u and F_d) for the upslope and downslope sliding masses, the piles in a row must at least provide a resistance force (P_D) corresponding to the net difference of the two effects.

Analysis procedure based on the Bishop method

Based on the Bishop method, this section derives the equations required to calculate the interactive forces. While the proposed method can be easily extended to more general cases of multi-rows of piles, the derivation here is conducted for the slope with a row of piles for convenience. We assume that, as shown in Fig.2, the interactive forces act in a horizontal direction neglecting the friction between the two faces of the piles in a row and the upslope and downslope soil masses. This is equivalent to ignoring vertical soil movement, e.g., Lee et al's approach (Lee, Hull & Poulos, 1995). It should be noted that such a simplified assumption is not always effective and therefore needs to be improved in future research. In addition, the location of the end E of the downslope slip surface BE is always constrained to be lower than or equal to the location of the end D of the upslope slip surface AD. Note that the thickness of the piles is not considered at this stage.

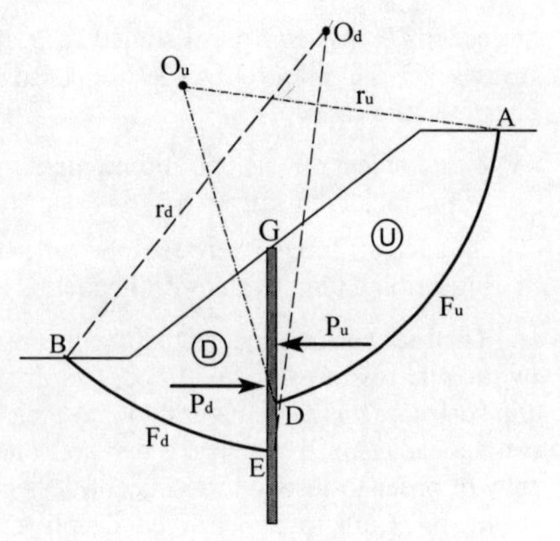

Fig. 2 Two circular slip surfaces

Derivation of interactive forces

Fig.3 shows the forces involved in the derivation of the factor of safety for a circular slip surface located in the upslope and downslope soil masses of the pile row, respectively. The moment equilibrium for the sliding mass on the upslope slip surface is described by summing moments about the center O_u of rotation:

$$\sum_u Wx = r_u \sum_u T + l_u P_u \qquad [1]$$

In Eq.[1], W is the total weight of a slice, P_u is the force from the upslope soil mass,

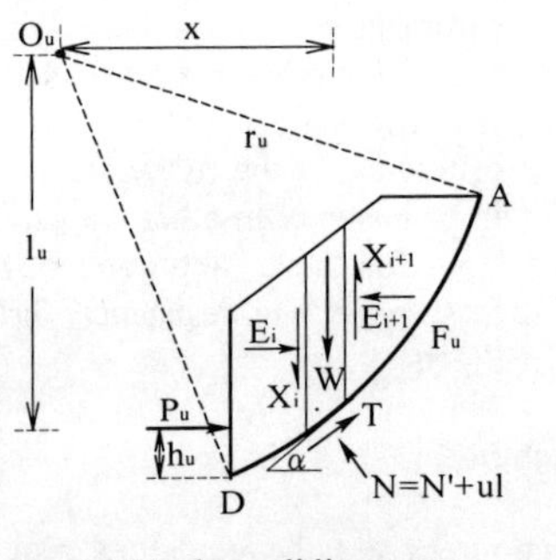

(a) Upslope sliding mass

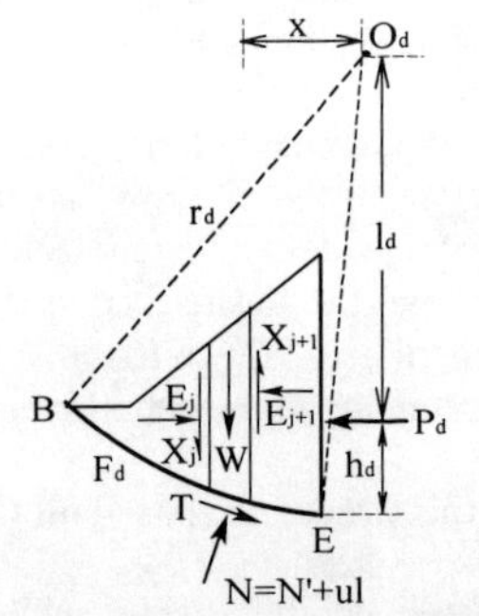

(b) Downslope sliding mass

Fig. 3 Derivation of interactive forces

T is the shear force mobilized on the base of each slice, r_u is the radius of the upslope circular slip surface, and l_u is the moment arm associated with P_u. We may also write:

$$F_u = \sum_u R_f \Big/ \sum_u T \qquad [2]$$

Where F_u is the factor of safety of the upslope sliding mass and R_f is the shear strength on the base of each slice. See Fig.3 for other symbols.

From the above equations the simplified Bishop method for this case yields

$$F_u = \frac{1}{\sum_u W \sin\alpha - \frac{l_u}{r_u} P_u} \sum_u \left\{ \frac{c'l\cos\alpha + (W - ul\cos\alpha)\tan\phi'}{\cos\alpha(1 + \tan\alpha\tan\phi' / F_u)} \right\} \qquad [3]$$

Rearranging Eq.[3], the equation of P_u can be obtained.

$$P_u = \frac{r_u}{l_u}\left[\sum_u W \sin\alpha - \frac{1}{F_u} \sum_u \left\{ \frac{c'l\cos\alpha + (W - ul\cos\alpha)\tan\phi'}{\cos\alpha(1 + \tan\alpha\tan\phi' / F_u)} \right\} \right] \qquad [4]$$

Similarly, the factor of safety equation and the force P_d for the downslope sliding mass can also be derived as follows.

$$F_d = \frac{1}{\sum_d W \sin\alpha + \frac{l_d}{r_d} P_d} \sum_d \left\{ \frac{c'l\cos\alpha + (W - ul\cos\alpha)\tan\phi'}{\cos\alpha(1 + \tan\alpha\tan\phi' / F_d)} \right\} \qquad [5]$$

$$P_d = \frac{r_d}{l_d}\left[\frac{1}{F_d} \sum_d \left\{ \frac{c'l\cos\alpha + (W - ul\cos\alpha)\tan\phi'}{\cos\alpha(1 + \tan\alpha\tan\phi' / F_d)} \right\} - \sum_d W \sin\alpha \right] \qquad [6]$$

When the factor of safety is prescribed for the upslope and downslope sliding mass respectively, the forces P_u and P_d can easily be calculated from Eqs. [4] and [6]. In the present paper, P_u and P_d are assumed to act at 1/3rd the height of GD and GE (Fig.2), respectively.

Determination of critical slip surfaces

In this section, a novel definition of critical slip surface is introduced. Then a search procedure is described to determine the critical slip surfaces in the upslope and downslope sides of the pile row. In the conventional slope stability analysis, the critical slip surface is defined as a surface on which the factor of safety is smallest. However, this definition is not valid for the present analysis as the values of the safety factors are assigned in advance. It is therefore necessary to make a modification of the definition of critical slip surface for application of the proposed method.

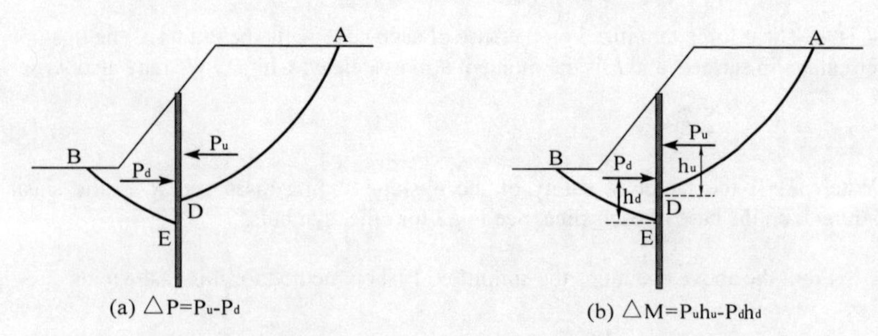

Fig. 4 ΔP_{max} and ΔM_{max} approaches for definition of critical slip surfaces

In this paper, the critical slip surfaces are defined as a pair of slip surfaces which are most dangerous (or critical) for the piles in row. And we have established the following two possible approaches to search for the critical slip surfaces:

1) ΔP_{max} approach: In this approach, the critical slip surfaces are defined as a pair of slip surfaces which give the maximum value ΔP_{max} of the difference in interactive forces: $\Delta P=P_u-P_d$ [Fig.4 (a)] subject to $P_u>0$ and $P_d>0$. Note that the piles in row should not be installed at a position where $P_d<0$.

2) ΔM_{max} approach: This approach defines the critical slip surfaces to be a pair of slip surfaces which provide the maximum value ΔM_{max} of the difference in moments: $\Delta M=P_u h_u - P_d h_d$ [see Fig.4 (b)], subject to $P_u>0$ and $P_d>0$.

In a theoretical manner, we can not conclude in advance which approach is more adequate and is better to use. It is thus suggested that two approaches are simultaneously applied and we chose the results in which stronger piles in row are required to meet the prescribed factors of safety.

It is the load the piles should sustain, mentioned in "Introduction", that is the one by which the stronger piles are required to meet the prescribed factors of safety.

A repeated trial procedure is used to find the most dangerous situation for the pile row and the associated critical slip surfaces. Fig.5 illustrates the search method schematically. A search region, indicated by a grid, is first specified for the upslope and downslope areas of the pile row, respectively. Then, for a trial center of the grid for the upslope side (e.g. point O_u

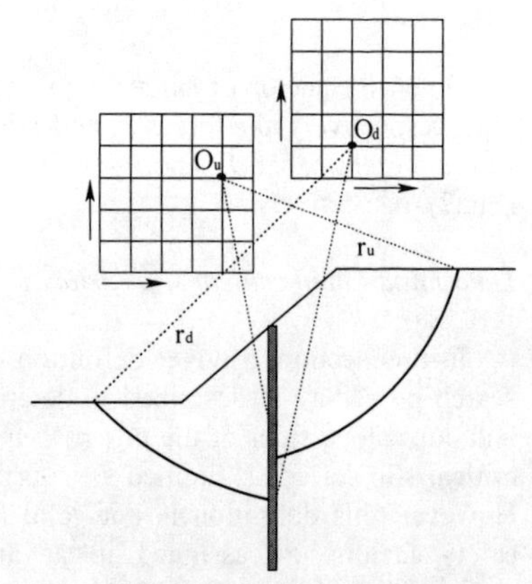

Fig. 5 Search procedure for a pair of critical slip surfaces

in Fig.5) and a given radius r_u, ΔP and ΔM are calculated separately for each grid point of the downslope side using various radii, and all calculated ΔP and ΔM are compared respectively. The above search procedure is repeated for all grid points of the upslope side, and finally ΔP_{max} and ΔM_{max}, together with the associated two pairs of critical slip surfaces can be determined.

Horizontal bearing capacity of passive piles

The Ito•Matsui model (1975, 1981 & 1982) is quite useful in evaluating the horizontal (allowable) bearing capacity, although this approach has a flaw that it is derived for rigid piles with infinite length. This model provides the lateral force between a pile and the surrounding soil in a state of plastic equilibrium, satisfying the Mohr-Coulomb yield criterion, due to movement of the sliding mass. Therefore, we may consider that the Ito•Matsui model gives the horizontal ultimate bearing capacity for each pile in row when subjected to lateral soil movement. What we need, however, is the horizontal allowable bearing capacity corresponding to the target factor of safety. It is thus obtained by modifying the original Ito• Matsui model with the following strength parameters:

$$\bar{c} = c / F_s \qquad \tan\bar{\phi} = \tan\phi / F_s \qquad [7]$$

Where F_s is the target factor of safety. For convenience, the following sections discuss only the case of F_u=F_d, though they may have different values.

The horizontal allowable bearing capacity of the pile should be calculated over the length from the top of pile to the middle between two points of intersection of the pile with the upslope and downslope slip surfaces. As will be shown in examples, however, two critical slip surfaces in downslope and upslope soil masses intersect at almost the same point with the pile. This is convenient for calculation of the horizontal bearing capacity of the pile based on the Ito• Matsui method.

Diameter and spacing of piles

Based on Ito•Matsui model, the required pile diameter and opening between

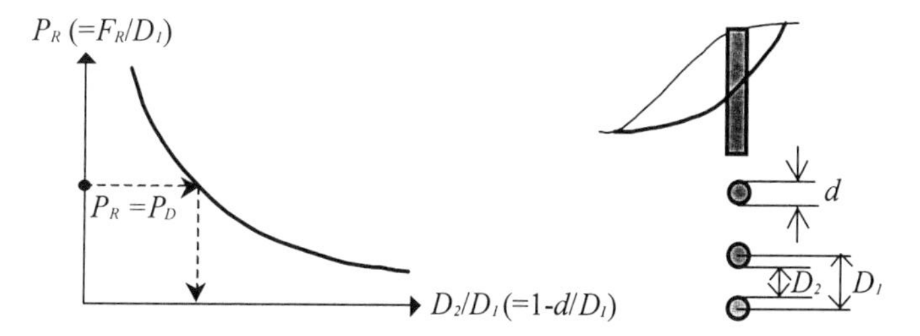

Fig.6 Schematical diagram showing relationship between P_R and D_2/D_1

piles can be determined, provided that the locations of the two end points D and E of the respective critical slip surfaces in upslope and downslope sliding masses do not differ much. Fig. 6 is employed to explain this context. A detailed diagram corresponding to Fig.6 will be given in an example problem in the next section.

We may first write down the following expression:

$$F_D = P_D D_1 \quad ; \quad D_2 = D_1 - d \qquad [8]$$

Where P_D is the load per unit length which the pile should bear to meet the target factor of safety for the pair of critical slip surfaces; this load is obtained from either ΔP_{max} or ΔM_{max} approach mentioned before; F_D is the same load as P_D but per pile; D_1 is center to center distance between piles; D_2 is opening between piles; and d is diameter of each pile.

It turns out from the Ito•Matsui model that the horizontal allowable bearing capacity per unit length P_R is expressed as a function of D_2/D_1 (=$1-d/D_1$), as shown schematically in Fig. 6 where

$$P_R = F_R / D_1 \qquad [9]$$

In Eq.[9], F_R is horizontal allowable bearing capacity per pile obtained from the Ito•Matsui theory.

In principle, we should design such that

$$F_R \geq F_D \quad or \quad P_R \geq P_D \qquad [10]$$

It is, however, the most economical design that meets the following equality:

$$F_R = F_D \quad or \quad P_R = P_D \qquad [11]$$

In order to meet this requirement, first we identify the point (mark ●)giving a value of P_R=P_D on the vertical axis, as shown in Fig. 6, then the corresponding value of D_2/D_1 (=1-d/D_1) is found on the horizontal axis. Consequently, as will be seen in example later, if a value for d is assumed, the associated value of D_1 is uniquely determined, and vice versa.

Examples and Results

Example-1

As an extreme case, a homogeneous slope with contiguously installed piles in a row (it may be regarded to be a wall) is solved here. In this example, the pile row is assumed to be installed at a location of *DIS*=9.0m, where *DIS* represents the horizontal distance from the toe of the slope, as shown in Fig.7(a). The slope in

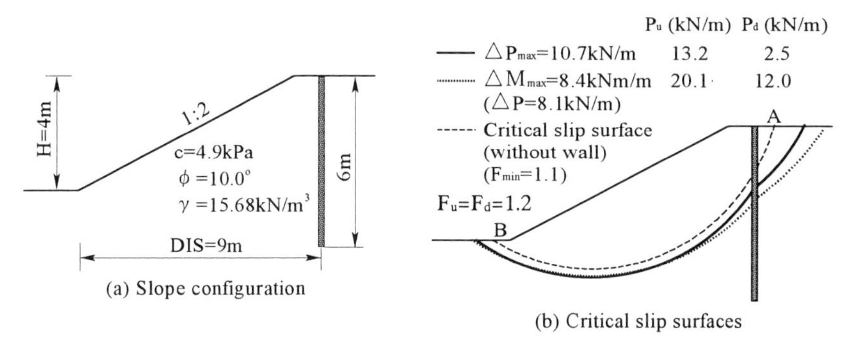

Fig. 7 Example 1

Fig.7(a) has a height of 4.0m, a slope inclination of 1:2 with the soil parameters of c=4.9kP$_a$, ϕ= 10.0° and γ=15.68kN/m^3. The original minimum factor of safety (without pile installation) was calculated to be 1.10 by the simplified Bishop method on the critical slip surface *AB* in Fig.7 (b). By prescribing the target value of the factor of safety to be 1.2 for both upslope and downslope sliding masses, the calculation results for the pile-installed slope in Fig.7(a) are illustrated in Fig.7(b). For achieving the required factor of safety, the ΔP value (i.e. P_D value), needed to be provided by the piles, was found to be 10.7kN/m from the ΔP_{max} approach and to be 8.1kN/m using the ΔM_{max} approach. The associated upslope and downslope critical slip surfaces together with the two cases are also illustrated in the same figure.

The influence of position of the pile row on interactive forces and critical slip surfaces are shown in Fig. 8. It can be seen from Fig. 8 that the locations of the critical slip surfaces differ significantly when the pile row is inserted at different

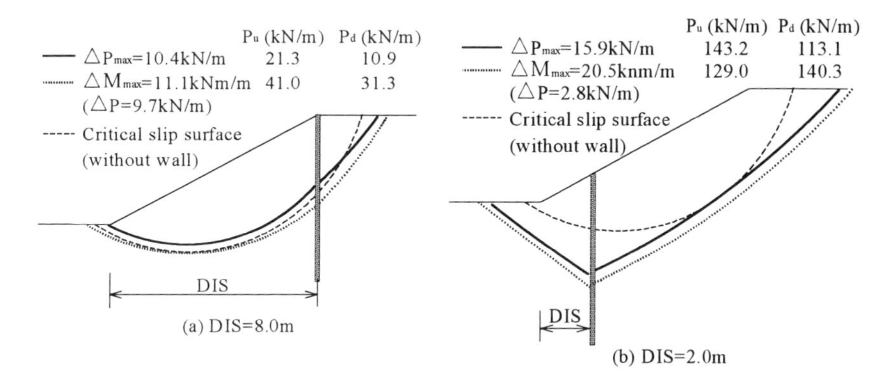

Fig. 8 Effect of pile row location on critical slip surfaces (for example 1)

positions. The values of ΔP $(=P_u - P_d)$ calculated in terms of ΔP_{max} are always larger than those in terms of ΔM_{max}.

It is of interest to note that the point of intersection of the downslope critical slip surface with the pile row almost coincides with that of the upslope critical slip surface [see Fig.7(b) and Fig. 8], regardless of the analysis in terms of the ΔP_{max} and ΔM_{max} approaches. These interesting results will be convenient for the detailed design of piles.

Example 2

The slope shown in Fig. 9(a) has the minimum factor of safety F_0 of 1.07 based on the simplified Bishop method with the strength parameters given in the figure when it has no piles. Then, the problem is to raise the factor of safety up to 1.20 by installing a row of piles at the location shown in the figure.

Fig.9(b) shows the critical slip circles obtained from the ΔP_{max} approach, together with that for no pile situation. Note the fact that at the pile location, the positions of end points of the two slip circles almost coincide despite the upslope and downslope slip circles are independent of each other. In this case, the load per unit length that the piles should bear is determined to be P_D=159.1kN/m.

In order to show how diameter and spacing of piles are determined, we assume, as an example, that piles with a diameter of d=0.5m are employed. Specifying several values of D_1 appropriately, the corresponding load F_D $(=P_D D_1)$ per pile can easily be calculated by using Eq.[8]. Therefore, the relationship between F_D and D_2/D_1 $(=1-d/D_1)$ can be obtained (Fig.10). On the other hand, from the Ito•Matsui model, the horizontal allowable bearing capacity, F_R, can also be determined for each of the specified values of D_1, and thus the F_R-(D_2/D_1) curve can be drawn, as shown in Fig.10. It can be seen that the intersection point of the two curves in Fig.10,

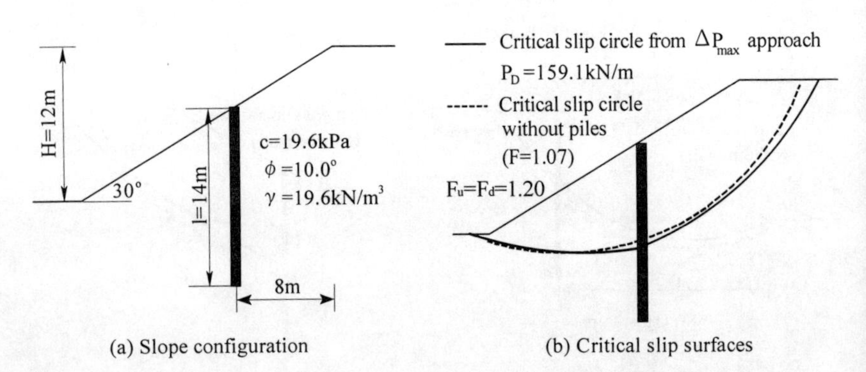

(a) Slope configuration

(b) Critical slip surfaces

Fig. 9 Example 2

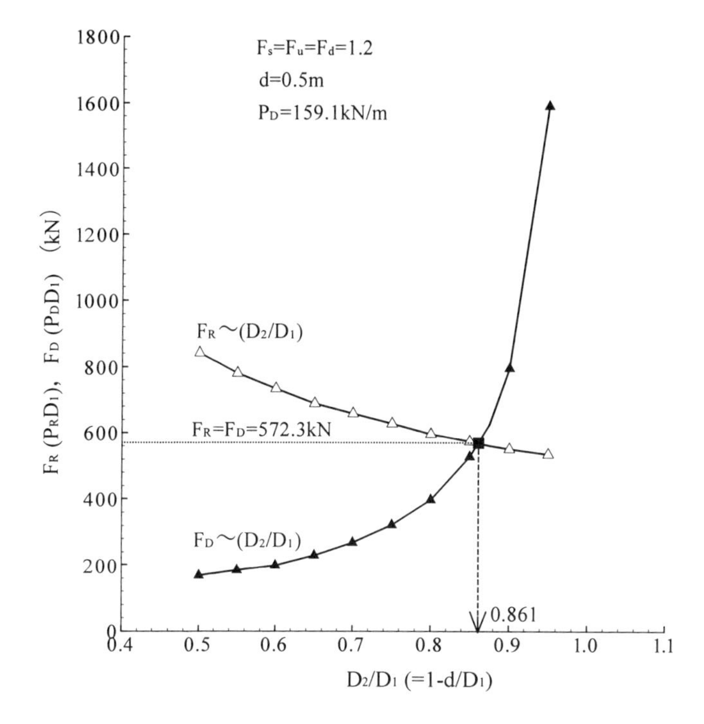

Fig.10 Relationships between $F_R - (D_2/D_1)$ and between $F_D - (D_2/D_1)$

satisfying the condition of F_R=F_D, corresponds to a value of 0.861 for D_2/D_1 (=1-d/D_1). Therefore, D_1, the opening between piles, becomes equal to 3.6m. This indicates that under d=0.5m, the target factor of safety of 1.2 is at least guaranteed when $D_1 \leqq 3.6$m. The problem yet to be solved is to design the piles themselves in order that they can really sustain the load per unit length P_D=159.1kN/m. But no further discussion on this matter is given here.

Concluding remarks

A methodology within the framework of limit equilibrium approach has been developed for the design of slopes reinforced by row of piles. The proposed method takes into account such a situation that a slip surface can occur separately in the upslope and downslope soil masses due to the presence of piles in a row. A unique feature of the method is that target values of the factor of safety are first specified by the designer and then stability analysis is carried out to determine the load (P_D) that the piles should bear to ensure the prescribed safety factors. A novel definition of critical slip surface has been introduced, and a repeated trial procedure based on

Bishop's method has been used to find a pair of critical slip surfaces which are most dangerous for the installed piles.

A horizontal allowable bearing capacity (P_R) as passive piles has been defined on the basis of the Ito•Matsui theory. This has been followed by a way to determine the necessary diameter and spacing of piles in a row on condition that P_D becomes equal to P_R.

The development of this study makes it possible to perform a satisfactory design for pile-reinforced slopes which not only meets the required factor of safety of the slope but also ensures the structural integrity of the piles. In future research, it is necessary to incorporate the structural analysis of the piles into the procedure of stability calculations. In addition, comparison studies with the results of both numerical simulations such as FEM and laboratory/field experiments should be conducted.

References

Chow, Y. K. (1996): Analysis of piles used for slope stabilization, *Inter. J. Nume. Anal. Meth. Geomech.*, Vol.20, pp.635-646.

Hassiotis, S., Chameau, J. L. & Gunaratne, M. (1997): Design method for stabilization of slopes with piles, *J. Geotech. and Geoenvir. Engrg., ASCE*, Vol.123, No.4, pp.314-323.

Ito, T. & Matsui, T. (1975): Methods to estimate lateral force acting on stabilizing piles, *Soils and Foundations*, Vol.15, No.4, pp.43-59.

Ito, T., Matsui, T. & Hong, W. P. (1981): Design method for stabilizing piles against landslide – one row of piles, *Soils and Foundations*, Vol.21, No.1, pp.21-37.

Ito, T., Matsui, T. & Hong, W. P. (1982): Extended design method for multi-row stabilizing piles against landslide, *Soils and Foundations*, Vol.22, No.1, pp.1-13.

Lee, C. Y., Hull, T. S. & Poulos, H. G. (1995): Simplified pile-slope stability analysis, *Computers and Geotechnics*, Vol.17, pp.1-16.

Popescu, M. E. (1991): Landslide control by means of a row of piles, Keynote paper, *Proc. Inter. Conf. Slope Stability Engineering*, Isle of Wight, pp.389-394.

Poulos, H. G. (1995): Design of reinforcing piles to increase slope stability, *Can. Geotech. J.,* Vol.32, pp. 808- 818.

Reese, L. C., Wang S.-T.& Fouse, J. L. (1992): Use of drilled shafts in stabilizing a slope, *ASCE, Geotech. Spec. Pub. No.31, Stability and performance of slopes and embankments – II,* Vol.2, pp.1318-1332.

Viggiani, C. (1981): Ultimate lateral load on piles used to stabilize landslides, *Proc. 10th Inter. Conf. SMFE,* Stockholm, Vol.3, pp.555-560.

Parametric Study and Subsurface Exploration Plan for Bluestone Dam

Greg Yankey[1], Rick Deschamps[2], Michael McCray[3], and David J. Bentler[4]

Abstract

Bluestone Dam is a concrete gravity structure founded on interbedded orthoquartzite and carbonaceous shale bedrock. Conventional analyses indicate the dam is unstable at pools approaching the probable maximum flood. Measurements of uplift pressures acting on the base of the dam at elevated pools are not available because 1) the dam has been operated at pool elevations significantly less than those used in design and 2) storms of record have not approached design levels. Numerical modeling was selected to aid in the extrapolation of uplift pressures operable at design pool levels. A preliminary modeling effort was undertaken using the distinct element program UDEC. The preliminary modeling effort had three primary objectives: 1) develop an approach for emulating drain performance and incorporating this within the numerical model, 2) determine the geometric and material parameters that have the greatest influence on uplift pressures for development of an efficient subsurface investigation program; and 3) assess whether numerical modeling can provide reliable information for use in the design process. The scope of this paper includes the parametric study and describes the developed exploration program. The reader is referred to a companion paper published in the 1999 ASDSO Dam Safety Proceedings, St. Louis, Missouri for descriptions of the developed drain model and related uplift pressure distributions.

Introduction

Bluestone Dam is located on the New River near Hinton, West Virginia. The dam forms Bluestone Lake, which covers 2,040 acres and has a drainage area of approximately 4,600 square miles. A site location map is shown in Figure 1. The project is operated and maintained by the United States Army Corps of Engineers, Huntington District (USACE-LRH).

---

[1] Project Manager, Fuller, Mossbarger, Scott, and May Engineers, Lexington, KY 40511-2050
[2] Senior Geotechnical Engineer, Fuller, Mossbarger, Scott, and May Engineers, Lexington, KY 40511-2050
[3] Geologist, Huntington District, U.S. Army Corps of Engineers, Huntington, WV
[4] Assistant Professor, University of Kentucky, Dept. of Civil Engineering, Lexington, KY 40506-0281

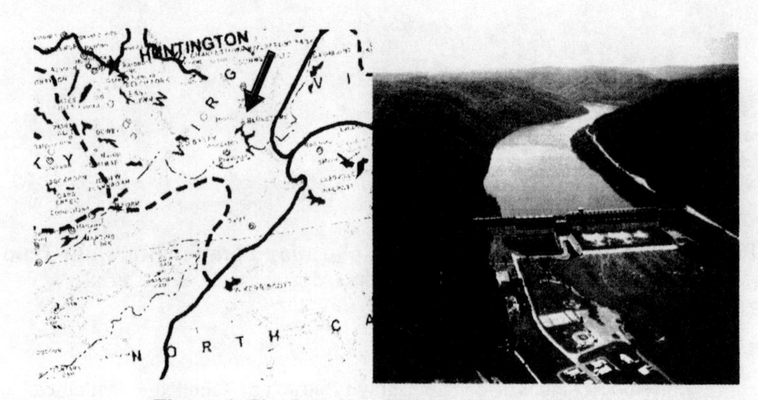

Figure 1. Site Location Map and Photo of Bluestone Dam

The structure is a concrete gravity dam 165 feet high and 2,050 feet long, consisting of 55 monoliths with an average width of 35 feet each. The elevation of the top of the dam is 1,535 feet above mean sea level. The dam was constructed in the 1940s. The town of Hinton, West Virginia lies directly downstream of the dam.

Bluestone dam was originally designed for flood control and to generate hydro-electric power; however, installation of the power generation unit was delayed indefinitely. The design pool for hydroelectric generation was 1,490 feet, however, as a flood control structure, the current operating pools are 1,406 feet in the winter months and 1,410 feet in summer. The pool of record for the project is elevation 1,506 feet, reached in 1960.

Recent hydrologic studies of the Bluestone Lake Basin indicate that the design storm event (100% PMF) is substantially more critical than that used in the original design, which will lead to overtopping of the structure. Overtopping is considered to lead to likely loss of pool because the abutments consist of non-durable shales and indurated clays. In addition, stability studies indicated that sliding along the near-horizontal bedding planes in the sedimentary rock foundation is likely when pool levels approach the top of dam. Accordingly, various remedial measures, including raising the dam and installing high capacity rock anchors, are currently being considered.

In the remedial measures currently under consideration, the pool elevation during the PMF would exceed the current pool of record by more than 40 feet. Extrapolation of uplift pressures from historical data for Bluestone dam to the new design pool is well beyond the limits suggested by regulatory agencies. Therefore, it was desired to do preliminary numerical analyses to assess whether or not numerical modeling would be beneficial during development of the exploration program and in consideration of final design alternatives. The objectives of the numerical modeling were to:

1) perform parametric studies to identify key geometric and material parameters for development of an efficient subsurface exploration program;
2) estimate uplift pressures acting on the base of the dam at pool elevations beyond the range of historical data; and
3) develop an approach to realistically model drain flow to allow assessment of drain performance at pool levels greater than the pool of record.

This paper describes the approach and findings of the parametric study (Objective No. 1), and the developed subsurface exploration program.

Foundation Conditions

The foundation beneath Bluestone Dam consists of near-horizontal orthoquartzite interbedded with carbonaceous shale. These bedding surfaces are typically undulating and very low tensile strength in the bedding joints between the orthoquartzite and shale exists. At depths greater than approximately 30 feet bellow the dam base, the bedrock becomes primarily orthoquartzite with occasional thin seams of carbonaceous shale. Close examination of the bedding planes between orthoquartzite and shale indicate small striations that likely developed during non-uniform strains occurring during erosional stress relief.

Parametric Study Methodology

The distinct element program UDEC, developed by Dr. Peter Cundall and supported by the Itasca Consulting Group, was used in this study. Some of the key features of UDEC are: 1) it is a distinct element model that is well suited to modeling discontinuous rock masses and large displacements along joints; 2) distinct blocks can be deformable with a specified failure criteria; 3) structural elements are available to simulate cables, beams, and grouted anchors; and 4) the program is capable of performing dynamic analyses. In addition, the model can simulate a fully coupled mechanical-hydraulic flow analysis for an intersecting joint system. Finally, UDEC includes an embedded programming language that allowed the development of an algorithm to model drain flow.

An idealized model of the dam and foundation conditions was generated using information available from the design, construction and operation information and from published references. Monolith 12 was selected for study because it had been included during historical analyses. Figure No. 2 illustrates an idealized cross-section of the dam.

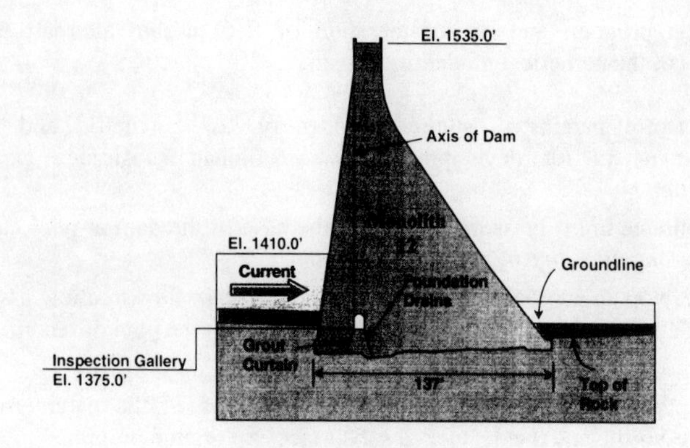

Figure 2. Typical Section Through Monoliths 11, 12 and 13 (not to scale)

The specific goal of the parametric study was to identify parameters that have a relatively large influence on dam stability so that they could be targeted in the planned subsurface exploration. A "baseline case" was developed in which a set of all input parameters was established. The relative influence of specific variables were then assessed by varying parameters over a range of values that were believed to encompass the possible conditions at the site. Except for one case, each parameter was adjusted while maintaining all other parameters at their baseline values.

Subsequent sections of this paper discuss the assumptions made during the performance of the parametric study, specific cases modeled, and the input parameters used. Finally, the results of the parametric study are presented followed by an overview of the developed subsurface exploration program undertaken to gather information for subsequent detailed modeling and design.

Assumptions, Cases Studied, and Input Parameters

Assumptions made during the modeling efforts included those related to the geometry of subsurface features, basic material behavior, and joint constitutive models. The opening of joints and bedding planes and the changes in size of these opening with changes in the stress field plays an important role in modeling results. To simplify discussion in the text the term “aperture” will be used synonymously with an open joint that is capable of carrying flow within the bedrock, regardless of the joint orientation. The following listing identifies specific base line assumptions made during the modeling effort.

1. Monolith 12 was utilized to develop the dam geometry. The foundation bedrock was idealized as containing a set of planar intersecting joints and is believed to provide a reasonable representation of a sedimentary environment (See Figure 5).

2. Near horizontal "bedding planes" and near vertical "joints" in the foundation bedrock were assumed to have the same aperture size-normal stress relationship. The apertures varied with the specific effective stress field within the foundation.
3. The bedrock foundation and concrete dam were modeled as being deformable blocks. The bedrock strength was characterized by a Mohr-Coulomb failure criterion, while the dam was assumed to be linear elastic. During numerical modeling, the dam was added in 12 increments and equilibrium was established at the end of each increment prior to progressing to the next increment.
4. Joints between blocks were characterized by the Coulomb model. Flow along the joints was assumed to be governed by the cubic law.
5. Each individual parameter was varied independently, except for one case as described herein.
6. Pool elevation at failure was selected as a common criterion to measure the importance of a particular variable. Although the onset of failure is difficult to define in numerical analyses, a combination of indicators were used including: out of balance forces, dam displacement, and the displacement rate. In all cases, failure occurred before the displacement of the dam toe exceeded two inches.
7. Strengths of the foundation bedrock and dam were not varied in the parametric study because sufficient site specific data were available.
8. The modeling was two-dimensional (plane strain) so that flow though transverse joints was not considered.
9. The potential change in aperture size associated with shear dilation was not considered.

Model validation efforts were undertaken prior to initiation of the parametric study. These included comparison of model results with simple limit equilibrium solutions, comparison with simple idealized closed-form solutions, and comparisons with previous reliable numerical analyses performed by Pace and Ebeling (1998). In addition, a series of analyses were undertaken to confirm that an adequate lateral extent was being modeled so that the location of the model boundaries did not impact results.

Cases Studied

The parameters that were varied can be divided into geometric parameters including initial aperture, bedding plane dip, spacing of vertical joints, etc., and material parameters such as the normal and shear stiffness of the joints. Figure 3 presents a flow diagram of cases analyzed during parametric studies and Table 1 describes the variation in parameters considered.

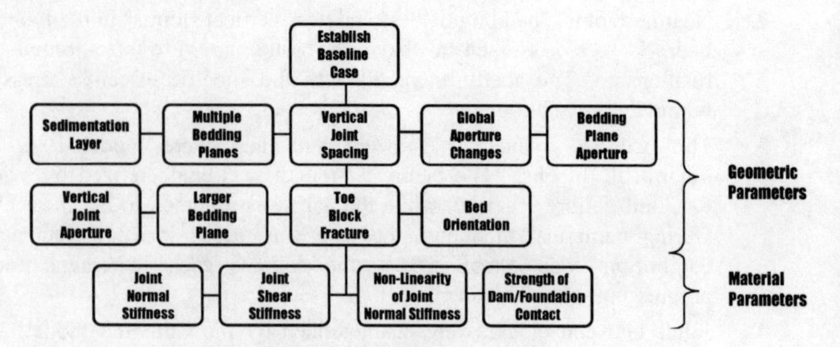

Figure 3. Parametric Study Cases

Table 1. Variation in Model Parameters

Geologic Details	Range
Headloss in sediment upstream	Assumed to be 0, 5 and 10 % of total headloss
Spacing and position of open bedding planes	Progressively eliminate open bedding planes at depth and reduce spacing below dam to 5 from 10 ft.
Spacing of near-vertical joints	Halve and double spacing
Global aperture size	Increase and decrease by a factor of 5.
Bedding plane aperture size	Increase and decrease by a factor of 5.
Near-vertical joint aperture size	Increase and decrease by a factor of 5.
Anomalous large bedding plane aperture	Position a single larger bedding plane aperture at different vertical positions within the baseline system.
Unfavorable inclined joint beyond toe	Include an inclined joint aligned with passive wedge beyond toe of dam
Dip of bedding planes	Vary between ± 4 deg. Downstream
Material Parameter Details	
Joint normal stiffness	Increase and decrease by a factor of 5.
Joint shear stiffness	Increase and decrease by a factor of 5.
Non-linearity of joint normal stiffness	Compare results for linear and hyperbolic joint normal stiffness aperture size relationships
Strength of dam/foundation contact	Reduce dam/foundation strength to a lower bound value.

Input Parameters

Although most input parameters were varied, some that are well documented were held constant for all cases investigated. The strength characteristics of the foundation rock and joints have been extensively investigated in previous studies. Accordingly, the strength, elastic moduli, and density parameters for the dam and foundation bedrock, and the strength parameters for the joints where not varied in the present work. Table 2 lists the bedrock and dam moduli; densities; and the bedrock, dam and joint strength parameters determined in previous studies. Table 3 illustrates the baseline values and ranges of values considered for each of the input parameters varied in the parametric study.

Table 2. Material Constants For Intact Blocks Utilized in this Study

	Orthoquartzite with Discontinuities	Orthoquartzite Interbedded with Shale	Concrete
Bulk Modulus (psi)	3,080,000	1,965,000	3,311,000
Shear Modulus (psi)	3,314,000	1,694,000	1,939,000
Unit Weight (pcf)	160	165	150
Friction Angle (deg.)	65	46	N/A
Cohesion (psi)	200	28	N/A

Table 3. Joint and Bedding Plane Properties for Mohr-Coulomb Area Model

Parameter	Value		
Bedding Plane Cohesion (psi)	Above Elev. 1335' = 6.0; Below Elev. 1335' = 11.0		
Bedding Plane Friction (deg)	Above Elev. 1335' = 32.0; Below Elev. 1335' = 30.0		
Vertical Joint Cohesion (psi)	0.0		
Vertical Joint Friction (deg.)	27.5		
	Baseline	**Upper**	**Lower**
Dam/Foundation Cohesion (psi)	33.8	N/A	18.3
Dam/Foundation Friction (deg)	47	N/A	40
Residual Conducting Aperture (in)	0.0012	0.006	0.00024
Conducting Aperture at Zero Stress (in)	0.0048	0.024	0.00096
Normal Stiffness (kip/ft^3)	40,000	200,000	8,000
Shear Stiffness (kip/ft^3)	40,000	200,000	8,000

Parametric Study Results

Subsequent sections present the results for variation of each parameter in the sensitivity study. The results of varying specific geologic parameters is shown first followed by influences of material parameters. To facilitate the following discussions, Figure 4 has been prepared which presents the primary conclusion or result (pool at failure) for each case in a concise form.

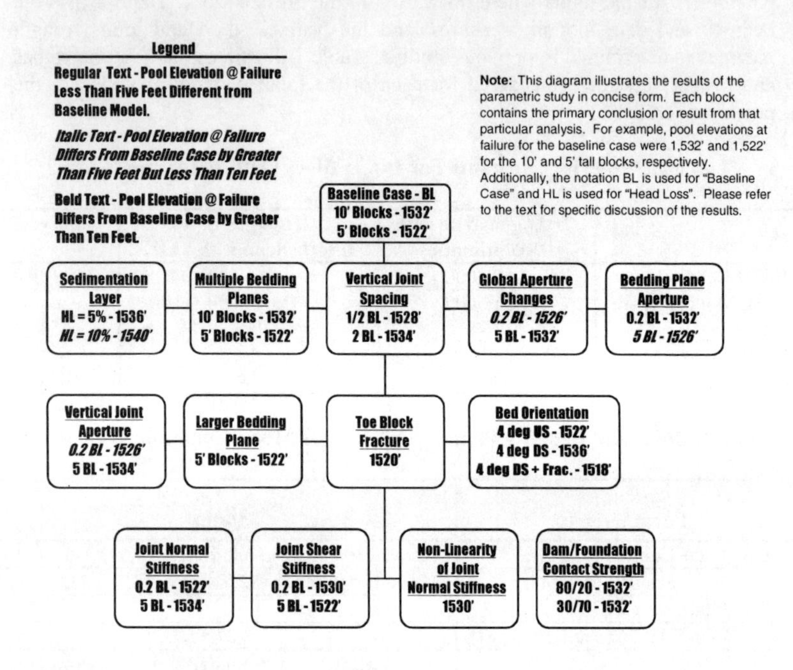

Figure 4. Summary of Parametric Study Case Results – Pool Elevation at Failure

Baseline Analyses. The pool elevation at failure for the baseline case was found to be between elevations 1,530 and 1,534 feet when using 10 foot high deformable blocks. The pool at failure was subsequently defined as 1,532 feet. During the numerical modeling process, the pool level was increased in two feet increments. However, the pool level at failure could not always be determined with this level of accuracy because of the difficult nature in defining the failure condition. When five feet high blocks were used the pool at failure was 1,522 feet. Figures 5 through 7 were generated to illustrate standard output that can be obtained from UDEC and represent results from the baseline case.

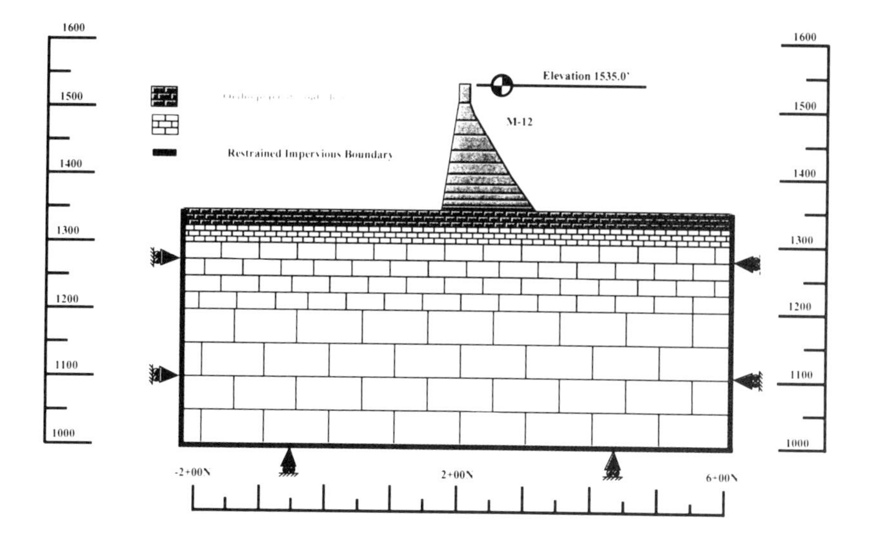

Figure 5. Baseline Case Mesh – 10' Blocks (not to scale)

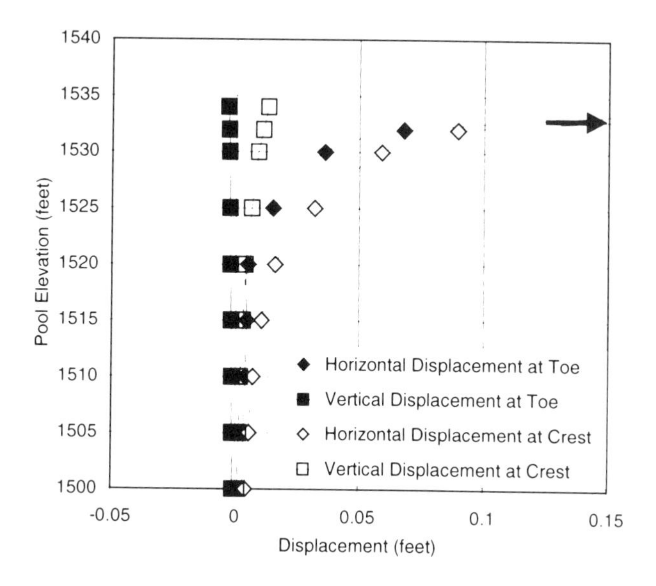

Figure 6. Displacements versus pool elevation for baseline case

The baseline lithology and model extents are shown in Figure 5. Figure 6 illustrates displacements versus pool elevation for the baseline case. This figure illustrates that the crest is moving horizontally downstream and vertically upward, and the toe is moving downstream and vertically downward. This pattern confirms the tendency for rotation of the dam when subjected to high pools, although sliding is the actual failure mechanism for this case. Between pool elevations 1,532 and 1,534 feet, the relative displacements increase dramatically indicating an unstable condition.

The results from the baseline case were further examined along the bedding plane below the dam between the heel and toe where sliding failure ultimately occurred at elevation 1,350 feet. Figure 7 a) shows the aperture, pressure head, and normal stress along the bedding plane for pool elevations of 1,510 and 1,532 feet. As the pool elevation increases, the aperture at the heel also increases while the aperture decreases slightly below the toe, again showing a tendency for dam rotation. Figure 7 b) shows the pressure head distribution below the dam. The pressure distribution becomes more non-linear at higher pool elevations indicating greater head loss in apertures below the toe. Figure 7 c) shows the normal stress distribution on the sliding surface below the dam.

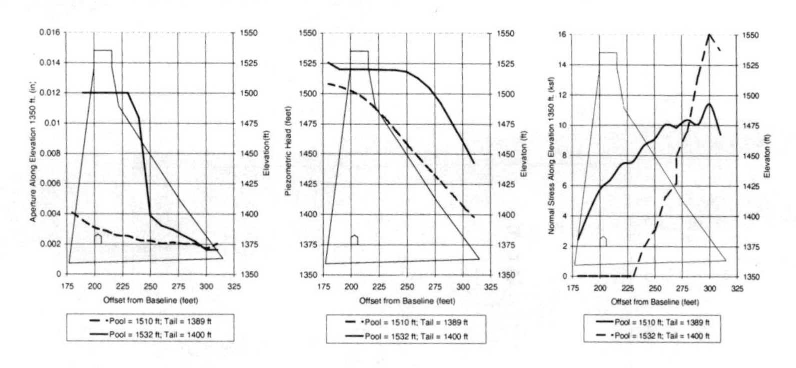

Figure 7. The variation in aperture, pressure head, and normal stress below the dam at two pools for the baseline case

Influence of Geologic Details

Sedimentation Layer. As expected, higher pools are necessary to cause failure when a sedimentation layer is present and this could have an important influence on stability and on interpreting historic uplift cell information. With head losses of five and ten percent, the pool elevations at failure were 1,536 and 1,540 feet, respectively. It should be noted that available data for Bluestone Dam does not indicate significant depths of sediment upstream of the dam (i.e. less than five feet), however, this may be a factor for other sites.

Number and Position of Bedding Planes. The influence of multiple bedding planes was investigated by deleting groups of horizontal bedding planes from the bottom of the baseline model upward. No change was noted in the pool elevation at failure from the baseline case, even when the only layer considered was the orthoquartzite interbedded with shale (elevation 1,360 feet to 1,330 feet). Consequently, this layer was broken up into five-foot tall blocks and the case re-analyzed. The results for this run indicated a pool elevation at failure of 1,522 feet. This indicates that as long as a sufficient bond exists between the base of the dam and the underlying bedrock, the vertical position of the first bedding plane has an important influence on stability and the total number is not critical. It should be noted that this conclusion may not be applicable to a condition with drains because the number of apertures will have an important impact on flow quantity.

Vertical Joint Spacing. Only modest changes in the pool elevation at failure were noted between the two cases investigated. Vertical joint spacing at one-half and two times the baseline case produced pool elevations at failure of 1,528 and 1,534 feet, respectively. The key factor appears to be when the combination of the number and/or size of the vertical joints are such that they limit the flow to the horizontal bedding planes.

Global Aperture Size. The overall or average size of joint openings has an influence on stability conditions with the condition of smaller size apertures being more critical. The smaller apertures tend to produce more non-linear pressure distributions because at the same joint normal stiffness a change in stress has a relatively greater variation in aperture size. The pool elevation at failure for the one-fifth and five times the baseline case were 1,526 and 1,532 feet, respectively. This is consistent with previous studies performed by Stone & Webster (1992) under the EPRI research program.

Size of Bedding Plane Apertures. The pool elevation at failure increases for global apertures greater than the baseline case, but was unaffected for the case of smaller bedding plane apertures. The pool elevation at failure for the one-fifth and five times the baseline case were 1,532 and 1,526 feet, respectively.

Size of Vertical Joint Apertures. This parameter was shown to have an important influence on stability. The pool elevations at failure for the one-fifth and five times the baseline case were 1,526 and 1,534 feet, respectively. As the vertical aperture decreases, stability decreases. This is an interesting result that would not be intuitively obvious without the benefit of the numerical modeling. For the case of smaller vertical apertures at low pools, there is relatively greater head loss that occurs during downward flow to horizontal bedding planes. This tends to reduce the uplift pressures acting below the base of the dam. However, as the pool elevation rises and lateral loads are transmitted to the dam, the imposed lateral dam movements tend to close the vertical joints beyond the toe and open the joints beyond the heel. This response reduces the head loss at the heel and increases the head loss at the toe, substantially increasing the uplift pressures acting on the dam.

Mixed Bedding Plane Apertures. A bedding plane with a larger aperture (3 times the baseline case) was placed within a system of bedding planes with baseline apertures. The bedding planes below the dam had a vertical spacing of five feet. The larger bedding plane was first placed in the position of the first bedding plane located five feet below the dam (Elev. 1,355 feet). Failure occurred along this bedding plane at a pool elevation of 1,522 feet. The larger aperture was then placed at the second aperture located 10 ft below the dam (Elev. 1,350 feet). Failure again occurred at a pool elevation of 1,522 feet along the aperture located five feet below the dam. This pool level at failure is the same as the previous case where all baseline apertures were used with five feet vertical spacing between bedding planes.

Inclined Toe Block Fracture. The presence of an inclined fracture located in the blocks at the toe of the dam was considered. This assumption was used in limit equilibrium stability analyses completed previously by the Huntington District. When a toe block fracture that rises 26 degrees downstream was included in the analyses, the pool elevation at failure decreased to 1,520 feet. This is a significant reduction from the baseline elevation of 1,532 feet because there is a substantial difference in passive resistance between a fractured wedge and intact rock blocks.

Bedding Plane Orientation. The pool elevations at failure for the upstream and downstream dipping cases were 1,522 and 1,536 feet, respectively. When the bedding dips upstream, the pool elevation at failure decreased because the bedding planes begin to daylight at the dam toe. This effectively eliminates passive resistance achieved with the baseline case, and with the case where beds dip downstream. A modification of this case with a favorably oriented inclined fracture at the toe of the dam was also considered for the bedding planes dipping four degrees downstream. When the toe fracture is included in the analysis for bedding dipping downstream this trend is reversed and the downstream dip becomes more critical. The pool elevation at failure for this case was 1,518 feet.

Influence of Material Properties

Magnitude of Joint Normal Stiffness. The pool elevations at failure for the one-fifth and five times the baseline case were 1,522 and 1,534 feet, respectively. The baseline case assumes a linear normal stiffness-normal displacement relationship and it can be seen that the magnitude has an influence on pool elevation at failure. Lower values of normal stiffness lead to a greater variation in aperture size, smaller apertures below the toe, and higher uplift pressures.

Non-linear Normal Stiffness. The normal stiffness-normal displacement relationship is recognized to be highly non-linear when viewed over a large stress range. However, the degree of non-linearity is expected to be less over the stress ranges applicable at Bluestone in the zones that control stability. The influence of non-linearity of the normal stiffness relationship was modeled by using a hyberbolic relationship based on the Barton-Bandis model (Barton et. al., 1985). The normal stress-normal displacement relationship utilized is shown in Figure 8. To facilitate comparison with the baseline case, the non-linear relationship has a comparable stiffness over the stress range of interest at the Bluestone site. The linear baseline case is shown tangent to the non-linear curve at the approximate average normal stress below the dam. The pool elevation at failure was 1,530 feet when using the non-linear relationship. This is only slightly smaller than the baseline case.

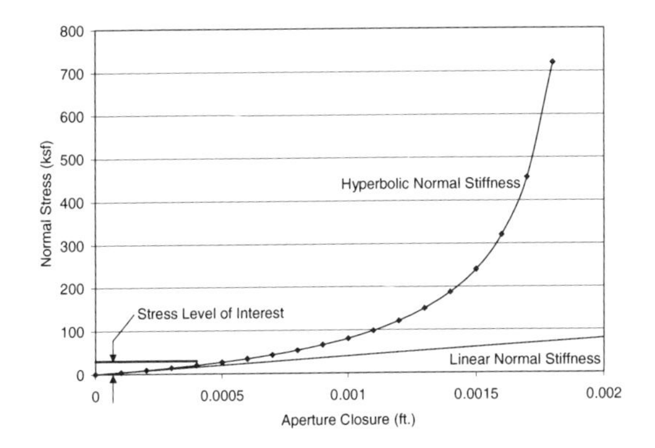

Figure 8. Comparison of Linear and Hyperbolic Normal Stiffness Relationships

Magnitude of Joint Shear Stiffness. The pool elevation at failure for the one-fifth and five times the baseline case was 1,530 and 1,522 feet respectively. The results of the direct shear testing completed previously indicate the shear stiffness at Bluestone Dam lies in the range between the baseline case and the lower bound case. Over this range of shear stiffness, the pool elevation at failure was not affected significantly (1,530 to 1,532 feet). Joint shear stiffness was varied in concert for all discontinuities (i.e. horizontal bedding planes and vertical joints).

Dam-Rock Interface Strength. Observations from recovered core samples indicate that in general, the bond between the dam and bedrock is intact although this is not true in all cases. The baseline case uses a constant strength that is weighted 80:20 between the peak strength and the sliding strength. For the present case, to study reduced strength along the interface, a weighting ratio of 30:70 was used between the peak and sliding strengths. In addition, the permeability of this joint was assumed to be the same as the other bedrock layers. For this strength assumption, there was no reduction in the pool elevation at failure because the underlying bedding plane was still the critical surface.

Critical Modeling Parameters

Based on the results of the parametric study, it is possible to rank the relative importance of the specific parameters investigated. Table 4 summarizes this ranking based on the possible influence of pool elevation at failure relative to the baseline case. The results summarized in Table 6 are important parameters in the generation of the subsurface exploration program.

Table 4. Relative Importance of Parameters Investigated

Important Influence on Stability (Potentially Influence Pool Elevation at Failure by >10 feet)
• Toe Block Fracture • Bed Orientation • Vertical Location of First Open Bedding Plane • Horizontal Bedding Plane Normal Stiffness • Horizontal Bedding Plane Shear Stiffness
Significant Influence on Stability (Potentially Influence Pool Elevation at Failure by > 5 feet)
• Sedimentation • Global Aperture Size • Gouge • Horizontal Bedding Plane Aperture • Vertical Joint Aperture
No Significant Influence on Stability (Small Influence on Pool Elevation at Failure)
• Vertical Joint Spacing • Non-Linearity of Normal Stiffness • Dam Foundation Bond Strength Reduction

It should be recognized that dependent combinations of parameters would produce variability in pool elevations at failure greater in magnitude than that presented in Table 4. It should also be noted that although a particular variable may have been identified as "critical", the specific case that illustrated this item may or may not be appropriate at Bluestone Dam. For example, sedimentation just upstream of the dam was identified in a recent sedimentation survey to be less than five feet. Consequently, although sedimentation may have a large influence on stability, it will not be an important item in the development of the exploration program.

Exploration Plan

The list of primary requirements from the exploration program and the approach that will be used to obtain the specific information follows.

1. Characterization of the predominant joint sets within the bedrock, including joint sizes and spacing, and possible presence of an inclined joint set at toe of dam.
   - Vertical and inclined logged rock cores.
   - Borehole camera imaging and interpretation.
2. Bedding plane orientation (dip).
   - Vertical rock cores upstream, downstream and within gallery of dam. Borings will extend to a reliable marker bed at depth.
3. Estimation of bedding plane aperture sizes.
   - Straddle packer tests in core holes: upstream of the dam (at lower confining stress), and through gallery below dam (at higher confining stress). Interpreted using numerical modeling.
   - Staged packer tests in core holes while measuring pressure and flow in adjacent core holes. Specific apertures were identified with the borehole camera. A shortly spaced straddle packer test was conducted on the aperture while the pressure or flow was measured in adjacent core holes spaced two feet and four feet away. Pressure and flow was measured at three different packer pressures. The purpose was to try to assess the changes in flow characteristics due to changes in aperture size with changes in the stress field. Variations in pressure and or flow will provide data to back calculate aperture sizes, and entrance losses.
4. Entrance loss coefficient and frictional resistance factor for flow in drains.
   - Entrance losses will be estimated with the staged straddle packer tests described in Item 3. The drain friction losses could be estimated by conducting an instrumented drain flow test in situ.

5. Joint normal stiffness.
   - The joint normal stiffness for bedding planes and other joints will be obtained from a combination of laboratory tests and correlation with joint roughness and hardness.

Summary and Conclusions

This paper summarizes a parametric study that was conducted in concert with other numerical modeling at Bluestone Dam. The primary goals included developing a better understanding of critical bedrock features that influence the stability of the dam. The distinct element program UDEC proved to be a very useful tool, and the modeling effort provided physically consistent data and much insight into the factors contributing to the applicable uplift pressures below concrete gravity dams founded on bedrock.

The simulation of real systems with the use of numerical models requires many simplifying assumptions. The number and importance of the assumptions made depend on the complexity of the system being modeled. In general, the modeling of geotechnical systems, which are inherently "data limited", is far more complex than modeling systems that contain fabricated materials of known geometry and physical properties. Philosophically, the objectives should focus on developing an understanding of the dominant mechanisms that affect the behavior within the system being simulated, not the specific magnitude of any one response. This was the primary objective of the present study consistent with the development of an efficient plan for subsurface investigation.

An important simplifying assumption made for the parametric analyses is the use of continuous, and planer, bedding plane apertures that are connected with a "regular" pattern of orthogonal joints, and that this structure can be represented by a two-dimensional system. This assumption is believed to be acceptable considering the geologic environment of sedimentary rock. Within this framework, the influence of most other geometric parameters was assessed in the parametric study.

Acknowledgements

The authors would like to thank Messrs. Robert Yost, Michael Keathly, Michael Szwalbnest, Robert Ebeling and Michael Klosterman of the U.S. Army Corps of Engineers for their valuable input on this project.

References

Barton, N., Bandis, S., and Bakhtar, K. (1985). "Strength, Deformation and Conductivity Coupling of Rock Joints," International Journal of Rock Mechanics, Mining Science, and Geomechanics Abstracts, 22(3), pp121-140.

Itasca Consulting Group (1996). UDEC Users Manuals, 3 Volumes, Thresher Square East, 708 South Third Street, Suite 310, Minneapolis, MN.

Pace, M. E., and Ebeling, R. M. (1998). "Interaction of a Gravity Dam, Rock Foundation, and Rock Joint with Uplift Pressures," Journal of Dam Engineering, Volume IX, Issue 3, pp. 265-305.

Stone & Webster (1992). "Uplift Pressures, Shear Strengths, and Tensile Strengths for Stability Analysis of Concrete Gravity Dams," EPRI, TR-100345, Research Proj. 2917-05.

Subject Index

Page number refers to the first page of paper

Author Index

Page number refers to the first page of paper